AS Science in Society

Project directors
Angela Hall
Robin Millar

Editor
Andrew Hunt

Contributors
Paul Bowers Isaacson
Angela Melamed
Jean Scrase

Nuffield Curriculum Centre

THE UNIVERSITY *of York*

Expert advisers
Suzanne Aigrain, University of Exeter; Charles Barclay; Emm Barnes, University of Manchester; Sean Beevers, Environmental Research Group, King's College London; Frances Fewell, Institute for Complementary Medicine; Sarah Gordon, Imperial College London; Shakoor Hajat, Public and Environmental Health Research Unit; Kate Harvey, Nuffield Council on Bioethics; Steve Hayes; Roger Highfield, Science Editor, *The Daily Telegraph*; Steve Jones, University College London; Frank Kelly, Environmental Research Group, King's College London; Randal Keynes, The Charles Darwin Trust; David Kirby, University of Manchester; Andy Longmore, UK Astronomy Technology Centre, The Royal Observatory, Edinburgh; Sivasegaram Manimaaran, Department for Innovation, Universities and Skills; Emma Newall, Science Learning Centre London; Michael Perryman, European Space Agency; Andrew Roddam, University of Oxford; Francoise Shenfield, Reproductive Medicine Unit, UCLH; Ian Stevens, University of Birmingham; Jo Tanner, The Coalition for Medical Progress; Bridget Fenn, London School of Hygiene & Tropical Medicine; The Association of the British Pharmaceutical Industry

www.heinemann.co.uk
✓ Free online support
✓ Useful weblinks
✓ 24 hour online ordering

01865 888080

Heinemann

Heinemann is an imprint of Pearson Education Ltd, a company incorporated in
England and Wales, having its registered office at Edinburgh Gate, Harlow, Essex CM20 2JE.
Registered company number: 872828

www.heinemann.co.uk

Heinemann is a registered trademark of Pearson Education Ltd

First published 2000 as AS Science for Public Understanding

12
10 9 8 7

British Library Cataloguing in Publication Data is available from the British Library on request.

ISBN 978 0 435654 64 1

Edited by Tim Jackson
Designed by Wooden Ark
Produced, illustrated and typeset by Wearset Ltd
Original illustrations © Pearson Education Ltd 2008
Picture research by Q2AMedia
Cover photo/illustration © dino4/iStockphoto (main image);
Vladimir Sazonov/iStockphoto (butterfly)
Printed in China (CTPS/07)

Websites
The websites used in this book were correct and up-to-date at the time of publication. It is essential for tutors to preview each website before using it in class so as to ensure that the URL is still accurate, relevant and appropriate. We suggest that tutors bookmark useful websites and consider enabling students to access them through the school/college intranet.

Contents

Acknowledgements

The editor and publisher would like to thank the following individuals and organisations for permission to reproduce photographs:

p1 CNRI/Science Photo Library; **p2** Jean-Loup Charmet; **p3** Mary Evans Picture Library/Photolibrary; **p4** Dr Kari Lounatmaa; **p5** Science Photo Library; **p7 T** Pasteur Institute; **p7 BR** Pasteur Institute; **p8 T** Mary Evans Picture Library/ Mary Evans ILN Pictures; **p8 ML** Michael Abbey/Photo Researchers, Inc.; **p10 T** Bernard Pierre Wolff; **p10 BL** Science Photo Library; **p11** National Library of Medicine/Science Photo Library; **p13** Science Photo Library; **p14** Popperfoto; **p16** Photo Researchers/Dinodia; **p21** CDC/Science Photo Library; **p22 TL** SIU/ Science Photo Library; **p22 ML** Erich Schrempp/Science Photo Library; **p22 BL** Sidney Moulds/Science Photo Library; **p23 TL** St. Mary's Hospital Medicine School/Science Photo Library; **p23 MR** Popperfoto; **p23 BR** Mary Evans Picture Library/Alamy; **p24** Science Photo Library; **p28** Rafiquar Rahman/ Reuters; **p30 TL** Science Source/Science Photo Library; **p30 B** USA Library of Medicine/Science Photo Library; **p31** Krista Kennell/ZUMA/Corbis; **p32** Peter Bowater/Alamy; **p37** NASA; **p40** Andrew Brown/Corbis; **p41** Rainer Raffalski; **p42** Cape Grim BA.P.S./Simon Fraser/Science Photo Library; **p45** Scott Barbour/Getty Images; **p50** Hattie Young/Science Photo Library; **p51** Richard Shiell/Photolibrary; **p52** Colin Cuthbert/Science Photo Library; **p53** D.Vo.Trung /Eurelios/LookatSciences; **p54** Geoff Tompkinson/Science Photo Library; **p56** Geoff Tompkinson/Science Photo Library; **p57 MR** The Lancet, 371, Copyright Elsevier (2008); **p57 BL** Associated Press; **p58** Tek Image/Science Photo Library; **p59 TR** Klaus Guldbrandsen/Science Photo Library; **p59 MR** Art Stein; **p60 TL** Dr P. Marazzi/Science Photo Library; **p60 M** Mark Clark/Science Photo Library; **p61 T** BSIP/Manceau; **p61 B** Cordelia Molloy/Science Photo Library; **p64** Associated Press; **p65** Pasieka/Science Photo Library; **p66** Elizabeth Hunter/The British Library; **p67** GlaxoSmithKline; **p70** R. Umesh Chandran, TDR, WHO/Science Photo Library; **p71 MT** BSIP, Hardas/Science Photo Library; **p71 ML** Sebastian Kaulitzki/Shutterstock; **p72** David Scharf/Getty Images; **p73 TR** Professor Stephen Proctor; **p73 BR** James King-Holmes/Science Photo Library; **p75** Eddie Gerald/Alamy; **p76** Professor Miodrag Stojkovic/Science Photo Library; **p77** Colin Cuthbert/Science Photo Library; **p78** Laguna Design/ Science Photo Library; **p79** Bernard Benoit/Science Photo Library; **p90** Lauren Shear/Science Photo Library; **p91** Dr Yorgos Nikas/Science Photo Library; **p92** Pascal Goetgheluck/Science Photo Library; **p95** Simon Fraser/RVI, Newcastle-upon-Tyne/Science Photo Library; **p97** Simon Fraser/Science Photo Library; **p100** AJ Photo/Science Photo Library; **p101 TR** Cordelia Molloy/Science Photo Library; **p101 B** David Parker/Science Photo Library; **p104** AFP; **p105** Kenneth Langford/Corbis; **p106 TL** Firefly Productions/Corbis; **p106 BL** Edward Shaw; **p118 M** Blair Seitz/Science Photo Library; **p118 B** Brian Harris/Alamy; **p120** Simon Fraser/Science Photo Library; **p121 TR** Image Science Médecine; **p121 B** Matthew Meadows, Peter Arnold Inc./Science Photo Library; **p124** PhotoDisc. Cole Publishing Group. Michael Lamotte. 1995; **p125 TR** Ace Stock Limited/Alamy; **p125 BR** Popperfoto/Alamy; **p131 TR** BSIP/Mendil; **p131 BL** Mark Sykes/Science Photo Library; **p133 TR** Rob Wilkinson/Alamy; **p133 MR** Aaron Kohr; **p134** Macduff Everton/Corbis; **p137** Carlos Goldin/Science Photo Library; **p138** Science Photo Library; **p140 TL** The Print Collector/Photolibrary; **p140 BL** Worldwide Picture Library/Alamy; **p141** Mary Evans Picture Library/ Photolibrary; **p142** William Ervin/Science Photo Library; **p143** The British Library; **p144** Keith Kent/Science Photo Library; **p146 TL** Mary Evans Picture Library/Photolibrary; **p146 BL** Robert Thompson; **p147** National Library of Medicine/Science Photo Library; **p148** National Library of Medicine/Science Photo Library; **p149** Paul A. Souders/Corbis; **p151** Tom McHugh/Science Photo Library; **p154 TL** David Parker/Science Photo Library; **p154 BL** W. M. Keck Observatory; **p155** Gordon Garradd/Science Photo Library; **p169** NASA Jet Propulsion Laboratory (NASA-JPL); **p173** Harvard College Observatory/Science Photo Library; **p174 M** Henri Boffin; **p174 BL** NASA; **p175** Doug Sanqunetti; **p180** Roger Ressmeyer/Corbis; **p184** Peter Menzel/Science Photo Library; **p185 TR** Adam Hart-Davis/Science Photo Library; **p185 B** Dr Ken Macdonald/ Science Photo Library; **p187** Dr Ian Stevens; **p188** Dr Suzanne Aigrain; **p189** CNES/D. Ducros; **p190** ESA 2001. Illustration by Medialab; **p191 TL** ESA 2002. Illustration by Medialab; **p191 BR** Dr Andy Longmore; **p193** The Royal Society

Thanks are due to the following for permission to reproduce illustrations:

Fig. 2.15: "TB Deaths, 2005", http://www.globalhealthfacts.org/topic.jsp?i=19, The Henry J. Kaiser Family Foundation, 2005. This information was reprinted with permission from the Henry J. Kaiser Family Foundation. The Kaiser Family Foundation, based in Menlo Park, California, is a nonprofit, private operating foundation focusing on the major health care issues facing the nation and is not associated with Kaiser Permanente or Kaiser Industries. **Fig. 3.3**: Royal Commission on Environmental Pollution, http://www.rcep.org.uk/studies/energy/98-6063/fisher.htm. Reproduced under the terms of the Click-Use Licence. **Fig. 3.5**: Department for Environment, Food and Rural Affairs, http://www.defra.gov.uk/environment/statistics/globatmos/gagccukem.htm. Reproduced under the terms of the Click-Use Licence. **Fig. 3.6**: National Statistics, http://www.statistics.gov.uk/downloads/theme_environment/transport_report.pdf. Crown copyright material is reproduced with the permission of the Controller Office of Public Sector Information (OPSI). **Fig. 3.14**: Air Quality Archive, http://www.airquality.co.uk/archive/monitoring_networks.php?n=aun. The Air Quality Archive is prepared and hosted by AEA Energy & Environment, on behalf of the UK Department for Environment, Food & Rural Affairs and the Devolved Administrations. **Fig. 3.16**: Parliamentary Office of Science and Technology, http://www.parliament.uk/post/pn188.pdf. Parliamentary material is reproduced with the permission of the Controller of HMSO on behalf of Parliament. **Fig. 3.17**: National Atmospheric Emissions Inventory, http://www.naei.org.uk/pollutantdetail.php?poll_id=24&issue_id=1. **Fig. 3.23**: National Statistics, http://www.statistics.gov.uk/cci/nugget.asp?id=24. Reproduced under the terms of the Click-Use Licence. **Fig. 4.10**: Downstate, http://library.downstate.edu/ebm/2400.htm. **Fig. 6.9**: American Cancer Society. *Breast Cancer Facts and Figures 2005-2006.* Atlanta: American Cancer Society, Inc. **Fig. 6.19**: Artwork originally created for the National Cancer Institute. Reprinted with permission of the artist, Jeanne Kelly. Copyright 2006. **Fig. 7.27**: Department for Environment, Food and Rural Affairs, http://www.defra.gov.uk/environment/statistics/radioact/radradon.htm. Reproduced under the terms of the Click-Use Licence. **Figs 8.6 and 8.7**: Cancer Research UK, http://info.cancerresearchuk.org/cancerstats/causes/lifestyle/tobacco/. Accessed March 2008. **Fig. 8.8**: Cancer Research UK, http://info.cancerresearchuk.org/cancerstats/types/lung/smoking/?a=5441#life. Accessed March 2008. **Fig. 8.9**: Artwork originally created for the National Cancer Institute. Reprinted with permission of the artist, Jeanne Kelly. Copyright 2006. **Fig. 8.10**: ABC of breast diseases: Breast cancer – epidemiology, risk factors, and genetics. Sept 9, 2000. K McPherson, C M Steel, J M Dixon. *British Medical Journal.* BMJ Publishing Group Ltd. **Fig. 8.13**: Reproduced with the permission of Biobank UK. **Fig. 8.19**: Cancer Research UK, http://info.cancerresearchuk.org/cancerstats/types/skin/incidence/. Accessed March 2008. **Fig. 11.4**: *Daily Telegraph*, 11 June 2007. **Fig. 11.8**: European Space Agency, http://www.esa.int/esaSC/SEMYZF9YFDD_index_0.html.

Every effort has been made to contact copyright holders of material reproduced in this book. Any omissions will be rectified in subsequent printings if notice is given to the publishers.

Foreword

Science in Society is a full A-level course. It replaces the one-year AS *Science for Public Understanding* course that was launched in 2000 after a two-year pilot. Like its predecessor, the AS units that make up the first half of the new course are built around the study of topical issues and key episodes in the history of science. In these contexts learners develop a more rounded and mature view of the major science explanations and ideas about how science works that they have met during their earlier studies.

The course is a means of broadening the curriculum for those whose interests lie mainly in the arts and humanities, or giving those who study science an opportunity to reflect on their specialist interests in a broader context.

We have developed the *Science in Society* specification in close collaboration with David Baker of the Awarding Body AQA, with advice from many teachers together with expert advisers from universities and other research centres.

We would like to thank our co-authors and advisers for their contributions to this book. We are also indebted to Liz Marchant, Claire Gordon and Sally Woods of Heinemann for their support and expertise.

This book covers the first year of the course leading to an AS qualification. It is complemented by the project website (www.scienceinsocietyadvanced.org). Both have been developed through collaboration between the Nuffield Curriculum Centre and the University of York Science Education Group.

We are very grateful to the Trustees of the Nuffield Foundation for their encouragement and financial support.

Angela Hall
Robin Millar
Andrew Hunt
Paul Bowers Isaacson
Angela Melamed
Jean Scrase

Introduction

The three strands of the course

As you work through this *Science in Society* course you study a series of topical issues and episodes from the history of science. In each topic you learn to apply your understanding of scientific explanations. You also reflect on the way that science itself works, and how it impacts on society.

The opening page for each of the first 11 chapters in this book shows you how the three strands of the course are interwoven: the issues, the science behind the issues, and what a study of the issues tells you about science and society.

Studying the course

During this course you will take part in discussions, debate issues and form your own opinions. You are not expected to recall all the information in this book. The details of the issues and case studies are not as important as the scientific explanations and the ideas about how science works that lie behind them. You should focus on the questions in each chapter, which indicate the kinds of things you are expected to be able to do with the information provided, and the general ideas you are expected to be able to bring into your discussion.

The coursework allows you to keep up to date and explore your own interests. You choose which topical scientific issue you will study and you select a piece of popular science writing to read, enjoy and analyse.

How this book is organised

This book matches the structure of the specification for the two units in the AS *Science in Society* course. There are 11 chapters covering the corresponding topics in Unit 1 and a twelfth chapter with advice on how to approach and carry out the critical reading and study of a topical science in Unit 2.

The course specification, together with specimen examination papers, are all available from the Awarding Body AQA. The latest information is available on the AQA website (www.aqa.org.uk). You will find a commentary on the questions in this book on the project website (www.scienceinsocietyadvanced.org), where you will also find links to other websites with up-to-date and authoritative information about the issues which feature in the course.

Chapter 1

The germ theory of disease

The issues

In many parts of the world millions die each year from infectious diseases such as cholera, tuberculosis (TB) and malaria. In some parts of the world people are too poor to ensure that their water is safe to drink or to pay for the treatments which can prevent or cure diseases. Even in richer countries food poisoning makes many people very ill and can kill. People everywhere, even when they have been taught the germ theory of disease, ignore basic rules of hygiene and lay themselves open to the risk of infection.

Preventing and curing disease costs money, and across the world there are real issues about how money for healthcare is spent. In order to make the right decisions, and to fight infectious diseases whenever and wherever they occur, it is necessary to understand the causes of such diseases and how they are passed from one individual to another.

The science behind the issues

We now know that the 'germs' which cause infectious diseases are small organisms including bacteria, viruses and fungi (Figure 1.1). These microorganisms are present in the environment and can be passed from one infected individual to another. Under ideal conditions the body defends itself against these invading germs using the immune system, but often not before unpleasant symptoms are experienced. In some cases the patient is dead before the immune system has done its work. By developing an understanding of how diseases are caused and spread we can move towards preventing or curing them – though this is not always as easy as it sounds.

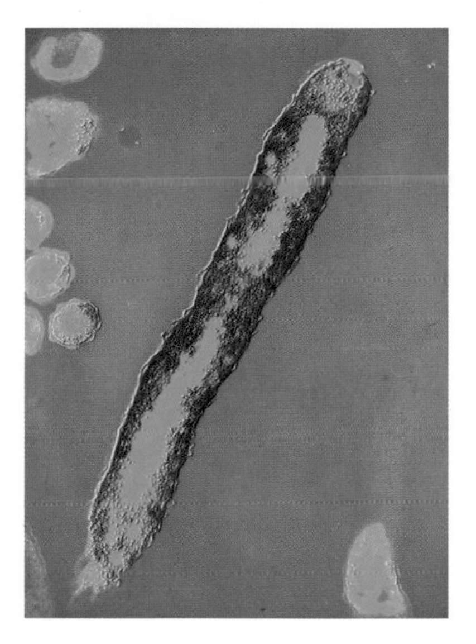

Figure 1.1

Vibrio cholerae *is the bacterium which causes cholera in humans. This false colour image from an electron microscope is magnified about 18 000 times.*

What this tells us about science and society

People have known about infectious diseases for centuries. But it was only in the nineteenth century that they began to understand how they are transmitted and what causes them. So having information (data) is not enough to guarantee that we will find a theory which can explain things. Coming up with a new theory needs creative imagination – to see patterns in the data and think up possible explanations. More than one explanation can account for the same data – so there is always room for disagreement and people's past experiences and wider commitments can influence their judgement. A scientist has to persuade the whole scientific community of his or her idea before it becomes accepted as 'reliable knowledge'.

The story of Ignaz Semmelweis

After giving birth, women are vulnerable to infections, but in the days when most women had their children at home attended by their female relatives, serious infections after delivery were relatively rare. In the late eighteenth and early nineteenth centuries, however, changes in the frequency and type of medical interventions in childbirth began to cost the lives of an increasing number of women.

Doctors became increasingly involved in the delivery of babies at home, and hospitals were set up with maternity wards for the 'safe' delivery of infants (Figure 1.2). These hospitals were used mainly by the poorer women who could not afford medical care in their own homes. The women began to die a horrible death. Within days or even hours of giving birth mothers developed a range of symptoms including pain and tenderness of the abdomen, together with a rapid pulse and high fever. Severe pain, inflammation of the womb, vomiting, convulsions and death followed within five days.

Figure 1.2

A women's ward of the Bridewell hospital in London in 1808. The beds are piles of straw. In the worst hospitals at this time the death rate from puerperal fever was as high as one new mother in every three.

This dreadful illness, known as puerperal fever or childbed fever, regularly claimed the lives of about one woman in every five who gave birth to a baby in the hospitals of European cities. Even when women delivered at home, the risk of fever rose – but it depended on who delivered the baby. Some doctors lost most of their maternity patients, while others never had a case of puerperal fever.

There were fierce debates about the causes of this mystery disease. In Britain, Charles White and, in America, Oliver Wendell Holmes tried hard to persuade people that it was doctors and nurses who were spreading the disease from patient to patient. However, the person we now remember for gathering the scientific evidence to show how the disease was spread was Ignaz Philipp Semmelweis, an Hungarian physician.

Semmelweis's evidence and idea

Ignaz Semmelweis was born in Buda, Hungary, on 1st July 1818. Soon after he qualified as a doctor he became an assistant at the maternity clinic of the Vienna General Hospital. The hospital had two delivery rooms, one staffed by female midwives and the other by medical students. Over 12% of the women delivered by the young doctors died of puerperal fever. This was over three times the percentage of women dying from the other delivery room. Semmelweis realised that the medical students were often dissecting a dead body as part of their training and then moving straight on to delivering a baby, without washing their hands first. He wondered if they were carrying the cause of disease on their hands from the corpses to their patients.

Then a colleague of his, Jacob Kolletschka, cut himself whilst carrying out an autopsy and subsequently died from symptoms identical to those of puerperal fever. For Semmelweis this confirmed his idea that puerperal fever was caused by some kind of infectious agent. He immediately insisted that his medical students wash their hands in chlorinated lime before they entered the maternity ward, and eventually he insisted that they should wash between each patient. Within six months the mortality rate of his patients had dropped to a quarter of the original figure, and after two years the death rate was down to only 1.3% of the women who gave birth in his wards.

Semmelweis presented his findings to other doctors. He was sure that they would recognise from his evidence that puerperal fever was spread from patient to patient by doctors. Yet in spite of the compelling evidence Semmelweis met with strong opposition. Eventually, in 1850 he left Vienna for the university hospital in Pest where, as professor of obstetrics, he was responsible for the care of mothers during childbirth. Again he enforced what are now recognised as antiseptic practices and the number of women dying from puerperal fever after having a baby in Pest fell to 0.85%. But yet again his findings and publications were resisted, not just in Hungary but also abroad.

Semmelweis's experience shows that experimental tests of a new theory may not be enough to convince others that the explanation is correct. Scientific data alone, however convincing it may seem to some, may not be enough to convince others of the truth of a new idea.

Questions

1. What were the 'factors' which Semmelweis was considering during his investigations?
2. What links between factors and outcomes did Semmelweis observe?
3. How did Semmelweis explain his observations?
4. Suggest, using your twenty-first century knowledge, why washing in chlorinated lime was effective in controlling the spread of puerperal fever.
5. Give two everyday examples of 'antiseptic practices' which people use at home to keep themselves healthy.

Opposition to Semmelweis

Pain and suffering during childbirth was an accepted part of European culture in the eighteenth and nineteenth centuries (Figure 1.3). It was hard for doctors to admit that they themselves had spread the disease and killed their patients instead of curing them. To change their point of view would mean accepting that the deadly disease was caused by a transferable agent.

Another factor might have been that hand washing probably seemed rather an odd practice at the time. There was no indoor plumbing, so getting water to wash in was not easy. Water brought in would have been cold, and the chemicals used to wash with (such as chlorinated lime) would have eventually damaged the skin of the hands. From the viewpoint of the twenty-first century it is difficult to imagine just how difficult such a simple procedure must have seemed in the nineteenth century. In those days surgeons operated in their outdoor clothes and did not wash their hands or their patients' wounds to prevent infection.

Even outside of hospitals, women died of puerperal fever. However, as the American Oliver Wendell Holmes pointed out, only the patients of certain doctors died. So if the doctor delivering a baby was clean, washed his hands and changed his clothes regularly, then the mother would probably live. If not, she had a high chance of developing puerperal fever.

Figure 1.3

An engraving by Holbein showing an angel, accompanied by Death, chasing Adam and Eve from the garden of Eden. In Genesis, the first book of the Bible, God says to Eve, 'I will increase your trouble in pregnancy and your pain in giving birth.' Some doctors preferred to accept this as an explanation of puerperal fever, rather than Ignaz Semmelweis's explanation in terms of infectious agents which he demonstrated clearly in two different hospitals.

Questions

6 Identify two reasons why doctors were unwilling to accept Semmelweis's ideas.

7 Why did the disease linger on in the UK, Europe and the USA until the middle of the twentieth century?

8 Suggest reasons why thousands of women in developing countries of the world still die each year from puerperal fever. How could these deaths be prevented?

Semmelweis found the rejection of his work unbearable, because he recognised that simple hygiene measures held the key to saving thousands of lives. By the 1860s he suffered a major breakdown and went to a mental asylum in Vienna. In 1868, aged only 47, he died from an infection picked up from a patient during an operation – an ironic twist of fate.

A lingering threat

Relying on hand washing alone leaves a great deal of room for human error. Dying of fever after childbirth remained a feared outcome until after the Second World War, which was when antibiotics became widely available (Figure 1.4). However, in some parts of the world even today, where knowledge of microorganisms is scanty and antibiotics are rarely available, mothers are still dying of puerperal fever in the days immediately after they have given birth.

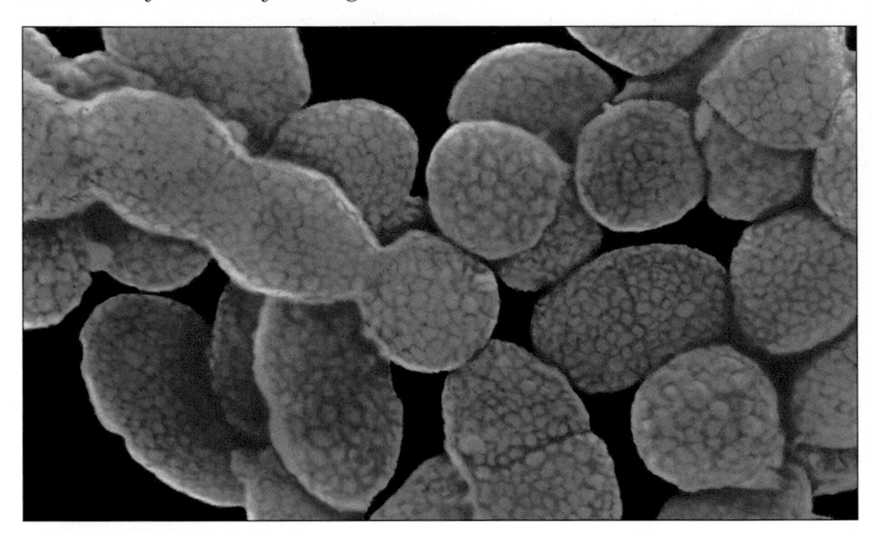

Figure 1.4

By the mid-twentieth century it was shown that puerperal fever was caused by the Streptococcus pyogenes *bacterium and that antibiotic drugs could destroy it. In developed countries the fear of death from puerperal fever was finally lifted from women and their families. This picture shows cells of the bacterium dividing. This strain is resistant to common antibiotics. The cells are magnified about 21 500 times.*

Cholera in London

A new disease

A number of severe cholera outbreaks gripped London and other parts of the country in the 1830s and 1840s – at the same time as Semmelweis and others were trying to convince people that puerperal fever was infectious.

Cholera was a new disease in Britain. People with cholera suffer severe stomach pains; they produce vast amounts of very watery diarrhoea and sometimes vomit as well. The disease spreads very quickly through a community. The cholera epidemic of 1831–32 killed 32 000 people in three months.

The people who suffered most from cholera were the poor in large towns who lived in overcrowded homes with poor sanitation. At the time there was no system for removing household rubbish other than throwing it into the streets.

Most doctors then thought that diseases such as cholera were spread either by touch or by 'bad air' (miasma). According to this theory it was the vapours or mists from rotting matter that could cause illnesses. One campaigner, Edwin Chadwick, used the miasma theory as the basis for a campaign to clean the rubbish from the streets and to build sewers (Figure 1.5).

Figure 1.5

The campaign to build sewers led to major engineering projects in London. This engraving shows the main northern sewer under construction in 1859. A programme to build 156 km of sewers was finally completed in 1865.

In this period John Snow was working as a doctor, first in Newcastle where he saw the terrible effects cholera on miners in the 1831 epidemic, and then in London. He became very interested in what was happening and made careful observations of all the cholera cases he came across. Snow traced one outbreak of the disease to the arrival of one sick seaman from Hamburg.

Snow came to the conclusion that cholera was actually caused by a poison which reproduced in the human body and was found in the vomit and diarrhoea of cholera patients. His hypothesis, as he stated it, was that 'disease is communicated by something that acts directly on the alimentary canal. The excretions of the sick at once suggest themselves as containing some material which being accidentally swallowed might ... multiply itself.'

In 1849 Snow published a pamphlet explaining his theory that the main way in which the disease was passed on was through water contaminated with an infectious agent. However, he was not the only person trying to come up with an explanation for the killer disease at the time. At this stage most people did not believe him and his pamphlet was largely ignored.

The Broad Street pump

In 1854 another major cholera outbreak hit London. Snow undertook a detailed record of all the cholera cases he could find. His meticulous documentation showed up a number of different and important features.

In one small neighbourhood under John Snow's surveillance the number of cholera deaths was terrifying. Within little over 200 metres of the junction of Cambridge Street and Broad Street, 500 men, women and children lost their lives in 10 days. By plotting the homes of all the cases on a local map he saw that they all got their drinking water from the same source – the Broad Street pump (Figure 1.6). For Snow this confirmed his ideas. He persuaded the panic-stricken local officials to have the handle of the pump removed. Once this was done the epidemic was contained and began to subside.

Key term

An **epidemic** is an outbreak of an infectious disease when many people are infected at the same time.

Questions

9 Which aspects of the cholera epidemic could be explained by the 'bad air' theory?

10 Edwin Chadwick argued that sewers should be regularly flushed with water to get rid of the smells. In London this washed the sewage into the Thames.
a) Why did Chadwick think that this would improve public health?
b) In fact it may have increased the incidence of cholera. How does John Snow's theory explain this?

11 Snow introduced the idea that the infectious agent might be able to multiply itself. Why was this idea an important feature of his theory?

12 In the 1830s, microscopes were powerful enough to see tiny organisms in water but they were not good enough to detect the infectious agent that causes cholera. What influence do you think this had on the early rejection of Snow's theory?

Figure 1.6

When John Snow plotted all the cholera deaths in a neighbourhood he found that they clustered around one water source – the Broad Street pump.

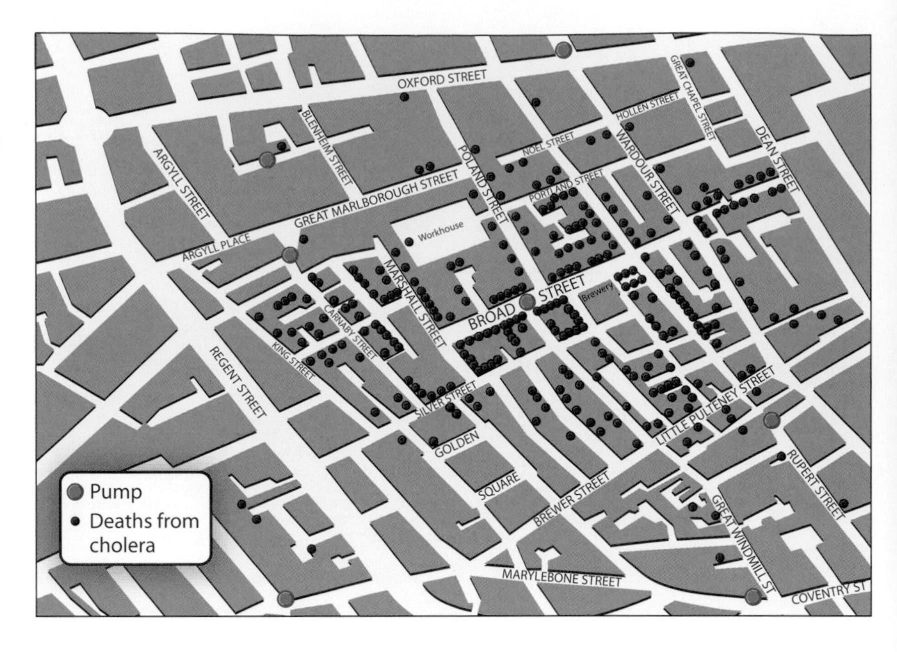

Snow's records showed him something else as well. In the same year he had studied a region of South London where houses in the same streets were supplied by two separate water companies. The Lambeth Water Company took water from the upper reaches of the Thames, before it flowed through the city. The Vauxhall Water Company took water from the lower Thames, after it had passed through London and received most of the city's sewage. People supplied with water by the Vauxhall Water Company were far more likely to develop cholera. As John Snow wrote of the Vauxhall supply in the *Medical Times* of 1854: 'Part of the water has passed through the kidneys and bowels of 2 million and a quarter of the inhabitants of London.'

We now know that John Snow's description of the disease was very near the truth. His way of working was an early example of epidemiology, which is the study of the patterns of incidence of a disease. Epidemiology is the branch of medical research which unravels the causes of disease by gathering data from large samples. Epidemiologists test hypotheses suggesting connections between the incidence of the disease and aspects of lifestyle (see Chapter 8).

Confirmation of John Snow's explanations emerged during 1883 in a cholera epidemic in Cairo, when Robert Koch identified the cholera bacterium in the victims of the disease, in water and in food.

Questions

13 Identify two ways by which Snow demonstrated a link, or correlation, between the type of water supply and the incidence of cholera.

14 Why was the removal of the handle of the Broad Street pump so effective in persuading some people that Snow's theory was valid?

15 Why did Snow's studies of water supplies in South London support his theory but not the miasma theory?

16 Why did Koch's discovery of the cholera bacterium help to convince people that Snow's theory was correct?

Pasteur and microbes

The work of Semmelweis and Snow led gradually to the acceptance of the idea that infectious diseases are brought about by an infectious agent (or 'germ') which is transferred from one individual to another and causes disease. But entrenched views of the causes of disease were hard to shift. It took many years of work by many people working in many countries before there was widespread acceptance of the theory. Two people who are now remembered for their part in establishing the theory were a French scientist, Louis Pasteur, and a German doctor, Robert Koch.

Figure 1.7

Pasteur dictating notes to his wife. Louis Pasteur was a major figure in the development of the germ theory of disease. His wife recorded his dictation, wrote up his notes and discussed his ideas with him.

Pasteur was born in 1822 and lived until the age of 72. Pasteur was always ambitious to make scientific discoveries. As a professor of chemistry at Lille he became interested in fermentation because of problems encountered by local vinegar makers. There was a new theory that fermentation was caused by microscopic yeast cells and was not simply a chemical reaction. Pasteur studied fermentation in great detail and produced the evidence to persuade people that the yeast theory was correct. As a result he became very interested in microbes and where they come from.

At the time there was a widespread belief that living things could arise from dead things. This was the theory of spontaneous generation. A number of scientists had studied the problem, but there was no consensus. This was despite the work of the Italian biologist Lazzaro Spallanzani, who had carried out a series of experiments in 1768 which he claimed showed that microbes must develop from other microbes and not by spontaneous generation.

In 1859 Pasteur decided to join the debate. He realised that it would be difficult to prove that spontaneous generation never happened but he intended to show that there is no evidence that it happens. This led to his series of classic experiments with swan-necked flasks, showing that the microorganisms which grew in broth, turning it cloudy and mouldy, did not appear by spontaneous generation but were already present in the air (Figure 1.7).

In 1866 Pasteur was asked by the Department of Agriculture in France to investigate killer diseases that were destroying large numbers of silkworms in the silk industry. Pasteur's careful observations allowed him to identify silkworms that were infected and should not be used for breeding. He devised a system for culturing eggs only from healthy moths. By keeping the healthy eggs away from all contact with living silkworms he made sure the eggs would produce moths free of disease (Figure 1.8).

This work had an important influence on Pasteur's thinking about infectious diseases. He saw that healthy worms became infected when allowed to nest on leaves used by infected worms. He also found that some worms died shortly after infection, others some weeks later and some not at all.

Key terms

Microbes (or microorganisms) are minute living beings which are only visible with the help of a microscope. We now know that yeasts are the microbes which cause fermentation and bacteria are the microbes which cause diseases such puerperal fever and cholera.

A **germ** is a microbe which can cause disease.

Questions

17 Why did Pasteur realise that it would be difficult to prove that spontaneous generation *never* happened?

18 What is the significance of the observation that healthy silkworms become infected if they nest on leaves previously used by infected worms?

Figure 1.8

Drawing of a healthy silkworm from Pasteur's book Diseases of Silkworms.

Figure 1.9

Robert Koch, German bacteriologist (1843–1910).

Koch and bacterial diseases

Pasteur had established the germ theory of disease but he did not know how to identify the different kinds of germ. In 1866, when Pasteur was already in late middle age, the young German doctor Robert Koch became interested in examining microbes with the more powerful microscopes which were becoming available (Figure 1.9).

Koch devised the techniques for growing cultures of bacteria on agar jelly. He also worked out how to study bacteria by using dyes to stain them on glass slides so that they could be seen and recognised under a microscope. With these methods, he and his fellow workers discovered the causes of 11 diseases – including anthrax (in 1863), tuberculosis (TB) (in 1882) and cholera (in 1883) (Figure 1.10).

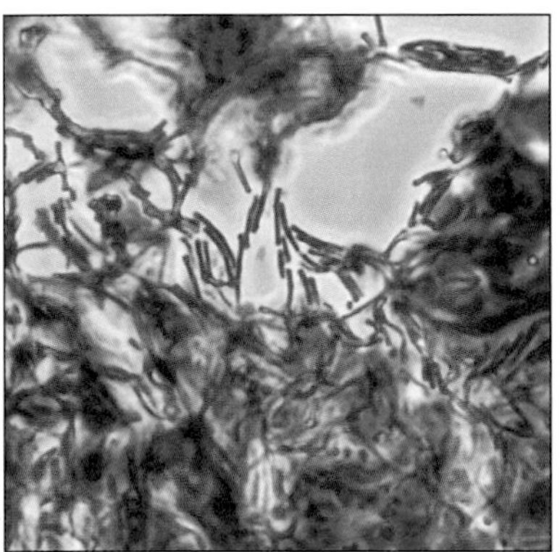

Figure 1.10

Photo taken through a light microscope showing tissue infected with the bacterium Bacillus anthracis. *The dye added to the specimen dyes the bacterial cells blue. Robert Koch was the first person to isolate this bacterium, which causes anthrax. Magnified about 400 times.*

Koch observed the germs multiplying and investigated the conditions which would stop them reproducing. In many ways he was the father of the science of bacteriology and in 1905 he was awarded the Nobel Prize for Medicine. By the end of the nineteenth century the idea that infectious diseases were caused by germs was almost universally accepted.

Resistance to infection

Doctors and medical science have played their part in the conquest of disease, but even now many diseases cannot be cured. Doctors are often only able to speed recovery or relieve discomfort.

In John Snow's time most deaths, especially deaths in childhood, were caused by infectious diseases such as typhoid, diarrhoea, whooping cough, measles, scarlet fever and tuberculosis. These diseases flourished in big cities where the children of the poor were underfed, lived in crowded homes with poor sanitation and drank polluted drinking water.

Question

19 Suggest why finding a way to see and distinguish different kinds of bacteria helped to convince people that the germ theory of disease was correct.

At first, remedies arose from a better understanding of the germ theory of disease leading to policies to limit infection. Success depended on social action by politicians and others to relieve poverty and improve housing. Sanitary engineers played their part by installing piped water supplies and constructing enclosed sewers. Nutritionists helped too, by means of research which helped to develop guidelines for the diet of children. All this activity greatly cut down the deaths from infectious diseases long before vaccination and antibiotics were available.

When people are healthy they have considerable resistance to disease. As Figure 1.11 shows, the human body can defend itself against infection.

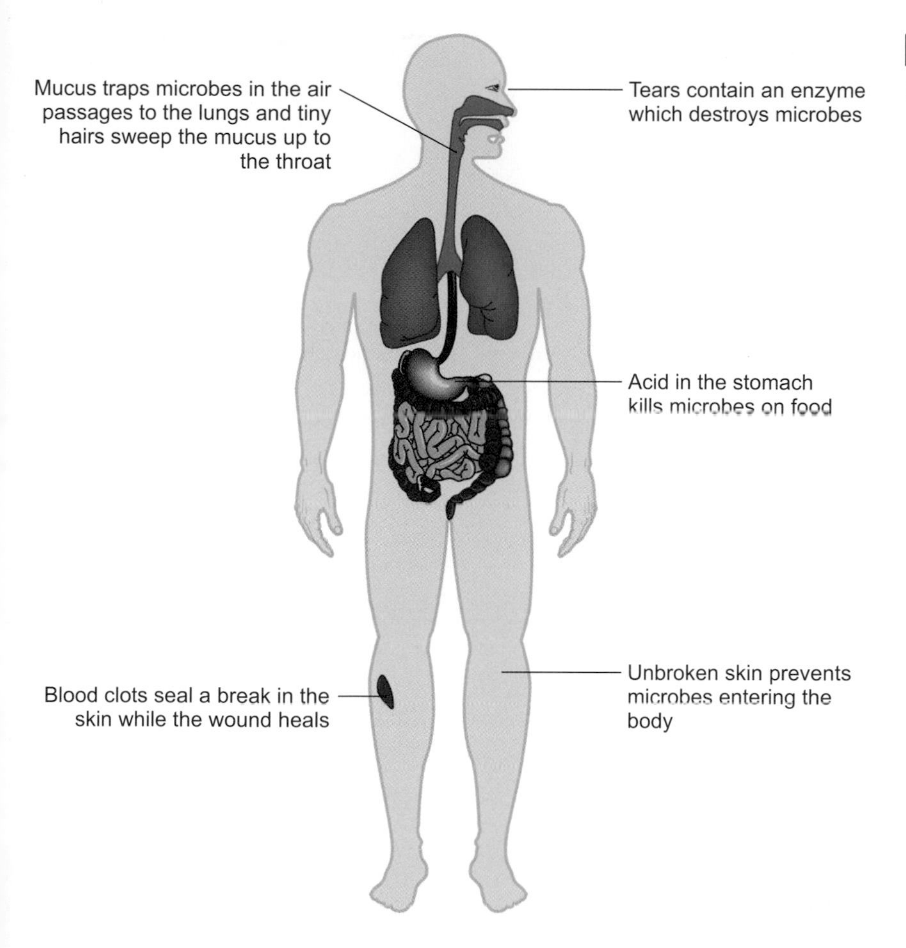

Figure 1.11

Some of the barriers to prevent infection.

People do not necessarily fall ill even if microorganisms manage to break through and invade their bodies. The immune system provides the next line of defence.

Immunisation

Immunisation helps to prevent sickness by using the body's own immune system to fight infection. The first disease to be tackled in this way was smallpox. Many people died if they caught smallpox and even if they survived they were disfigured with pox scars (Figure 1.12).

It has been known for hundreds of years that once someone has had smallpox they cannot catch the disease again. This led to experiments to explore the possibility of immunising people with a mild attack of the disease. As long ago as 1718, Lady Mary Wortley Montagu came across this approach to prevention in Turkey and returned to England to try out the idea. The first people tested were criminals from Newgate prison under

Questions

20 Give two examples to explain why the body's defences against infection sometimes fail (see Figure 1.11).

21 Cholera and other diseases that cause diarrhoea are still the main killers of children throughout the world. Why do people still die of cholera, and why do you think children are particularly vulnerable?

Figure 1.12

A photo taken in 1973 showing a young Bangladeshi child with smallpox in a relief camp. Smallpox was caused by the variola virus. The patient's body was covered by large spots. Survivors of the disease were usually scarred for life.

Key terms

Immunisation is the procedure for making people immune, or resistant, to a specific disease. The term is often used interchangeably with vaccination.

The **immune system** is the system of organs and cells that help to defend the body against infection and fight disease.

sentence of death. They were offered their freedom if they agreed to be immunised with smallpox. All survived and were set free. But it turned out that the technique worked with some people but not with others, who instead developed a serious attack of the disease and died or infected others. As a result this method of immunisation was made illegal in England by 1840.

Jenner and cowpox

The story of how smallpox was defeated began in the county of Gloucestershire, England, where the country doctor Edward Jenner (1749–1823) practised medicine. His ideas were based on his everyday experiences of visiting the sick among the farming community.

The local farmers probably set Jenner's mind to work on the problem of smallpox. They told him that the girls who milked the cows often suffered from the mild disease cowpox but rarely caught the more severe and often fatal smallpox. Cowpox comes from handling cows and causes spots, especially on the hands.

Spots are also a symptom of smallpox, and Jenner wondered if deliberately infecting people with cowpox would protect them from smallpox. Jenner did not test his theory until 1794, which was a year in which one in five of *all* reported deaths were the result of smallpox.

He took pus from cowpox spots on the hands of Sarah Nelmes and, with a needle, scratched it into the arm of James Phipps, a healthy boy (Figure 1.13). The boy recovered from the cowpox, but the experiment was not yet complete. Two months later Jenner went one risky step further and scratched pus from the spots of someone suffering from smallpox into the boy's arm. Fortunately, the boy did not develop smallpox. His survival helped confirm the idea that exposing a person to a mild dose of a disease stimulated the person to resist its more serious form.

Jenner had shown that infecting someone with a mild disease could give protection against a fatal illness. The pus from cowpox spots was the first example of a vaccine – Jenner called the process vaccination, from the Latin name for a cow (*vacca*).

Figure 1.13

Drawing of a statue of Edward Jenner, showing him infecting James Phipps with cowpox.

Figure 1.14

This Gillray cartoon illustrated people's nightmares about vaccination, suggesting appalling results.

Key term

Today the term **vaccination** means the injection into the body of killed or weakened organisms to give the body resistance against disease. The word comes from the Latin for cow, and reminds us of the importance of cowpox in giving medicine such a powerful weapon with which to prevent disease.

Pasteur and chicken cholera

Following his successful investigations into fermentation and the disease of silkworms, Pasteur had established a substantial reputation for solving practical problems and farmers sought his advice. He knew about Jenner's work and wondered whether or not vaccination could be used to prevent other diseases.

In 1879 Pasteur was studying a disease called chicken cholera. He had identified the microbe which causes the disease and succeeded in getting it to grow as a cell culture in a special medium in his laboratory. He and his assistants were injecting healthy chickens with the bacteria and showing that microbes grown outside the body did infect healthy chickens so that they rapidly became ill and died.

One day, by mistake, an assistant injected a batch of chickens with a culture of microbes which they had grown but had then left standing for several weeks during a holiday. The chickens were ill for a time but then recovered. The assistant was about to ignore this experiment; Pasteur, however, told him to give the chickens a second injection – this time using freshly grown microbes. The birds stayed healthy.

Further experiments confirmed Pasteur's idea that the microbes which had stood over the holiday had become weakened. They were only able to give the chickens a mild form of the disease. Once chickens had recovered from the mild infection they became immune to the fresh injection. Pasteur had discovered the basis of vaccination using a weakened (attenuated) form of the infective agent.

Questions

22 What fears are illustrated by Figure 1.14? How do they compare with the worries people have about vaccination today?

23 a) What was Jenner's theory?
b) How did he test his theory?
c) What part did prediction and experiment play in his discovery?

24 Pasteur once famously said that; 'In the field of experimentation, chance favours only the prepared mind'. Show that this is illustrated by his work with chicken cholera.

Pasteur and anthrax

Following his work on chicken cholera, Pasteur set out to develop a vaccine for anthrax. Anthrax is a disease of animals which is highly infectious. In the 1870s up to 50% of all the sheep and cattle in France were dying of the disease. Animals have a strong chance of catching the disease if they graze above the burial sites of animals that have died of anthrax. Pasteur showed that an infectious agent was being brought up from the buried corpses by worms.

Pasteur enjoyed publicity and was willing to make enemies in the medical profession. He enjoyed convincing his opponents that they were wrong. His quest for a vaccine for anthrax gave him a famous opportunity to do so. By this stage in his life Louis Pasteur had suffered a very severe stroke. Although he recovered and his mind was unaffected, his speech, gait and ability to use his hands were never the same again – for the rest of his life Pasteur relied heavily on his team of trusted fellow scientists to carry out the experiments he dreamed up.

Pasteur and his team were confident that they could find a way of beating anthrax – but it proved more difficult than they thought. The anthrax germ proved very hard to grow in the laboratory.

Meanwhile Robert Koch had developed a way of culturing anthrax spores. Pasteur immediately used this new technique to grow anthrax bacteria and then tried to make a vaccine. None of the methods he attempted seemed to give reliable results.

Then Toussaint, a young vet, claimed to have produced a successful vaccine using a different method. Pasteur claimed that it was unreliable, and quickly announced his own vaccine was on the way (produced in a very similar way to that described by Toussaint). What Pasteur and his assistants did was to keep anthrax bacteria warm at just over 40 °C. After eight days the microbes were much weakened and no longer able to cause the fatal disease. They could, however, be used for vaccination.

As a senior and highly respected scientist, Pasteur's version was accepted and Toussaint retired, a broken man. But Pasteur's steady work on anthrax was then interrupted. Hippolyte Rossignol, a vet who had little time for the germ theory of disease, threw down a very public challenge to Pasteur. Rather than appear unsure of his work, Pasteur accepted the challenge, although he was not at all certain his vaccine would work.

The trial took place on Monsieur Rossignol's farm at Pouilly-le-Fort (Figures 1.15 and 1.16). A crowd gathered to watch the start of the experiment. The farmers, vets and doctors returned on 2nd June and applauded Pasteur as he arrived to find all the vaccinated sheep alive and well while all the unvaccinated sheep were dead or dying.

After the success of the trial at Pouilly-le-Fort, Pasteur completed the development of his vaccine against anthrax. The vaccine had a major effect on farming for generations to come – but the confirmation of the germ theory of disease was to have even greater implications for the health and wellbeing of people all over the world.

Questions

25 Why was the trial at Pouilly-le-Fort such an effective demonstration of the germ theory of disease?

26 Pasteur's version of an anthrax vaccine that was accepted rather than the one proposed by the young vet Toussaint.
a) Why do you think this was the case?
b) Suggest possible errors which might arise from taking into account the reputation of scientists when deciding whether or not to accept their ideas.

Figure 1.15

Louis Pasteur performing his anthrax vaccination experiment at Pouilly-le-Fort in 1881.

5th–17th May 1881

25 sheep given Pasteur's anthrax vaccine

25 sheep left unvaccinated

31st May 1881

All sheep injected with virulent anthrax spores

Pasteur predicts vaccinated sheep will survive

2nd June 1881

Vaccinated sheep all alive

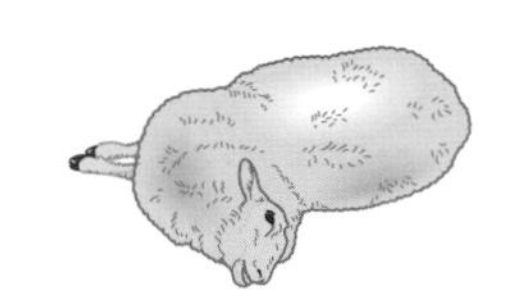

Unvaccinated sheep all dead or dying of anthrax

Figure 1.16

Pasteur's success at Pouilly-le-Fort silenced almost all critics of his germ theory – the evidence was there for all to see.

Questions

27 Identify similarities and differences between Jenner's vaccine for smallpox and Pasteur's vaccines for anthrax and rabies.

28 Pasteur's opponents argued that Pasteur was risking giving people rabies by his injections and this was not justified because most people bitten by rabid dogs did not die. What is your view of the argument that Pasteur should have spent more time developing the treatment before trying it on people?

29 At the time that Pasteur was studying rabies there was no other research centre in Europe that could repeat his work. Why did this make it harder for Pasteur to convince other scientists and doctors that his rabies vaccination was effective and safe?

Pasteur and rabies

Soon Pasteur was to face the challenge of showing that his methods were a safe and successful way to protect people, as well as animals, from disease. He and his assistants had developed a technique of vaccination to protect dogs from rabies. This took a long time but eventually they discovered that they could prepare an effective vaccine by drying out parts of the spinal nerves from rabbits which had been infected with rabies.

When a rabid dog bites people they develop a fatal disease which leads to an agonising death. This human form of rabies is called hydrophobia. In 1885 a young boy in Alsace was badly bitten by a dog with rabies. His mother had heard of Pasteur and his work. She travelled with her boy to Paris as quickly as possible at the time to ask Pasteur to help her son.

Pasteur and his assistants were very nervous of trying their treatment on a human being, but they knew that if they did nothing the boy would die. Fortunately he survived. News of this success brought many others to Paris in search of a cure (Figure 1.17). Not all those who were treated survived, often because they were vaccinated too long after they were bitten. In the early months only 10 people died out of the 1726 people treated, and this was at a time when 16 out of every 100 people died if they were bitten by dogs with rabies.

This was the first example of a laboratory-based medical breakthrough which became a major news story. Pasteur's success established microbiology as a scientific discipline. His work on rabies led, in the late 1880s, to the creation of the Pasteur Institute as a centre for research.

Figure 1.17

Pasteur with English children who were sent to him after they had been bitten by dogs.

Review Questions

30 Use examples from this chapter to show how the scientific explanations (a)–(d) helped to make sense of the causes of disease or the effectiveness of immunisation to treat disease.

a) Living things are composed of cells.
b) Many diseases are caused by microbes.
c) The human body can defend itself against infections with its immune system.
d) An individual who survives an infection by a particular microbe is protected against future invasion by that organism.

31 Give at least one example from this chapter to illustrate each of these ideas about how science works:

a) If something happens only when a factor is present, we say there is a correlation between the factor and the outcome.
b) To investigate the relationship between a factor and an outcome, it is important to control all other factors that might affect the outcome.
c) We are more likely to accept that a factor causes an outcome if we can identify a plausible mechanism to link them.
d) Many scientific theories involve objects that cannot be directly observed.
e) Scientific theories do not 'emerge' from data by a process of logical deduction; proposing an explanation always involves imagination and conjecture.
f) Scientists test an explanation by seeing if predictions based on it are in agreement with data from planned observations or from an experiment.
g) The topics that scientists choose to work on are strongly influenced by previous work of other scientists.
h) Scientists are motivated by the desire to be the first to a new discovery.

Chapter 2

Infectious diseases now

The issues

Governments have to consider how best to deploy the resources available for healthcare. The answers differ from one country to the next. Primary healthcare and programmes to prevent disease can be more cost-effective than building hospitals and setting up expensive specialist services, but political pressure from those with money and influence in society may mean that the interests of the better off dominate over the needs of the majority.

The science behind the issues

If we know which microbes cause a disease, and where these can be found, then the best way of preventing disease is to eliminate the causes of infection. Vaccination is also a very effective way of preventing disease because it stimulates the immune system to produce antibodies which protect against future infection. Vaccination, however, has its limits against diseases such as influenza because new forms of the virus keep appearing and each variety needs its own vaccine.

The discovery of antibiotics gave doctors the power to cure as well as treat the symptoms of bacterial diseases. A growing worry is that tried and tested antibiotics are beginning to fail as more and more bacteria develop resistance. Overuse of antibiotics has accelerated the evolution of resistant bacteria that are no longer killed by the drugs.

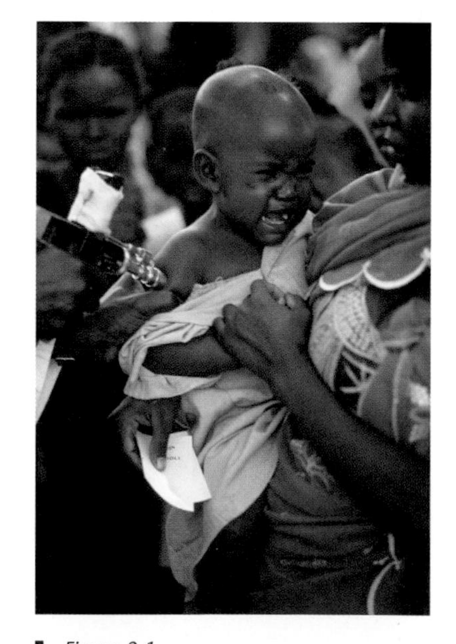

Figure 2.1

A child being vaccinated.

What this tells us about science and society

Decisions about whether or not to apply scientific knowledge can highlight the tension between the rights of individuals and the interests of society as a whole (see Figure 2.1).

Vaccination programmes to prevent diseases such as measles or whooping cough are only fully effective if nearly all babies are immunised. Yet there are parents who feel strongly that the risks of vaccination are too high. If too many parents ignore the advice of doctors they put at risk the health of many babies because they increase the chance of an epidemic. Science alone cannot resolve such issues.

The UN World Health Organization (WHO) applies science to monitor diseases that threaten health and lives worldwide, such as tuberculosis, influenza and HIV infection. The WHO works with governments to develop strategies to protect people from infection and fight the spread of disease.

Microbes that cause disease

The work of scientists such as Semmelweis, Snow, Pasteur and Koch led to general acceptance of the germ theory of disease (see Chapter 1). This theory has moved much further forward since then, helped by many developments in science and technology.

Infectious diseases are still widespread all over the world and affect not only humans but also all other animals and plants. They range from minor inconveniences such as the common cold and mild sickness through to devastating and fatal illnesses such as AIDS, yellow fever and malaria. People catch these diseases when an infectious agent (or microbe) invades their bodies. Microbes cause tissue damage as they reproduce themselves in or on the body, producing the symptoms of disease. In some cases, the damage and symptoms are not caused directly by the microbe, but by a toxin which the microbe produces. For example, the symptoms of cholera are caused by a toxin which affects the lining of the gut.

> **Key term**
>
> **Toxins** are poisons produced by living cells or organisms.

Cells as the basic units of living things

Figure 2.2

All living organisms are made up of cells. This was discovered in the very early days of the microscope. Cells carry out all the basic functions of life, such as making new proteins and materials for growth, transferring energy from food in respiration, replicating to form new cells, and getting rid of waste material.

The mechanisms by which these processes take place are similar in all living organisms.

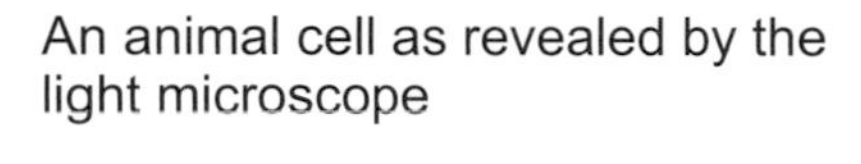

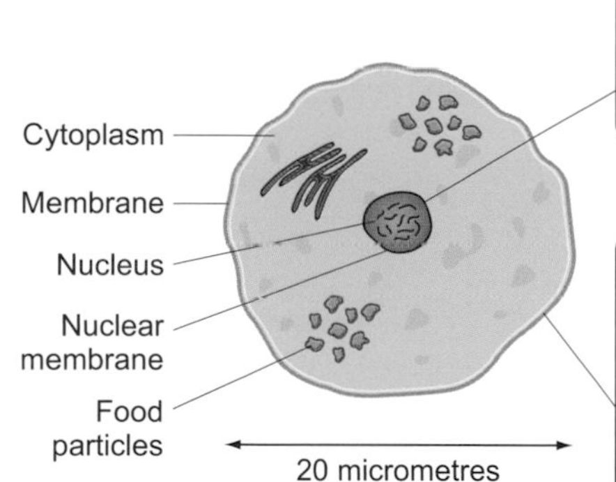

All cells have features in common, and ways in which they differ. Animal cells have a nucleus which contains the genetic material (in the form of DNA), which controls how the cell works.

Chemicals can enter and leave the cell through the membrane. Incoming chemicals are the ones the cell needs for respiration and growth, including oxygen and glucose. Outgoing chemicals are waste products such as carbon dioxide.

Bacteria, viruses and fungi

The microbes that cause disease include bacteria, fungi and viruses. Bacteria are single-celled organisms (Figure 2.3). Most bacteria are either harmless or beneficial. Some bacteria invade human tissues and reproduce, causing disease. Examples of bacterial diseases include tonsillitis, tuberculosis (TB) and cholera.

Figure 2.3

Very small they may be, but some bacteria can wreak havoc on the human body and cause death in a matter of hours.

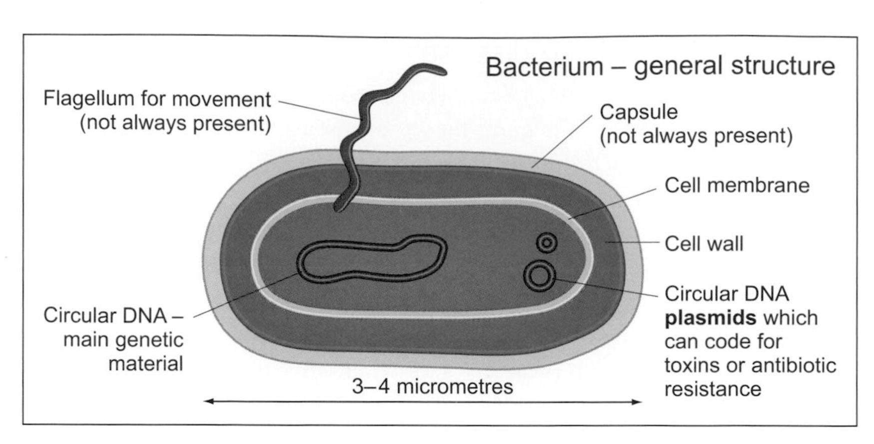

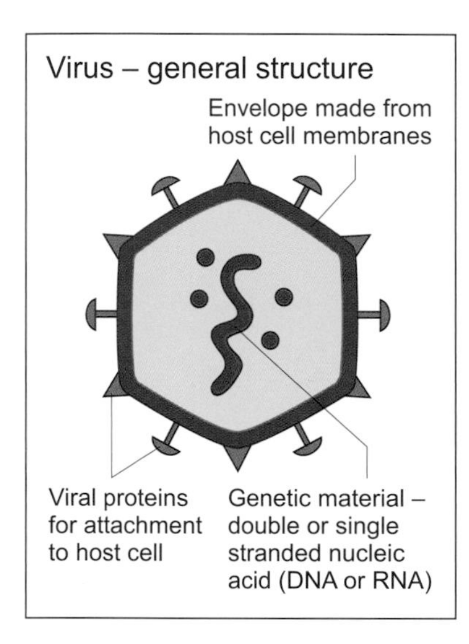

Figure 2.4

Even smaller than bacteria, viruses nevertheless cause some of the deadliest diseases to affect the human race. Viruses range in size from 10 to 50 nanometres (a nanometre is 10^{-9} m).

Pasteur, Koch and others could see bacteria with light microscopes, but it wasn't until the electron microscope was developed that scientists could identify viruses. It turns out that viruses are not really independent organisms. They are 'packets' of DNA, along with some enzymes, contained in an outer case of protein (Figure 2.4). Viruses can only multiply by invading a healthy cell. Once inside the host cell the virus takes over the cell's biochemistry and uses it to produce new copies of itself (Figure 2.5). This process continues until the host cell structure completely breaks down, releasing new viruses to infect other cells. It is this cell destruction and the body's reaction to it which gives rise to the symptoms of viral diseases such as polio, influenza and HIV infection.

Fungal infections can be superficial, such as athlete's foot, ringworm and thrush. They can also cause serious diseases of internal tissues and organs. *Aspergillus*, for example, is a fungus that causes lung infections.

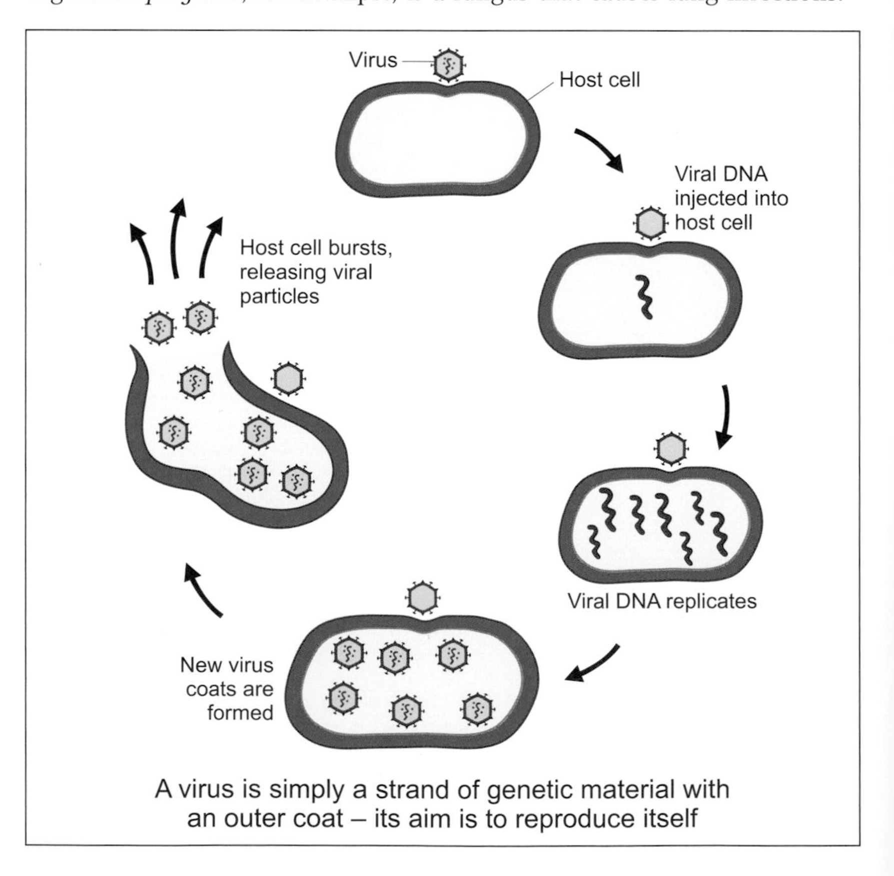

Figure 2.5

A virus multiplying inside an infected cell.

Aspergillus lung infection is an occupational disease (see Chapter 8) associated with horticultural workers who inhale peat dust. Fungal infections which invade the body can threaten the lives of people who are already weakened by diseases such as leukaemia or the late stages of HIV infection.

The cells of bacteria and fungi and the structure of viruses are quite distinctive. Often a doctor can diagnose a disease simply from the symptoms, but sometimes pathologists have to culture the infected tissue to identify the microbe causing the infection. (*Culture* means growing the microbe in a laboratory.) Because the different kinds of microbes are so different it is relatively easy to tell them apart with microscopes.

Preventing the spread of disease

Once the germ theory of disease was accepted, people could really begin to understand how infectious diseases spread from one person to another. There are all sorts of ways in which microbes can be passed on – and knowing this can help people avoid spreading disease.

- Living creatures can transmit infection from one person to another. Mosquitoes, for example, spread malaria as insect vectors.
- Materials such as clothing and bedding can carry microbes from one person to another. For example, *Staphylococcus* infections are often caught from hospital bedding.
- Direct contact is often important in the spreading of skin diseases such as impetigo and sexual diseases such as syphilis.
- Whenever someone coughs, sneezes or talks millions of droplets containing microbes are expelled from their respiratory tract, to be inhaled by someone else. Diseases such as influenza, measles and tuberculosis are spread like this.
- Many of the microbes which cause gut diseases are transmitted by contaminated food, drinking water or by the hands of infected people if they are not careful about hygiene. Most diseases causing diarrhoea and vomiting are spread in these ways, including cholera.
- Microbes can also enter the body directly through cuts in the skin or wounds – for example, hepatitis B, HIV, rabies and tetanus.

Questions

1. Suggest why the development of more powerful microscopes led to the final acceptance of the germ theory of disease.
2. In a disease caused by bacteria the symptoms tend to come on gradually, steadily getting worse. In diseases caused by viruses the symptoms come and go – for example, the temperature goes right up, then falls a bit, then shoots up again. Suggest an explanation for these differences.
3. Use the information about the way diseases are spread to explain:
   a) Semmelweis' observations about childbed fever (see pages 2–4);
   b) Snow's observations about cholera (see pages 5–6).
4. Outline, for one of the ways of spreading disease, how you might set about showing that there is a causal link between the method of transmission and people getting a disease. How, for example, could you show that people catch skin infections from infected clothing?
5. For each of the six ways that diseases can spread, suggest one public health measure that can help to prevent infection.

Key term

An **insect vector** is an insect species which transmits human disease. The mosquito is an important insect vector, transmitting malaria and dengue fever. Flies are vectors for sleeping sickness.

Before the discovery of antibiotics and other effective drugs that could cure diseases, the only effective ways of stopping infection were through public health measures. There was a long tradition of policies to prevent epidemics of diseases such as the plague, based on the theory that infections happened through the air. The public health authorities believed that good ventilation and fresh air were important. They tried to avoid contact between people by closing public meeting places and putting people with highly infectious diseases into quarantine.

Once people understood better the various ways that diseases can spread it became possible to plan and implement other more effective public health measures. These included better housing, better sewage disposal and the supply of treated drinking water.

Preventing infection by immunisation

Vaccination

The human body has an immune system which helps to fight disease. When microbes enter the blood, white blood cells recognise these cells

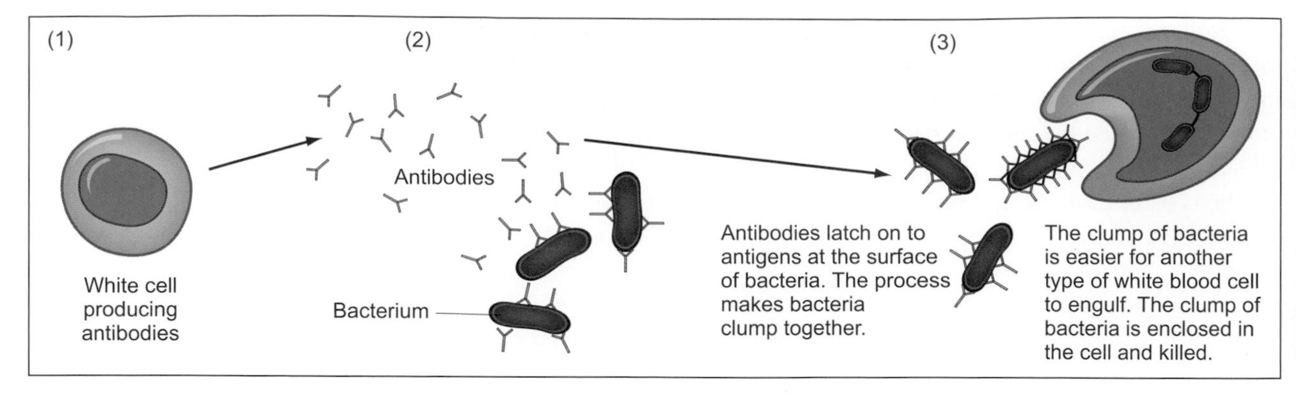

Figure 2.6

How white cells destroy invading bacteria. This diagram is not to scale. Antibodies are Y-shaped protein molecules which bind to the cell surface.

Key terms

Tears, stomach acid, white blood cells and the antibodies made by some white cells, are part of the body's **immune system** to prevent and fight infection.

White cells produce proteins called **antibodies**. Antibodies bind to a specific bacterium or virus so that it can be destroyed by other cells of the immune system. Each infection by a new microbe stimulates production of a new antibody for that particular microbe.

as 'foreign' and set about destroying them. Some white cells attack the microbes directly. There are also white cells which produce antibodies that bind to the invading microbes. This makes it easier for the other white blood cells to recognise and kill them (Figure 2.6).

The white blood cells which responded to a particular microbe stay in the body for months or years. This means that, after infection, people become immune to further infection by the same disease. The body can respond quickly to a second infection by a particular microbe, making the white blood cells that produce the specific antibodies to fight the infection.

These natural defences against infectious diseases are much weaker in young children and older people. Diarrhoea and respiratory illnesses can kill undernourished children. Others vulnerable to infection include old people, people who are stressed or weak from recent infections and people who damage their body through drug and alcohol abuse.

Moderate exercise has been shown to boost the immune system, although extreme training can have the opposite effect. A good diet and healthy lifestyle is perhaps the best protection against infectious diseases.

The defence mechanisms of the immune system can be stimulated artificially by immunisation (see Chapter 1). This is also called vaccination. A vaccine gives immunity to a specific infection.

Vaccination policy and safety

The elimination of smallpox

The development of vaccines has made a major contribution to public health programmes designed to promote health and prevent disease. Vaccination gradually became more popular in the UK after Jenner's demonstration that the procedure could be safe and effective (see page 10).

By 1853 the smallpox vaccination had become widely available in the UK and was made compulsory for infants before they were three months old. Smallpox was such a fearful disease (Figure 2.7) that most people were willing to overcome any fears of vaccination, and of the serious side effects of the vaccination. The proportion of children vaccinated, however, was never close to 100%. Compulsory vaccination ended in 1946.

By 1971 the chance of infection in Britain was so low that routine vaccination for children was no longer recommended. In May 1980 the WHO was able to declare that the world was free of the scourge of smallpox. The eradication of smallpox is the greatest triumph so far for vaccination programmes. In the early 1950s, smallpox infected 50 million people a year, disfiguring many of those who did not die. Mostly through the efforts of the WHO, many millions of dollars were spent in a worldwide vaccination programme between 1967 and 1977. Even so, the cost of the vaccination programme was less than the cost of treating people and imposing quarantine restrictions to help prevent the disease spreading.

Vaccination against other diseases

Vaccination programmes mean that diseases such as polio, measles, mumps, chickenpox, rubella and typhoid are now much less common than they were in countries such as the UK. Immunisation helps to prevent the spread of disease so long as a high enough proportion of people are immune. A small number of vulnerable people are effectively protected if they only meet people who cannot catch the disease and pass it on. The benefits of mass vaccination programmes are, however, put at risk if too many parents are unwilling to have their children vaccinated.

The necessary level of vaccination depends on how infectious the disease is. Measles, for example, is highly infectious. Nearly everyone in a community catches measles if most people have not been immunised. At least 90% of children must be vaccinated year by year to stop the disease spreading.

Compared with measles, meningitis is much less infectious. Up to a quarter of young adults may carry the microbes in their noses and throats but the number of actual cases is quite low.

Risks of vaccination

There is no such thing a perfect vaccine which protects everyone who receives it and is entirely safe. There are three questions which a parent might ask before allowing a child to be vaccinated:

- Is the vaccine really effective in preventing disease?
- What are the possible side effects and what is the chance that my child will be affected?
- What will the authorities do to compensate if my child suffers lasting damage from a vaccination recommended as part of public policy?

Some effective vaccines can produce side effects which are not serious and which clear up quickly. In some cases the side effects are not caused by the vaccine but are the result of human error or are just coincidences. Where there is a possibility of more serious side effects it is seldom possible to predict for sure which children are likely to be affected.

Medicines to treat and cure disease

The development of the modern pharmaceutical industry has been based on remarkable developments in techniques of analysis and synthesis, which now allow chemists and biochemists to model and manipulate the detailed structure of complex molecules (see Chapter 4).

Until the mid-1930s the pharmaceutical industry was small, producing mainly simple chemicals and extracts from plants. Local pharmacists mixed the ingredients to make liquid mixtures, pills and ointments. Most of these drugs were not cures. Doctors prescribed them to relieve the symptoms while relying on the immune systems of their patients for the cure. Today we continue to take mild painkillers, cough mixtures and apply ointments in a similar way.

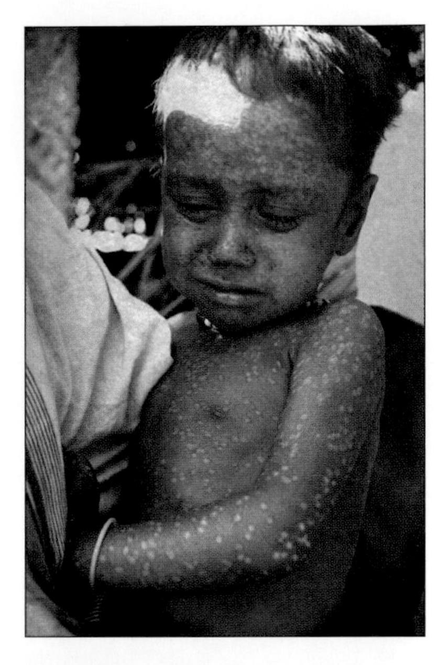

Figure 2.7

This case, one of the last known infections of smallpox occurring naturally in humans, was reported on 16th October 1975. The infected patient is Rahima Banu, a 2-year-old girl from Bangladesh. She has the pustular rash characteristic of smallpox. The pustules usually leave permanent scarring on the skin.

Question

6 Do the benefits of mass vaccination for the community as a whole ever justify compulsory vaccination for all children? State your opinion on this question, giving your reasons.

Key terms

Drugs are the active ingredients in medicines used for the treatment, relief or prevention of disease. People also take drugs for pleasure, stimulation and relaxation.

Medicines normally consist of one or more drugs, mixed with other inert materials, and combined in a way which makes the treatment available as pills to swallow, ointments to rub onto the skin, powders or vapours to inhale, solutions to inject or drops for the eyes or ears.

The **pharmaceutical industry** is the part of the chemical industry which makes drugs and medicines.

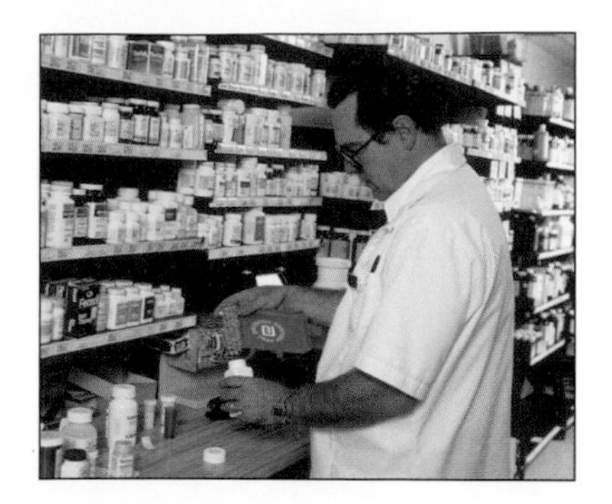

Figure 2.8

A pharmacist preparing a patient's prescription in a hospital pharmacy.

Figure 2.9

An array of modern medicines might include pills, capsules, ointments, drops, injections, inhalers and powders.

Key term

An **antibiotic** is a chemical produced by a microbe that kills or limits the growth of another microbe.

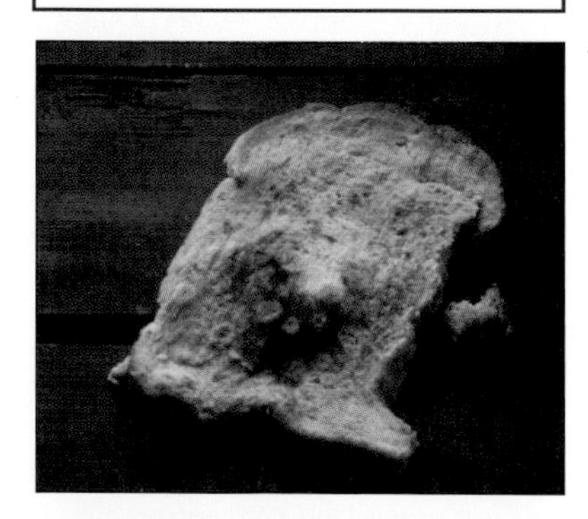

Figure 2.10

Different species of bacteria and mould growing over the surface of a piece of bread. Moulds are a type of fungus. The fruiting bodies of Penicillium are blue while those of Aspergillus are green or yellow. The pin moulds such as Mucor form black fruiting bodies.

One or two of the older medicines were effective. Doctors could prescribe vitamins for patients suffering from deficiency diseases, for example vitamin C for scurvy. Digitalis from foxgloves was available to treat heart failure. Quinine was an anti-malarial drug, first isolated over a 100 years ago from the cinchona tree in the tropical rainforests of the Amazon – where its effects had been known by the peoples of the forest for centuries.

In 1922, the Canadian doctor Frederick Banting and a young medical student, George Best, successfully isolated and tested insulin from the healthy pancreases of dogs and then cows. This drug was being used worldwide to treat diabetes within a few months of their first successful tests.

However, the revolution in the use of chemicals to treat disease really started in 1935 with the launch of sulfonamide drugs. These were followed in the 1940s by the discovery of antibiotics. These discoveries transformed the pharmaceutical industry. Companies began to build up research teams engaged in systematic studies to develop new drugs. The companies have grown so that most of them are now large, international businesses producing a wide range of products (Figures 2.8 and 2.9) (see Chapter 4).

Antibiotics

The discovery of penicillin

A slice of bread left open to the atmosphere for a few days becomes a battleground for microbial warfare. In Figure 2.10, notice the splashes of colour spread over the bread's surface. Each colour is an area covered by the growth of a particular species of bacterium or mould. Once one species has established itself it prevents other species from trespassing on its territory by releasing a chemical which kills the invaders. These chemicals are antibiotics.

One summer's day in 1928, the British bacteriologist Alexander Fleming was examining some Petri dishes in which were growing *Staphylococci* – the kind of bacteria which causes boils and sore throats (Figure 2.11). The growth on the dishes was several days old, and Fleming noticed that one of the dishes was contaminated with a mould. He was surprised to see that the areas of *Staphylococci* growing near to the mould were dying. Fleming realised that there was a possibility that the mould was producing a substance which killed the bacteria.

Fleming grew the mould in a special solution. He found that the mould released chemicals into the solution which could kill bacteria that cause human diseases. The solution was effective even when diluted. Fleming also injected the solution into mice and showed that it did not harm them. The mould was later identified as *Penicillium notatum*, and Fleming named the mysterious anti-bacterial substance it produced penicillin. By chance he had found the first antibiotic, but he did not fully understand the importance of his discovery. At the time other scientists were also indifferent to his work. They were concentrating on other ways of treating human disease.

Figure 2.11

Alexander Fleming at work in his laboratory at St Mary's Hospital, London, in 1909. It was nine years later that he discovered that secretions from the mould Penicillium notatum *destroy the bacterium* Staphylococcus.

Large-scale production

There matters rested until 1938 when Ernst Chain and Howard Florey at Oxford University took up the investigation (Figures 2.12 and 2.13). They quickly established penicillin's effectiveness against different bacteria and that it was harmless when injected into mice. By 1941, the scientists had developed methods to produce enough penicillin for clinical trials (see Chapter 4). The mould was grown in a special solution using all kinds of vessels (including milk bottles). About 100 litres of the mixture were needed to produce enough penicillin for one day's treatment of one patient. The supply was so short that penicillin was saved by extracting the drug from the urine of treated patients.

The trials proved that penicillin was very effective for treating infections, but by then the Second World War (1939–45) had begun. There was an urgent need to treat wounded soldiers. Untreated wounds often became infected with bacteria, causing fatal diseases. Production of penicillin had to increase, and the work moved to the USA where large-scale manufacturing processes were developed. By 1944, the production of penicillin was sufficient to treat all of the British and American casualties that followed the invasion of the European mainland to defeat Hitler's Germany. Today, the massive demand for penicillin worldwide is met by the development of new high yielding strains of *Penicillium* mould.

Improvements in yields have also come through advances in genetic engineering. Penicillins are produced in huge containers (fermenters) which hold up to 200 000 litres of mould and culture solution. At the end of a production run, the mould is filtered off and the penicillin extracted from the solution.

Biochemists have now discovered how penicillin destroys bacteria. The drug weakens the cell walls of the bacterial cells so that they burst and die. Human cells do not have cell walls so they are not affected by the antibiotic.

Figure 2.12

Ernst Chain checking the formulas for penicillin.

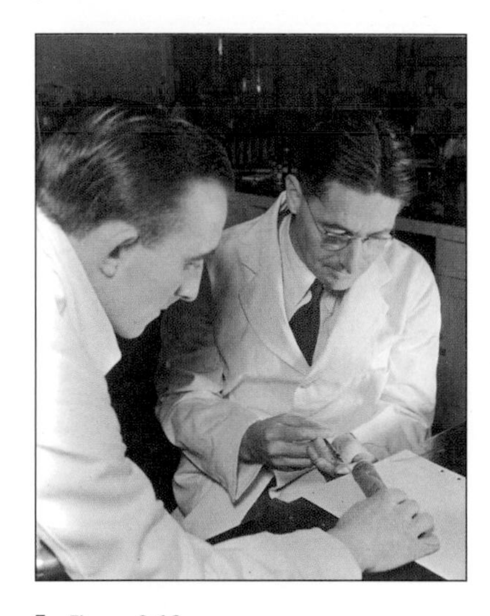

Figure 2.13

Howard Florey injecting penicillin into the tail of a mouse in the 1940s.

Questions

7 Suggest reasons why Fleming is famous for the discovery of penicillin while the achievements of Florey and Chain are much less celebrated.

8 a) What aspects of the behaviour of doctors and their patients have contributed to the development of antibiotic resistance?
b) Do you think that anything could have been done to delay the onset of resistance to penicillin and other antibiotics?
c) What could be done in future to prolong the effectiveness of highly effective new drugs?

Antibiotic resistance

With time, antibiotics become less effective because bacteria develop resistance. Following the introduction of penicillin in the 1940s, strains of bacteria have emerged which produce the enzyme penicillinase. The enzyme breaks down the penicillin, making the drug ineffective.

In the case of penicillin, populations of bacteria always contain a few individuals with genes for penicillinase, making them resistant to the antibiotic. These individuals survive the onslaught of penicillin when a patient is treated with the drug, and reproduce new individuals which inherit the genes for penicillinase. The offspring are resistant and, because bacteria multiply very quickly (in some species a new generation is produced every 20 minutes), resistance spreads quickly. As resistance becomes more widespread the dosage of drug has to be increased step by step until the drug becomes ineffective or so poisonous to the patient that an alternative has to be found.

Doctors are now aware they must limit the use of antibiotics to people who really need them. Incorrect diagnosis, patients not finishing a course of antibiotics (so that not all bacteria present are killed) and the use of antibiotics in animal feed all contribute towards development of resistance.

Newspapers often report on 'superbug' infections in hospitals. These infections involve bacteria which have developed resistance to a wide range of antibiotics. Using terms such as 'superbug' in headlines may be emotive, but these microbes are a real threat to life in situations where antibiotics are used constantly and extensively. Keeping hospital-acquired infections at bay is possible through careful use of antibiotics and strict hygiene procedures when caring for patients. Hospitals routinely test new patients to identify those carrying dangerous bacteria; these people can transfer the bacteria to others even though they may not have symptoms themselves.

Tuberculosis

The disease

Tuberculosis (TB) has always been a very common human infection. The disease still kills many people each year around the world.

TB comes in many forms, but it is most commonly caused by infections of the bacterium *Mycobacterium tuberculosis*. When infectious people cough, sneeze, talk or spit they spread TB bacteria into the air. An untreated person with active TB is likely to infect 10 to 15 other people each year. Crowded living or working conditions add to the ease with which it spreads from person to person. Infection, however, does not necessarily lead to sickness with the disease. The immune system gives protection and the bacteria may lie dormant in the body (this is latent TB).

The WHO estimates that about one-third of the world's population is currently infected with TB – that is about 2 billion people. Only a minority of infected people fall sick or become infectious at some time during their life. Even so, more than 8 million develop active TB each year and about 2 million die, mostly in developing countries.

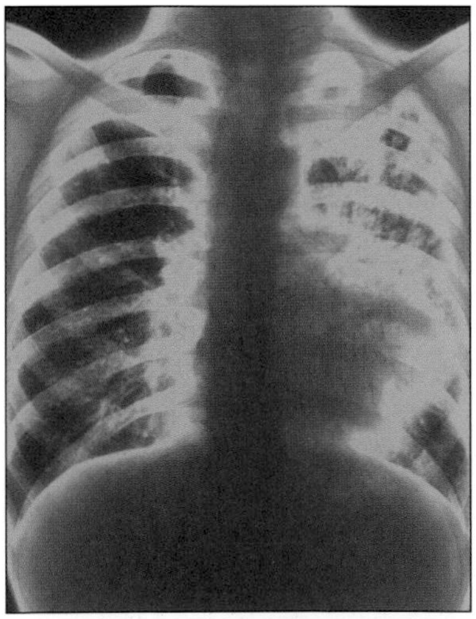

Figure 2.14

Coloured X-ray photograph of the lungs of someone infected with TB – note the fluffy yellow areas in the left lung of the patient.

Symptoms of TB

Tuberculosis can affect many areas of the body, including the lungs and the bones, but the most common forms of TB affect the respiratory system (Figure 2.14). The bacteria not only damage and destroy the lung tissue, they also suppress the action of the immune system, making the body less able to fight the disease.

Typical symptoms of TB are fever, night sweats, inability to eat and loss of weight. In response to the damage in the lungs there is often a cough which produces mucus from which *M. tuberculosis* can be cultured. In severe cases the mucus is blood-stained.

Improving living standards is the most effective way of controlling TB. Less crowded housing and working conditions mean people are less likely to pass on the disease. Generally healthier and better-fed people are less likely to develop debilitating TB even if they meet *M. tuberculosis*. Vaccination also has a part to play in reducing the numbers suffering from the disease.

Immunisation

Every bacterium is different and it has taken years of research since Pasteur's time to develop safe and effective vaccines for a wide range of diseases.

No vaccine yet exists which is fully effective against TB. The BCG (Bacillus Calmette-Guerin) vaccine was invented in 1921. It is made from a weakened form of a bacterium closely related to human TB. It is now the most widely used vaccination in the world and is still the only vaccination available against the disease.

The BCG vaccine is used because it reduces the likelihood and severity of TB in infants and young children. That is especially important in areas of the world where TB is highly prevalent (Figure 2.15) and the chances of an infant or young child coming into contact with an infectious person are high.

In the UK, a targeted vaccination policy offers BCG for infants (aged 0 to 12 months) living in areas where the incidence of TB is 40 cases per 100 000 people or greater. It is also offered for infants with a parent or grandparent who was born in a country where the incidence of TB is 40 cases per 100 000 people or greater.

Key term

The frequency of a disease in a particular population or area is called the disease's **incidence**. Incidence is expressed as the number of newly diagnosed cases during a specific time period.

Figure 2.15

Global TB deaths per 100 000 population in 2005. Drugs for TB are unlikely to bring the sort of profits needed to recoup the cost of research, because it is mainly a disease of impoverished developing countries (see Chapter 5). Anyone can get TB, but you are more likely to if you already have another disease, don't eat well or live in overcrowded or sub-standard housing.

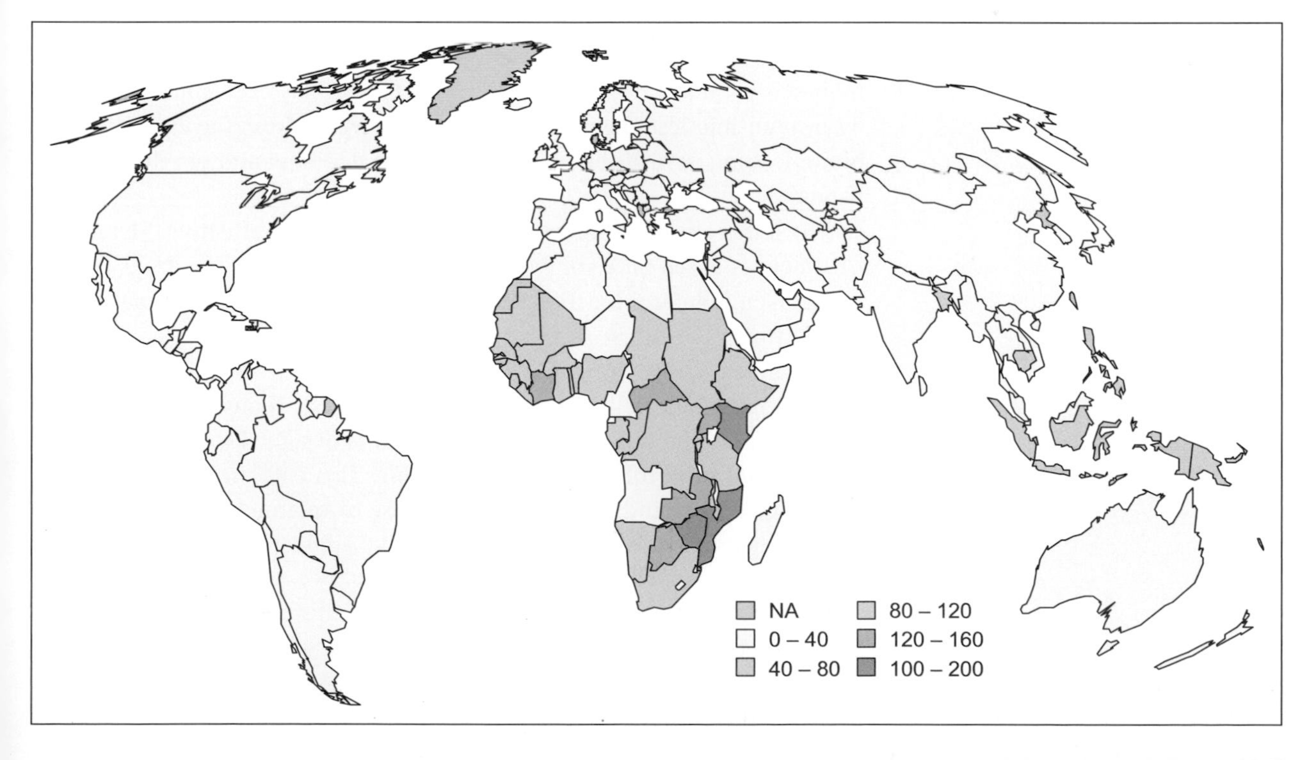

Questions

9 Give examples of ways in which improved social conditions can help to reduce the spread of TB. Explain why the changes are effective.

10 Why is TB no longer common in countries such as the UK but still widespread in other parts of the world?

BCG is also recommended for people such as healthcare workers who may be exposed to TB at work, and immigrants from, or people going to live in, countries with a high prevalence of TB, if they have not previously been immunised.

BCG protects infants and young children from the most serious forms of TB. Unfortunately, the vaccine does not continue to protect people when they become adults. As a result, many people develop tuberculosis even though they received BCG when they were younger.

Also BCG does not stop the reactivation of latent TB in the lungs, which is the main way that the bacteria spread in a community where the disease is common. So the overall impact of BCG vaccination on the spread of TB is limited.

Drug treatments for TB

The discovery of penicillin stimulated a search for other new antibiotics. Soon scientists in the USA discovered streptomycin, which promised to provide an effective treatment for TB. The British government bought enough of the drug to treat up to 200 patients for a cost of £1 500 000 – an enormous sum in the 1940s.

Not many patients could be treated and it was important to find out just how effective the new drug could be. So Austin Bradford Hill, a distinguished statistician, was brought in to design a randomised clinical trial (see Chapter 4). One hundred and seven patients were monitored in hospitals in London, Wales, Scotland and Yorkshire. Streptomycin was given to 55 patients and the other 52 received the only other available treatment, bed rest. A particular patient's treatment was decided at random by the equivalent of tossing a coin. At the end of six months, 14 patients given bed rest had died and four had shown a considerable improvement. Of those given streptomycin only four had died and 28 were very much better.

For a while, streptomycin was widely used in the treatment of tuberculosis, but it has harmful side effects so it has been superseded by new antibiotics. The drug can affect the nerves from the ears to the brain, leading to disturbed hearing, possible deafness and problems with balance.

The antibiotics used to treat TB today can be very effective. The usual treatment is based on a combination of four antibiotics. The drugs have to be taken regularly for 6 to 12 months for a complete cure. If patients do not finish the course of treatment, their TB can come back and then it is very much harder to treat.

Drug resistance develops rapidly in tuberculosis if any one drug is used alone, so the initial phase of treatment is designed to bring the disease under control as quickly as possible. This cuts the danger of resistance developing. A continuation phase of treatment starts once tests have discovered which drugs are most active against the particular strain of TB bacterium affecting the patient.

The treatment of TB using combinations of drugs depends on the notion that bacteria resistant to one of the drugs are unlikely to be resistant to the others in the drugs cocktail. So far most deaths from multi-drug resistant tuberculosis have been of patients who were also infected with HIV. However, the threat of multiple resistance to people with tuberculosis but without the complications of HIV infection is a growing problem. There are strains of bacteria resistant to all of the known anti-TB drugs.

In England and Wales the effects of better living conditions, vaccination and drug treatment led to a dramatic fall in the number of cases of TB during the twentieth century (Figure 2.16).

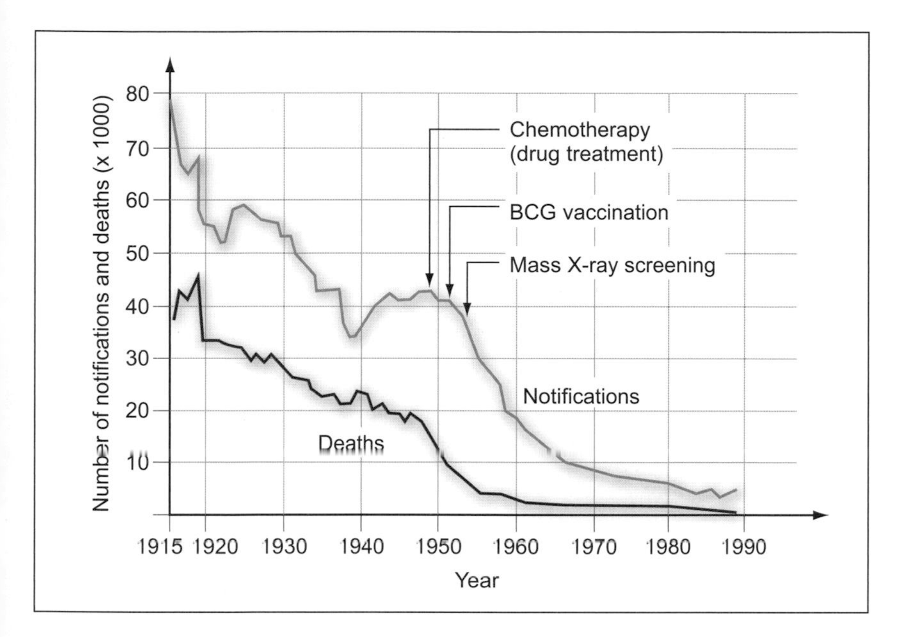

Figure 2.16

The incidence of respiratory tuberculosis in England and Wales from 1913 to 1990.

Question

11 Study Figure 2.16.

a) Which of the lines of the graph would you expect to be based on more reliable data and why?

b) Which important methods of controlling the spread of the disease are not indicated on the graph?

c) How do you account for the decline in the incidence of TB from 1913 to 1940?

d) Suggest reasons for the rise in the number of notifications of the disease between 1940 and 1950.

e) What role did mass X-ray screening of the population play in helping to lower the incidence of the disease? Why has the mass X-ray programme ended in the UK?

f) Has prevention or cure played the greater part in the reduction in the number of notifications of the disease?

TB on the rise

The number of cases of TB in England and Wales started to rise again in the early 1990s, but the disease remains quite rare. There are about 8000 new cases a year, mainly in the major cities. About 40% of all cases are in London. In the UK, those at risk are people in close contact with a person with TB, or people who have lived in places where TB is still common. Those most susceptible to infection are those whose immune system is weakened by HIV or other medical conditions, people with chronic poor health as a result of homelessness, alcoholism or drug abuse, and the very young and very old.

HIV weakens the immune system (see Figure 2.17). This means that someone who is HIV-positive is very much more likely to develop the symptoms of TB if they are infected. TB is now the leading cause of death for people who are HIV-positive. One-third of deaths associated with HIV infection worldwide are the direct result of tuberculosis.

Badly managed TB treatments are threatening to make TB incurable through the development of drug resistance. This process is accelerated if doctors prescribe the wrong treatment or patients do not take the full course of drugs and stop taking their medicine once they feel better.

TB worldwide

The WHO (World Health Organisation) estimates that more than 90% of cases of TB and deaths from TB happen in the developing world (Figure 2.18). This is economically very serious because three-quarters of cases affect people in the 15–54 age range, i.e. when they are most productive.

HIV and AIDS

Figure 2.17

Acquired immune deficiency syndrome (AIDS) is a collection of symptoms and infections resulting from the damage to the immune system caused by the human immunodeficiency virus (HIV) in humans.

AIDS (*acquired immunodeficiency syndrome*) is a new human disease caused by HIV (*human immunodeficiency virus*).

Acquired	– Something gained or caught
Immune **D**eficiency	– Destruction of the immune system so that the body cannot defend itself against infection
Syndrome	– The set of symptoms associated with a disease

AIDS was first recognised in the USA in 1981 among homosexual men who had previously enjoyed good health. HIV infection was already widespread in Africa and is now found worldwide.

HIV attacks the white cells in the blood that control the response of the immune system to microbes (see page 20). Destruction of the immune system by HIV exposes infected people to a variety of fatal diseases, including TB, pneumonia, skin cancer and brain damage.

Infection with HIV does not necessarily lead to full-blown AIDS. People infected by the virus who do not display the symptoms of AIDS are described as being HIV-positive. This is because tests on their blood detect antibodies to HIV.

Figure 2.18

A doctor visiting a tuberculosis patient in Dhaka, Bangladesh.

An adult with TB misses three to four months' work, with serious consequences for the family income. Children have to leave school if their parents have the disease and women may be abandoned by their families if they have the disease.

There has been a large increase in the number of TB cases in countries where HIV infection is common, especially in sub-Saharan Africa. At the same time, multidrug resistance caused by poorly managed TB treatment is a growing problem.

At the start of the 1990s, the WHO developed a treatment strategy for detecting and curing TB – which can be very successful if applied systematically. The key elements of the strategy are:

- political commitment
- microscopy services to detect the TB bacteria in samples of saliva
- steady drug supplies
- a drug regime which has proved effective
- monitoring of patients by a health worker to observe and record patents swallowing the full course and the correct doses of drugs.

Countries that have adopted the strategy on a wide scale have achieved impressive results. The spread of the disease has fallen sharply in some countries. In Peru, for example, the number of new cases has dropped by about 6% per year over 10 years. The number of deaths from the disease have fallen too. In China, the districts that have adopted the strategy have seen the total number of deaths fall by 30 000 a year. The strategy can also help to control drug resistance. This has been shown in New York, where an intensive programme to ensure that patients completed the treatment meant that new cases of drug resistance fell by three-quarters.

When the WHO introduced this strategy in the early 1990s, it set ambitious targets for its campaign – including a target of 70% detection of TB cases and a target of 85% for successful treatment of the detected cases. A report in 2007 showed that overall the targets were missed – with a detection rate of 60% and a cure rate of 84%. There were very wide variations across the world. Both targets were met in the Western Pacific Region and in 26 countries including China. Overall there are some grounds for optimism. The incidence of TB has levelled off and may even be starting to fall in all WHO regions, including Africa and Europe.

A crucial problem in some parts of the world is the lack of laboratory services with the capacity to confirm diagnosis and monitor treatments. This is a particular problem in many African countries.

Influenza

Hundreds of millions of people suffer from an influenza infection every year. Between half a million and a million infected people die because of the disease. A worldwide epidemic of influenza could kill millions more.

The disease

Influenza – more commonly referred to as flu – is a relatively common respiratory disease caused by the influenza virus. There are several strains of the virus. The disease is highly infectious and has a very short incubation period.

The symptoms of influenza include fever, often accompanied by shivering and sweating, feeling very unwell and unable to do anything, loss of appetite, aching muscles and painful joints. Simple influenza lasts for about 5–7 days before the fever goes down and convalescence begins but the exhaustion which follows can last from 6–12 weeks, even in patients without secondary bacterial infections.

Influenza infects the cells lining the tubes leading to the lungs, causing them to die. This leaves the airways open to infection, and many of the deaths associated with influenza are from severe secondary bacterial infections on top of the original viral invasion. This makes influenza more hazardous than other respiratory diseases such as the common cold. The people most likely to die of the disease are the elderly and anyone who is prone to asthma or heart disease.

Influenza spreads very quickly and there are often major outbreaks affecting thousands of people – these are known as epidemics.

The treatment for influenza is rest, warmth, plenty of fluids (so as to avoid dehydration) and mild painkillers. Although there are drugs which will ease the symptoms, there is no drug yet which will cure the disease. If secondary bacterial infections set in, then antibiotics can be used to combat them.

Questions

12 Suggest reasons why combating TB is seen as an important part of programmes to reduce poverty.

13 Explain the importance of political commitment in any countrywide programme to control TB.

14 Patients have to be watched while taking their drugs during the TB treatment programme recommended by the WHO. Why is this necessary? Do people responsible for medical care have the right to insist that patients are kept under surveillance during treatment?

15 a) What aspect of human nature and what pressures on people have contributed to the development of resistance to the drugs effective against TB?
b) What might have be done, worldwide, to prevent resistance developing? What are the lessons for the future?

16 From the point of view of public health, it can be argued that poorly supervised treatment of TB is worse than no treatment at all. Do you agree?

17 The WHO monitoring of its campaign to combat TB depends on reliable data. WHO studies show that in many countries there is great uncertainty about TB incidence and detection rates. Suggest reasons for this.

18 More and more people are travelling internationally on business and for holidays. There are many people who have been displaced from their homes by war, famine, natural disasters or the search for work. Why does movement of people help the spread of TB and make it more difficult to control the disease?

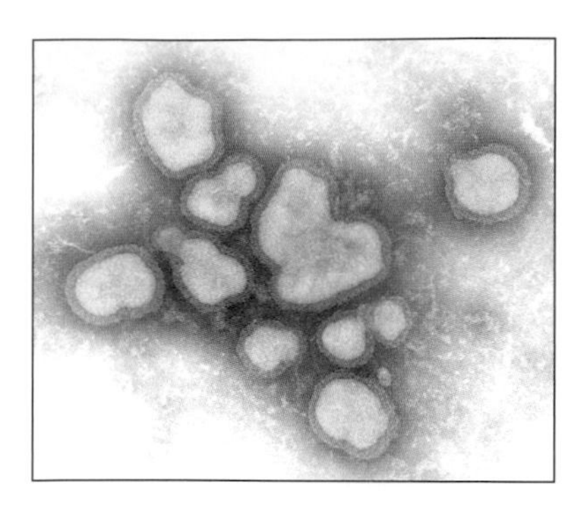

Figure 2.19

The protein coat of the influenza virus changes as the virus reproduces in its host. This makes it impossible for humans and other animals to keep up their immunity to flu year after year. Antibodies produced after a flu infection one year may not recognise the mutated version of the virus the following year.

Immunisation

Each year the various strains of the influenza virus are subtly different; the change is usually quite small, so having influenza one year leaves people with some immunity against infection for the next. But every so often there is a major change in the virus (Figure 2.19), and this heralds a major influenza outbreak as no-one's immune system is prepared.

The influenza vaccine manufactured each year has to be different to allow for the changing nature of the virus. A cocktail is made up of the strains of the virus thought most likely to cause disease in any one year, and this vaccine is then made available to those in high risk groups such as the elderly and medical workers. The great fear is that a new and very different strain of influenza will appear again, and without an effective vaccine millions of people worldwide could once again die (Figure 2.20).

The WHO is responsible for a worldwide surveillance network with over 100 centres for monitoring influenza outbreaks. This network keeps track of the strains of the influenza virus circulating in the world and recommends the appropriate composition for the vaccine each year.

Antiviral drugs

At present there is no drug treatment available for viral diseases which would be as powerful in its effects as antibiotics are against bacterial diseases. In other words, the antiviral equivalent of penicillin has still to be discovered. However, new substances which are at the research stage or just becoming available may hold the key to an antiviral future.

Two antiviral drugs used to treat influenza patients are oseltamivir (trade name *Tamiflu*) and zanamivir (trade name *Relenza*). These drugs are designed to halt the virus spreading in the body. The main effects of the drugs are to reduce the symptoms of the disease and make it less likely that the patient will suffer from complications.

Resistance has not yet been a serious problem with these antiviral drugs. However, different strains of influenza virus vary in the extent of their resistance to the drugs so it is not possible to predict what degree of resistance a future pandemic (global epidemic) strain might have.

Developing effective antiviral drugs is proving to be very difficult. Viruses constantly change as a result of mutations in the viral genetic code (see page 18). This means that antiviral drugs effective against

Questions

19 Why is it possible to catch influenza more than once despite the body's immune system?

20 Flu vaccines are widely given only to certain groups of the population, particularly the elderly, rather than to everyone.
a) Why do you think this is?
b) Is it acceptable to limit access to a vaccine which could save lives?
c) On what basis should such a decision be made?

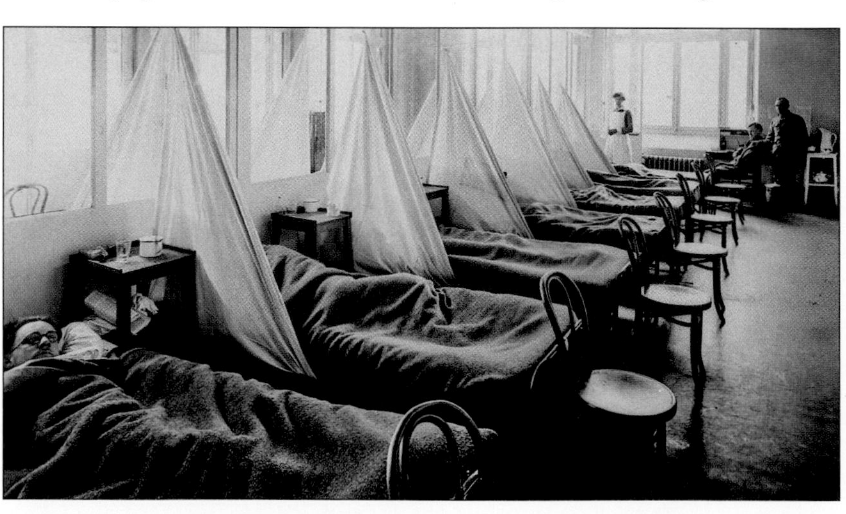

Figure 2.20

The 1918 outbreak of flu hit soldiers and civilians who were exhausted by the rigours of the war. It caused more deaths in a few months than had occurred throughout the whole of the war.

a particular form of a virus can quickly become useless against the different mutated version. New drugs may be developed to cope with the most recent version, but the threat of future mutations keeps research scientists hard at work trying to stay one step ahead (see Chapter 4).

Pandemics of influenza

If a new influenza virus appears against which the human population has no immunity, there is a risk of simultaneous epidemics of the disease all over the world, leading to a huge number of deaths. This is a pandemic.

Three massive influenza pandemics happened in the 20th century – in 1918, 1957 and 1968. During the 'Spanish flu' pandemic in 1918–20 at least 20 million people died from influenza (Figure 2.20).

Influenza infects many animals including birds, dogs, horses and pigs as well as humans. Poultry farming is particularly vulnerable to some forms of 'bird flu' which can spread very rapidly where chickens or turkeys are crowded together on intensive farms. The disease can kill all birds in just a couple of days (Figure 2.21).

Past outbreaks of bird flu have originated in South east and East Asia, where birds, pigs and humans live together closely. In these conditions, a virus infecting birds could potentially mutate into a form that can infect humans. If this were to happen, and if the infection could then be passed between humans, there could be a major pandemic. With this in mind, some governments are spending large amounts of money on research into vaccines while also stockpiling antiviral drugs such as Tamiflu.

Millions of birds have been killed in attempts to control the disease, but scientists believe that an influenza pandemic is inevitable in the near future.

The WHO reports that fewer than 10 countries have pharmaceutical companies that are involved in research into pandemic vaccines. The bleak assessment is that neither vaccines nor antiviral drugs will be available in adequate quantities worldwide at the start of a pandemic, or for many months afterwards. Also, these medical resources will not be fairly distributed; poorer countries will not be able to afford the necessary drugs.

Questions

21 What aspects of modern life are likely to mean that a new influenza virus will cause a pandemic of the disease?

22 Why is it not possible to start manufacturing vaccines to fight a pandemic before it starts?

Figure 2.21

A woman with a basket of chickens in an Asian market.

Review Questions

23 As more and more people are immunised by vaccination, the threat of the disease becomes much less so that the disease effectively disappears. At this point people start to worry about safety because of possible side effects of the vaccine. So people stop being immunised and the disease starts to spread again. Suggest ways of dealing with this problem.

24 Make summaries of the information about the bacterial disease TB and the viral disease influenza. Include in each summary: the cause of the disease, how the disease spreads, methods of preventing infection, possible treatments and factors which make the disease hard to control worldwide.

25 Use examples from this chapter to illustrate the way that people making decisions about public health policy and medical care take into account:

- technical feasibility
- benefits expected
- economic cost
- risks.

Chapter 3

Transport issues

The issues

People now travel much more than they once did (Figure 3.1). Driving to the shops, commuting to work and flying off on holiday is only possible thanks to fuels. As a result, people in many countries use more fuel today than they did in the past. Carbon dioxide, produced when fuels are burnt, is a greenhouse gas that can change the global climate. Burning fuels also causes local pollution which can harm people's health. People worry enough about air pollution for weather forecasters to include reports about the quality of the air we breathe. Air quality matters directly to people with heart conditions or lung diseases such as asthma and bronchitis. In the longer term polluted air affects us all and may contribute to chronic health problems.

The science behind the issues

Fuels are valuable because they are concentrated energy sources. When we use a fuel, we transfer the stored energy to other places. Energy is not destroyed in the process but it is spread out and becomes less concentrated – and so is less useful for doing anything more.

Understanding chemical reactions helps us to understand why motor vehicle engines and power stations pollute the air. When fuels burn at a high temperature the waste fumes include gases, such as oxides of nitrogen, carbon monoxide and sulfur dioxide, and sooty particles. These gases do not just disappear. They affect the quality of the air we breathe.

Figure 3.1

Traffic on a busy motorway.

What this tells us about science and society

Our high standard of living is largely the result of technological developments, many of them based on scientific ideas. But the demands this makes on fuels, particularly fossil fuels, has rapidly depleted a precious finite resource, and led to emissions of gases which damage the environment. The global inequalities in fuel use (and in standards of living) raise serious moral questions. While technical developments can contribute to tackling these problems, they cannot provide a complete solution. Economic, environmental and ethical considerations are often involved, and decisions involve balancing different considerations against each other.

Fuels for transport

As fuels burn, chemical reactions give out energy. Sometimes we use the energy directly to make a car engine run or power a jet engine. But we can also use it to make steam to drive turbines in power stations, which turn generators to generate electricity. Electricity is then used by the motors in trains and trams and also to charge the batteries in electric buses and cars (Figure 3.2). Electricity is a secondary energy source. It has to be generated using a primary energy source (such as a fossil fuel, nuclear fuel or one of the renewable energy sources).

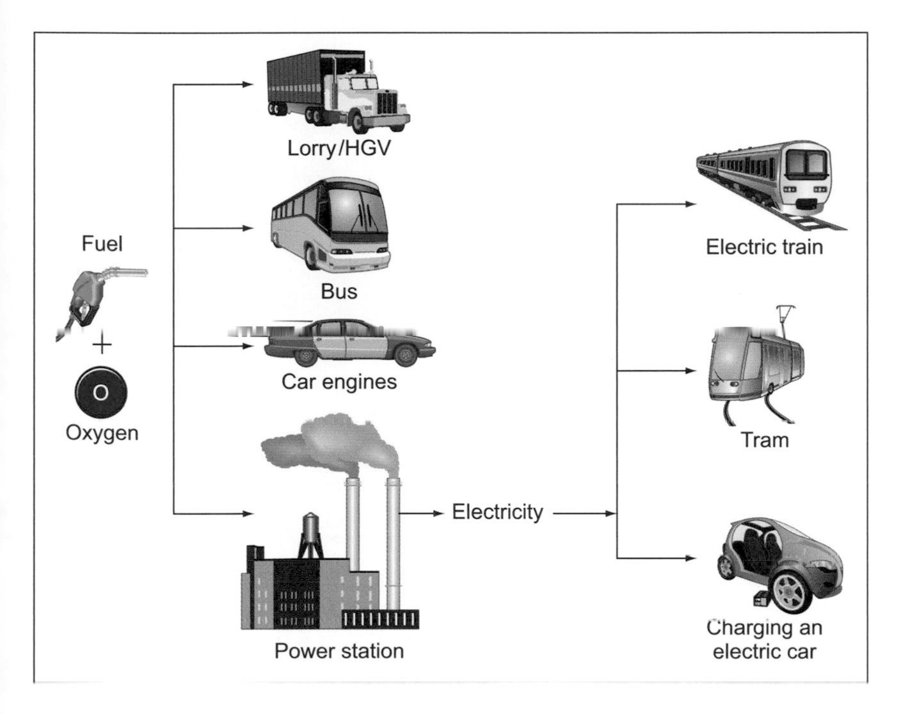

Figure 3.2

Pathways from primary energy resources to transport.

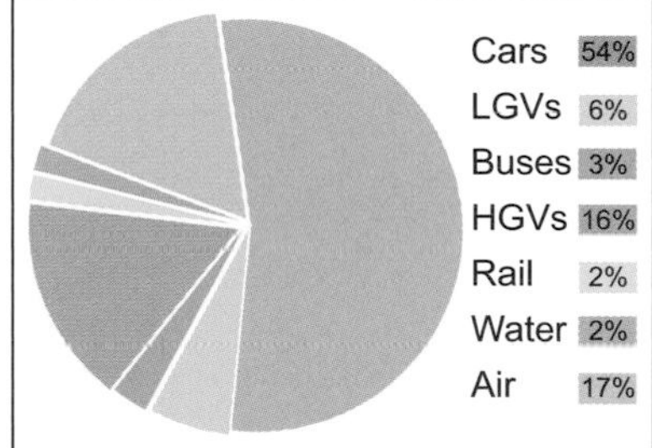

Figure 3.3

Chart showing energy for transport in the UK.

Fuels are useful because they are concentrated energy sources. Most of the energy for transport comes from burning fossil fuels – coal, oil and natural gas. At present these fuels are readily available, but they are a finite resource which will eventually run out. Some of our energy comes from nuclear fuel which is used in nuclear power stations to generate electricity. In some parts of the world, renewable biofuels are becoming increasingly important.

Burning fossil fuels has done more than anything else to increase the concentration of greenhouse gases in the atmosphere. When fuels burn the chemicals in them do not simply disappear. The products of burning, which pour out of chimneys and exhaust pipes, spread out into the air. During the past 200 years, since the start of the Industrial Revolution, the concentration of carbon dioxide in the air has risen from about 280 parts per million to 377 parts per million by volume in the air.

There is now a strong scientific consensus that the carbon dioxide from burning fuels is changing the climate. An international panel of scientists, called the Intergovernmental Panel on Climate Change (IPCC), has confirmed that global warming is a real problem and that urgent

Key terms

The original meaning of a **fuel** was 'a material for burning'.

More recently, the term has been extended to include **nuclear fuel**, which is used in nuclear power stations, though it is not literally 'burned'.

Fossil fuels are the fossilised remains of prehistoric rainforests or tiny marine animals, and have formed over millions of years.

A **renewable energy** source is one which is being (or can be) replaced as it is used.

Questions

1. With the help of Figure 3.4, give examples to show that you understand what these terms mean:
   a) element
   b) compound
   c) atom
   d) molecule
   e) hydrocarbon
   f) chemical reaction
   g) combustion
2. Draw a diagram in the style of Figure 3.4 to show what happens when the hydrocarbon called propane burns. Propane is a molecule with three carbon atoms and eight hydrogen atoms (C_3H_8).
3. a) What do you understand by the term 'energy industries' in Figure 3.5?
   b) What are the implications of Figure 3.5 for the commitment to make significant reductions in greenhouse emissions?

Burning: a chemical reaction

Figure 3.4

Petrol and diesel fuel consist almost entirely of a mixture of hydrocarbons. Hydrocarbons are compounds of just two elements: hydrogen and carbon. When petrol burns in an engine the main reactions involve hydrocarbons reacting with oxygen from the air. The hydrogen atoms in the fuel join with oxygen to form hydrogen oxide (water) while the carbon atoms join with oxygen to form carbon dioxide. This combustion reaction releases energy.

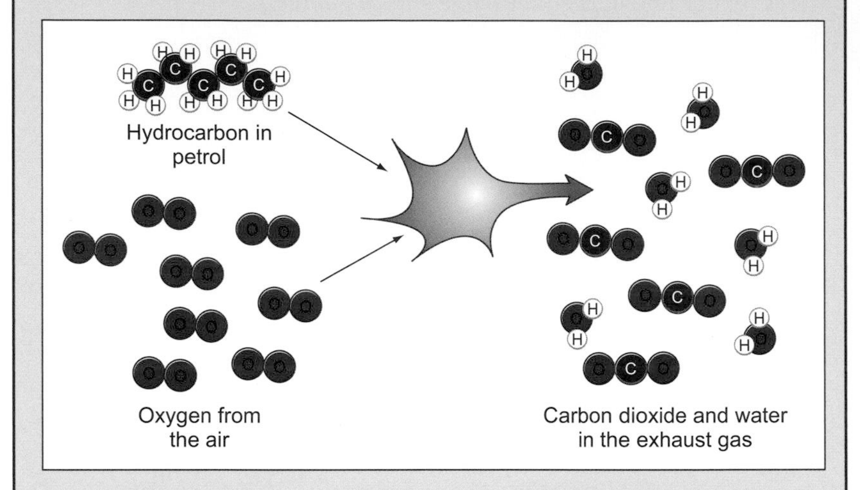

Chemists use models of molecules to describe what happens during chemical reactions. A hydrocarbon in a fuel reacts with oxygen to form water and carbon dioxide. The atoms of the elements do not change but they rearrange to make new molecules.

action is required by governments to cut emissions of greenhouse gases worldwide. This conclusion is based on the work of thousands of scientists all over the world.

Emissions of carbon dioxide in the UK fell by 6.4% between 1990 and 2005 (Figure 3.5). Carbon dioxide is the main greenhouse gas. It made up about 85% of the country's greenhouse gas emissions in 2005.

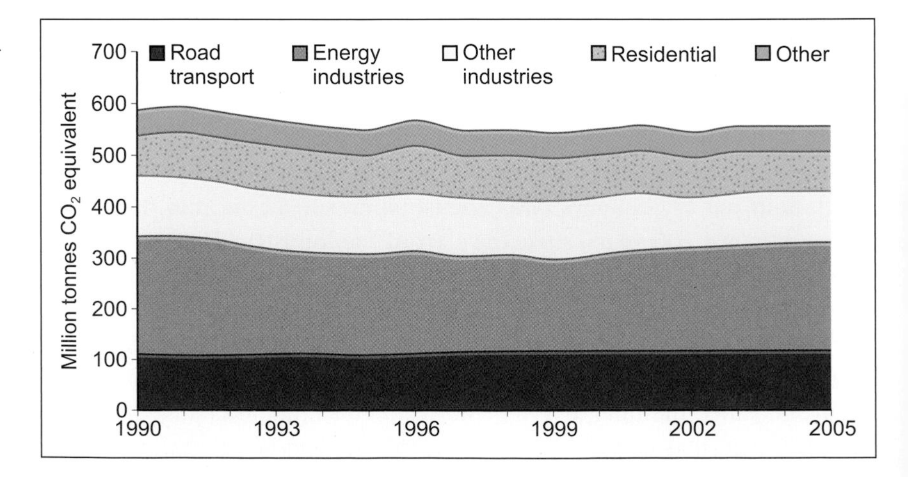

Figure 3.5

Greenhouse gas emissions from sources that burn fuels in the UK, 1990–2005.

The main exception to the downward trend has been transport. Greenhouse gas emissions from all forms of transport in the UK were 47% higher in 2002 than in 1990 (Figure 3.6). The upward trend suggests that emissions will double by 2045, despite a government commitment to reduce transport emissions of carbon dioxide to around 90 million tonnes by that date.

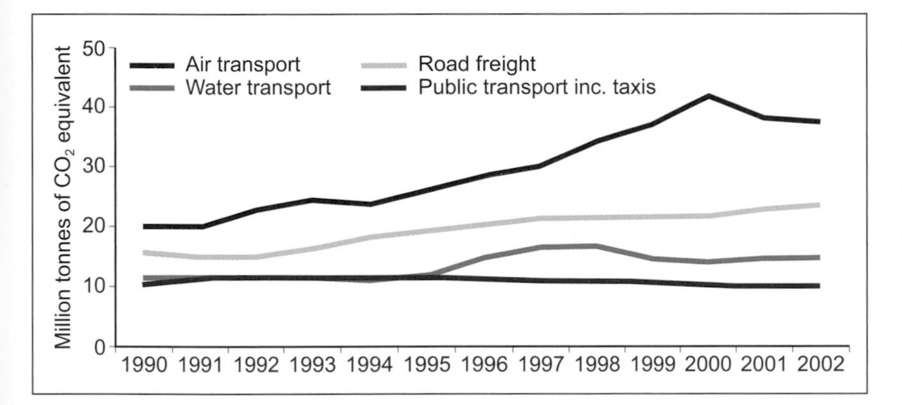

Figure 3.6

Trends in greenhouse gas emissions from different forms of transport in the UK, 1990–2002.

Questions

4 Estimate from Figure 3.6 the percentage increase in greenhouse gas emissions between 1990 and 2002 from:
a) road freight
b) air transport

5 Summarise in note form the main messages in Figure 3.6 about the contribution of transport to UK greenhouse gas emissions.

Alternatives to fossil fuels

Biofuels

Apart from less travel, there are not many ways to cut down the carbon dioxide emissions from transport. One possibility that can seem attractive is to make fuels from biomass. These biofuels come from plants: bioethanol comes from sugars and starches, biodiesel comes mainly from rapeseed and palm oil. Generally these fuels are blended with normal fuels, making up about 5% of the product.

In theory biofuels can be carbon neutral because the carbon dioxide released when the fuel burns is offset by the carbon dioxide taken in by photosynthesis as the plants grow. This possibility has led the UK government to introduce the Renewable Transport Fuel Obligation (RTFO) in April 2008. This sets a target of supplying 5% of transport fuels by 2010.

In Brazil, cars have been running on bioethanol for many years. The reason that the state set up a bioethanol fuel programme was to reduce the country's reliance on imported crude oil and not to cut costs or protect the environment.

In practice biofuels can cut emissions of carbon dioxide, but they are not carbon neutral because of the emissions during the stages of growing, harvesting and processing (Figure 3.7).

When assessing the benefits of biofuels it is important to compare the energy needed to produce the fuel with the energy released when the fuel burns. The energy from bioethanol can sometimes be less than all the energy needed to produce it from corn grown in North America.

The balance is more favourable for biodiesel from vegetable oil. Biodiesel from soybeans can reduce greenhouse gas emissions by 41% compared with conventional diesel fuel. This is mainly because no energy is needed for distillation during the making of the fuel. Also, far fewer fertilisers and pesticides are used in growing soybeans.

Key terms

In the energy industries, **biomass** is biological material that can be used as a fuel. Biomass can be converted into a **biofuel** such as bioethanol or biodiesel.

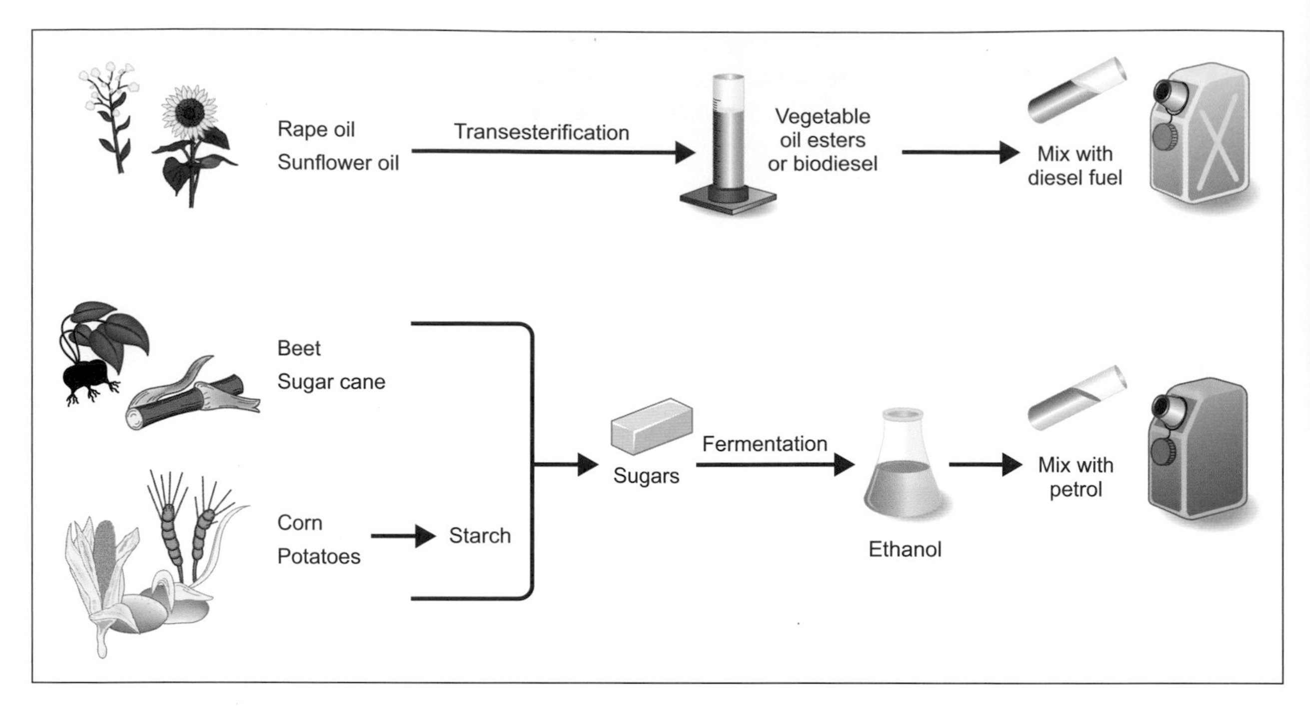

Figure 3.7

Sources and production of biofuels.

The *UK Biomass Strategy*, published in 2007, pointed out that turning plant products into biofuels is the least effective way of using biomass. The report said that best value comes from burning biomass for heating and to generate electricity. Experts estimate that we would need to plant a land area twice the size of the UK to produce enough biofuel crops to halve our carbon dioxide emissions.

The biofuels which offer the best value are those made from crops such as sugar cane grown in tropical countries. In Brazil ethanol is manufactured by fermenting sugars from sugar cane. The refineries can meet all their energy needs by burning the waste from sugar cane both for heating and to generate electricity. However, there are several negative aspects of this industry in Brazil where, among other things, large-scale deforestation has been carried out to make way for sugar cane plantations. Burning forests generates vast amounts of CO_2.

Questions

6 Where does the energy for the growth of plants come from?

7 With the help of Figure 3.7, explain why biofuels:
a) could, in theory, be carbon neutral
b) in practice do contribute to carbon dioxide emissions
c) are more energy efficient if made from biomass grown in the tropics.

8 Draw up a table to show the advantages and disadvantages of biofuels.

The idea of energy

Energy is an idea that was developed and defined over the 100 years or so from 1750 to 1850. It is not easy to give a simple and clear definition of 'energy'. In 1750 it was called *vis viva* (or living force). Although this name has not survived, it gives a sense of what 'energy' means. It is a measure of the maximum possible amount of change that an object (or a collection of objects) can cause to happen. An object can have energy because it is moving, or because it is hot, or because it has been raised to a height, or stretched or compressed.

Energy can be transferred from one object to another by a force, by heating, by radiation or by electricity. Energy is not just a vague concept: the amount of energy transferred in each of these ways can be calculated. The reason why energy is such a useful idea in science is that it is always conserved. There is always the same total amount of energy at the end of a process as there was at the beginning. So if you imagine a process, and then work out that it would involve increasing or decreasing the total energy, this means that it cannot occur. This is what scientists mean when they say that energy is needed for a process to occur.

Energy conservation has a rather different meaning in everyday usage (Figure 3.8). It refers to the fact that energy sources like fossil fuels are in limited supply and so are valuable – and we ought therefore to 'conserve' them in the sense of 'using them carefully and not wastefully'.

Scientists and others who work on energy policy use the word 'energy' in a rather different way from the formal scientific meaning discussed above. They talk about 'energy' as a commodity that can be 'supplied', 'used' and 'consumed'. This is the kind of energy language used in this chapter. Although it is not so formal and precise, it can be reconciled with the scientific idea of energy. When people talk about energy being 'used' or 'consumed', what they really mean is that energy has spread out from a concentrated source and dispersed – to different places. There is still the same total amount, but it is no longer useful for doing anything. It has effectively been 'used up' (Figure 3.9).

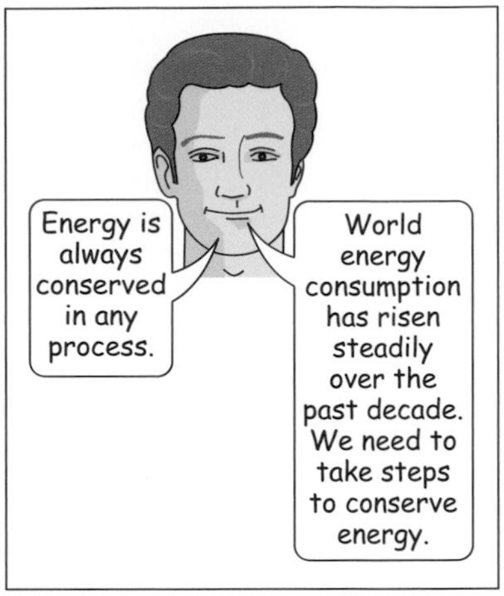

Energy is always conserved in any process.

World energy consumption has risen steadily over the past decade. We need to take steps to conserve energy.

Figure 3.8

Two kinds of 'energy language'.

Figure 3.9

After the launch, the energy from the burning fuel spreads out and warms up the surrounding air.

Key terms

Energy is measured in **joules, J**.
1 000 000 J = 1 MJ (megajoule).
Another widely used measure of energy from fuels is the **tonne of oil equivalent** (toe). This energy unit is the amount of energy released by burning one tonne of crude oil. 1 toe is equal to about 42 gigajoules (GJ). 1000 MJ = 1 GJ.

Why does burning fuels cause pollution?

Burning fuels in engines

As well as carbon dioxide, burning fossil fuels is the main source of many other pollutants. Carbon dioxide is a greenhouse gas, but it does not harm people's lungs. If the only gases in exhaust fumes were carbon dioxide and water, we would not have to worry about their effect on air quality and health.

Question

9 'Fuels can be used up, but energy cannot.' Explain this statement as fully as you can.

Questions

10 a) Draw a diagram in the style of Figure 3.10 to show the energy available from nuclear fuel used in a power station. There is 6500 GJ energy from the fuel. Of this energy, 2000 GJ is used to concentrate the fuel and produce fuel rods, 1000 GJ is left in the spent fuel rods, 2400 GJ energy is lost as heat from the power station. The powers station provides 1100 GJ electrical energy.
b) Calculate the efficiency of the power station.

11 Put these four words into the correct sequence to summarise the working of a four-stroke engine: bang, blow, squeeze, suck (Figure 3.11).

Efficiency

Figure 3.10

Because of the tendency of energy to spread out, an important thing to know about any process is how much of the energy goes where we want it to. For example, if we burn a fuel and use it to drive a car, some of the energy released from the burning fuel goes into making the car move; but some also goes into the metal of the engine and the surrounding air, heating them up. The efficiency of the process is the percentage of the energy released which goes where we want it to – into making the car move.

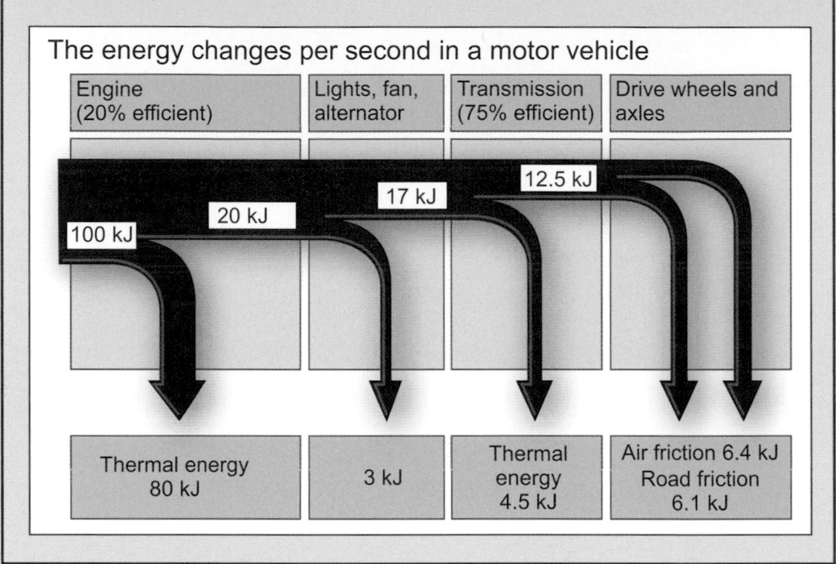

Engines pollute the air with gases that are harmful to health for three main reasons:

- they do not burn the fuel completely
- the fuel contains other chemicals as well as hydrocarbons
- they run at such high pressures and temperatures that nitrogen and oxygen in the air can react together to form oxides of nitrogen.

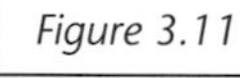

Figure 3.11

A schematic diagram of the four-stroke cycle of a petrol engine.

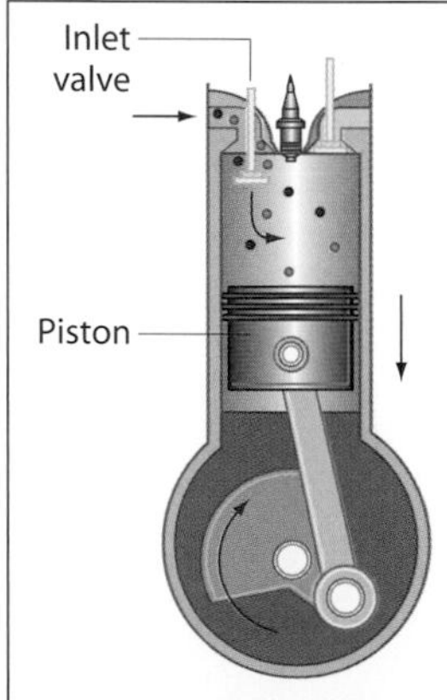

Intake stroke

As the piston moves down it draws a petrol/air mixture into the cylinder.

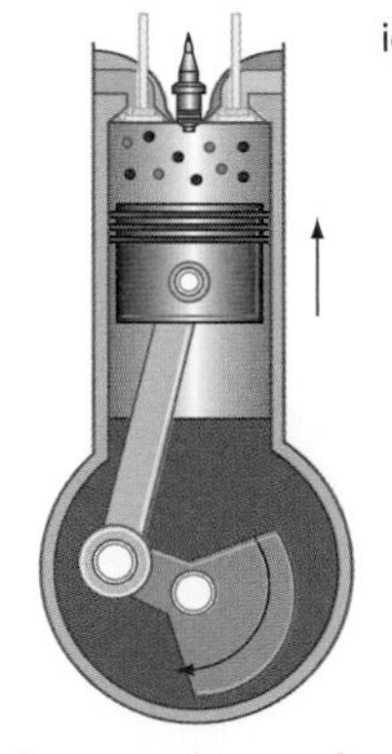

Compression stroke

The inlet valve closes and then the piston moves back up the cylinder, compressing the vapour.

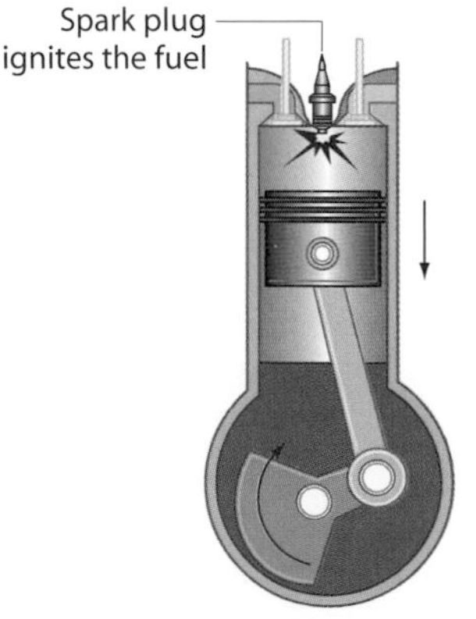

Power stroke

The petrol/air mixture burns explosively causing a huge expansion of the gases and pushes the piston down the cylinder. This is the 'power stroke' which propels the car.

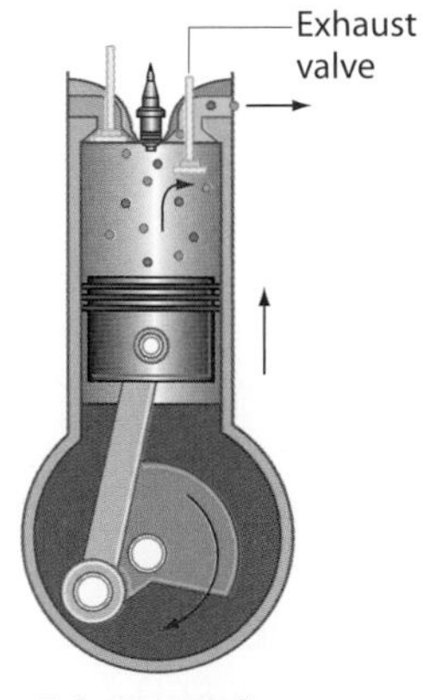

Exhaust stroke

The exhaust valve opens when the piston is at the bottom so that the piston can push out the burnt fuel as it moves up the cylinder.

Pollutant	*Main sources*	*Properties*	*Range of concentrations in the air we breathe*
Carbon monoxide	Incomplete combustion of fuel when there is not enough oxygen, which is often the case in engines.	Colourless gas with no taste or smell. It is highly toxic because it combines very strongly with haemoglobin in the blood. At low doses puts a strain on the heart and circulation. Fatal at high doses.	In the range of 4–20 mg/m^3, depending on the density of traffic and the weather. The UK target is that the mean level should not exceed 10.0 mg/m^3.
Oxides of nitrogen, NO and NO_2, which together are often referred to as NO_x	Mainly from engines and furnaces which burn fuels at a high enough temperature for oxygen and nitrogen in the air to combine.	Gases which oxidise in moist air to form nitric acid. Can make the symptoms of asthma worse.	20–90 μg/m^3 but can rise to as much as 700 μg/m^3 at the rush hour peak in city traffic. The UK target is that the annual mean value should not exceed 40 μg/m^3.
Sulfur dioxide	Burning fuels which contain sulfur in motor vehicles and power stations. Main source is burning coal. There is more sulfur in diesel fuel than in petrol.	Colourless acidic gas which oxidises in moist air to sulfuric acid. Makes the symptoms of asthma worse and causes wheezing and bronchitis.	Normally well below 90 μg/m^3 in urban areas but can rise to 1000 μg/m^3 or more. One of the UK targets is that 266 μg/m^3 should not to be exceeded more than 35 times per year.
Ozone	Formed when sunlight shines on polluted air containing oxides of nitrogen.	Highly reactive form of oxygen with O_3 instead of the normal O_2 molecules. Irritates the eyes, nose and throat. Can make asthma worse.	Background levels are usually below 32 μg/m^3 but in summer and in the south of England levels can rise to 300 μg/m^3 or more. The UK target is that 100 μg/m^3 should not be exceeded more than 10 times per year.
Particulates	The coarser particles include dust from roads and industry. Finer particles include soot (carbon) from diesel engines. Sulfur dioxide and NO_x react to form fine particles of sulfate and nitrate salts.	Of great concern are the PM_{10} particles with such small size (less than 10 μm) that they can penetrate deep into people's lungs. Diesel engines are the main source of PM_{10} particles in towns and cities.	The levels of pollution recorded depend on the method of measurement. In urban areas levels range from 10–150 μg/m^3. The UK target is that 50 μg/m^3 should not to be exceeded more than seven times per year.
Volatile organic compounds (VOCs) which are normally gases or liquids but evaporate so easily that their vapours stay mixed with the air.	VOCs include a wide variety of carbon compounds, which are mainly hydrocarbons from the evaporation of fuel and from unburnt fuel out of engine exhausts. Another source is leakage from pipelines and storage tanks.	The main VOC is usually methane from natural gas. People are worried about benzene, which makes up about 2% of petrol and can form in engines as fuels burn. Benzene can cause cancer (it is a carcinogen).	Measures often distinguish methane from other non-methane volatile organic compounds (NMVOC). The total NMVOC concentration normally reaches a concentration of several hundred parts per billion in urban air.

Figure 3.12

The key air pollutants which lower the quality of the air we breathe.

Questions

12 What does 1 mg of salt look like? Could you taste a salt solution containing 1 mg per litre?

13 Give examples of pollutants produced by car engines because:
- they do not burn their fuel completely
- the fuel contains chemicals other than hydrocarbons
- they run at high pressures and temperatures.

14 Why do car engines produce nitrogen dioxide even though there is no nitrogen in petrol?

15 Why is it important for a car engine to be adjusted so that the air/fuel ratio supplied to the cylinders is correct?

Key terms

Concentration units:

ppb = parts per billion by mass or volume

ppm = parts per million by mass or volume

$\mu g/m^3$ = micrograms per cubic metre (a microgram is a millionth of a gram)

mg/m^3 = milligrams per cubic metre (a milligram is a thousandth of a gram)

Ozone and its effects

Ozone often features in stories about environmental problems. The ozone layer in the upper atmosphere is a 'good thing'. This ozone, high in the stratosphere, protects living things by absorbing harmful UV radiation from the Sun. CFCs and other pollutants, such as the pesticide methyl bromide, tend to destroy the protective ozone layer.

In the lower atmosphere ozone is a 'bad thing' because it is harmful to living things. Ozone is a very reactive gas which irritates the eyes and causes breathing difficulties. Ozone is toxic and can affect the growth of crops as well as other plants including trees. The gas is reactive enough to attack materials such as fabrics and rubber, which gradually disintegrate if exposed to the gas.

Most of the pollutants shown in Figure 3.12 are primary pollutants. Ozone is a secondary pollutant which forms as sunlight shines on air containing nitrogen dioxide. The energy from sunlight splits up nitrogen dioxide molecules, NO_2, into nitrogen monoxide molecules, NO, and oxygen atoms, O. The free oxygen atoms then link up with oxygen molecules, O_2, to make ozone, O_3.

In still, sunny weather near cities the levels of ozone rise and mix with unburnt hydrocarbons from motor vehicles (VOCs, see Figure 3.12) converting some of them to compounds which together create a soup of eye-watering, irritant chemicals. A yellow haze appears. The action of sunlight on chemicals thus creates photochemical smog (Figure 3.13).

Figure 3.13

Photochemical smog blanketing the Hollywood district of Los Angeles.

Questions

16 In the UK ozone levels are higher in summer than in winter; they are also higher in the south than in the north. Suggest an explanation.

17 In the city of Los Angeles, on still sunny days, the levels of nitrogen dioxide and hydrocarbons reach a peak at 8 am. The peak formation of photochemical smog is between noon and 2 pm. How do you account for this?

18 a) Explain why the table in Figure 3.12 does not include carbon dioxide.
b) Why is there great concern about the carbon dioxide released from burning fossil fuels?

How do scientists monitor air quality?

The monitoring network

There are over 1400 sites in the UK where scientists measure air quality. The sites vary in the range of measurements and methods. The aims of monitoring are to:

- understand the scale of air pollution so that the authorities can devise cost-effective ways of dealing with the problems

- see if the UK and European standards for air quality are being achieved
- inform the public about air quality and warn people who may be at risk.

For each kind of information and measuring method there is a network of sites which work in the same way so that their data can be compared. Over 100 of the sites have automatic instruments which take measurements every hour and send the data to a central database (Figure 3.14). Some sites collect information less frequently using special apparatus to collect samples for chemical analysis (Figure 3.15).

Scientists measure particulates which are less than 10 μm across (PM_{10}s) (Figures 3.16 and 3.17). These are the particulates which people are likely to breathe in and inhale deep into their lungs. The particles are made up of soot mixed with other chemicals.

Figure 3.14

The network of automatic air quality monitoring sites in 2007. The instruments measure the level of pollution by oxides of nitrogen, hydrocarbons, carbon monoxide, sulfur dioxide, ozone and particulates.

Measurements and their accuracy

Errors

Any scientific measurement is affected by errors which are not mistakes but unavoidable differences between measured values and the true values. We can never be sure we know the true value of anything, but it is possible to assess how close to the true value a measurement is likely to be. So, analysts measuring air pollution can assess the uncertainty in their measurements.

There are random errors which cause repeat measurements to vary and scatter around a mean value. Averaging a number of readings helps to take care of random errors. The average value is the best estimate of the true value.

There may also be systematic errors which affect all measurements in the same way, making them all lower or higher than the true value. Systematic errors do not average out. Identifying and eliminating systematic errors is important for increasing the accuracy of data. Some systematic errors can be corrected by checking the reading with another instrument. If both agree, this increases confidence that the readings are correct. But it is still possible that something in the measurement procedure makes all measurements high or low. The better our understanding of the measurement process, the more likely we are to be able to spot, and remove, any causes of systematic error.

Question

19 Suggest reasons to account for the distribution of monitoring sites in the UK (Figure 3.14).

Sampling and analysis

Taking a sample is a vital first step in any measurement of air quality. It is important to try to make sure that the sample is representative of the air in the place where pollution is being monitored (Figure 3.18).

Much depends on where the sample is taken. One of London's air quality monitoring stations stands in Russell Square. The roads round the square are busy with traffic all day and much of the night, but the monitoring station stands a little way from the nearest road behind the café in the square where it is surrounded by trees and shrubs which protect the gardens from the full impact of the exhaust fumes. Moving the monitoring station to the roadside would certainly affect the readings. Even changing the height at which the sample is collected would have an effect.

Figure 3.15

A particulate analyser used for measuring the concentration of pollutant particles in a sample of air.

Figure 3.16

Trends in the UK annual emissions of pollutants from 1970 to 2000.

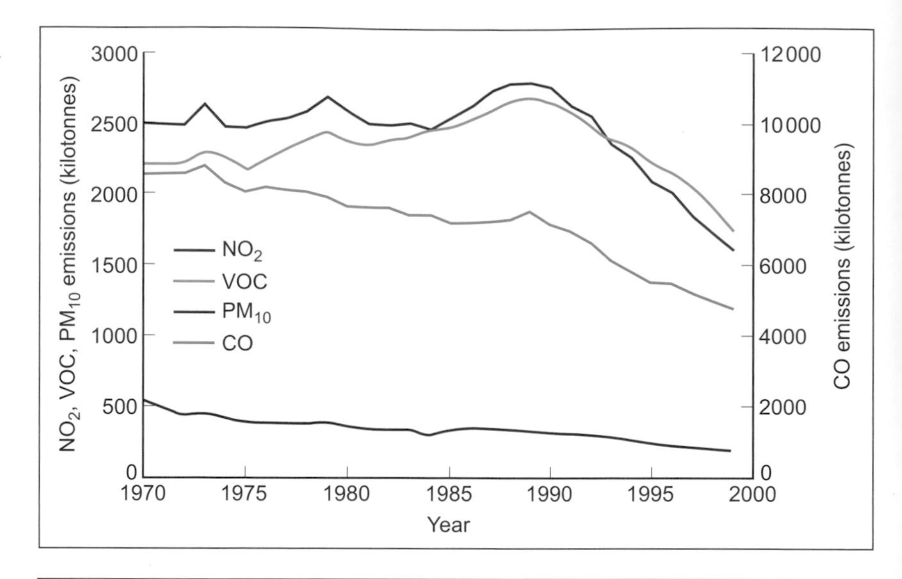

Figure 3.17

Trends in the UK emissions of PM_{10}s from a range of sources.

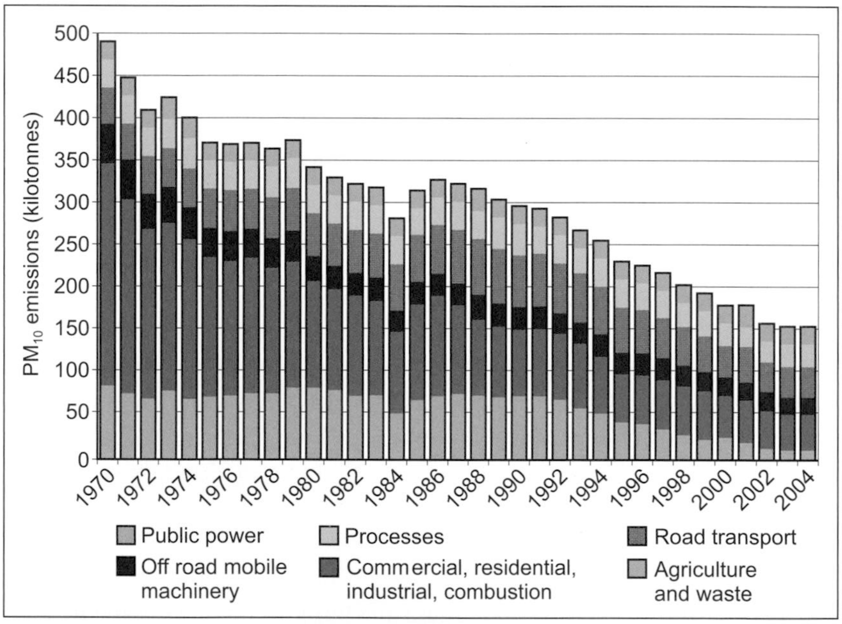

PM10 emissions (kilotonnes)

Public power | Processes | Road transport | Off road mobile machinery | Commercial, residential, industrial, combustion | Agriculture and waste

Questions

20 Summarise briefly what Figure 3.16 shows about the trends in air pollution by PM_{10}s in the UK.

21 Where in the UK would you expect the levels of these pollutants to be particularly high or especially low:

- nitrogen dioxide?
- carbon monoxide?

22 Which sectors have contributed most to the decline in PM_{10} emissions since 1970? Suggest reasons for the main trends.

23 Estimate the proportion of total PM_{10} emissions from road transport in 1970 and in 2004.

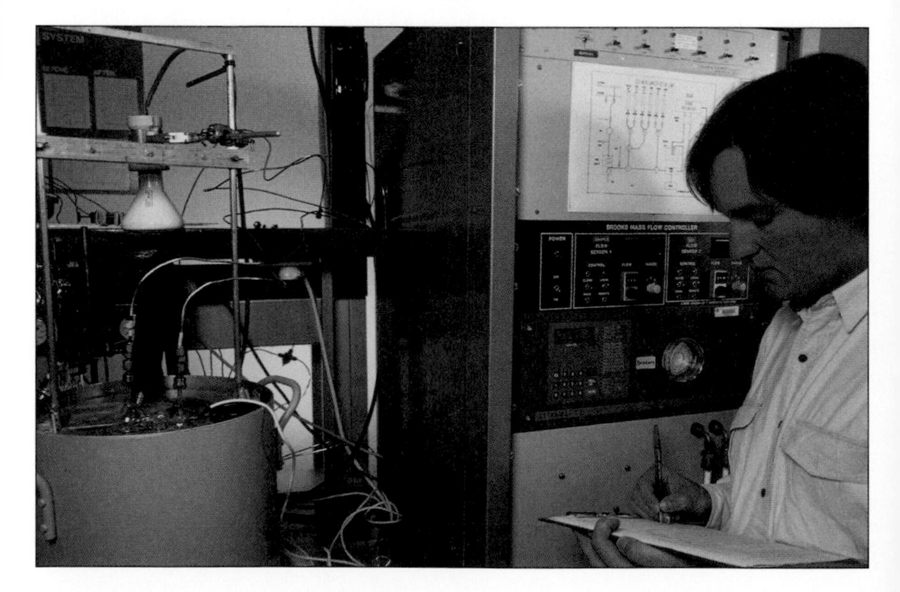

Figure 3.18

Air pollution monitoring. Air analysis: a scientist notes the level of carbon dioxide in a sample of air using a non-dispersive infrared gas analyser.

Another factor affecting the measurements is the sampling method. One common method is to use a pump to draw air continuously through an analyser. This method can introduce a range of errors. For example, some of the pollutants in the air may be absorbed onto the walls of the tubes as it flows through to the measuring instrument. Alternatively, some of the pollutant gases may react with each other once they enter the apparatus.

Another approach to sampling is to grab a sample of the air in a bag, syringe or evacuated bottle and then to take the sample to a laboratory for analysis. It is clearly very important that the containers are clean, free of leaks and made of an inert material. Otherwise, the composition of the sample will change between the time of collection and the time of analysis.

Given a sample, the analyst next has to choose a method of measuring the concentration of the pollutant which is specific to that pollutant, sensitive enough to cover the range of concentrations expected and reliable.

Uncertainty in the results

Analysts have a variety of techniques for checking the accuracy of their results. One method is to calibrate the instruments with samples prepared with a known concentration of the pollutant. Another technique is to collect several samples in the same place at the same time and to study the variation in the results.

The uncertainty in the measurement of gaseous pollutants is less than the uncertainty in the measurement of particulates. Uncertainties in the values for sulfur dioxide, for example, are in the range ±3%.

How do air pollutants affect human health?

It is one thing to measure the levels of pollution. It is quite another to determine the extent to which individuals are exposed to polluted air. On a day when air quality is poor, someone living in the suburbs and spending most of the day indoors experiences very different levels of pollutants from a commuter travelling into the city centre and walking along streets jammed with traffic.

The evidence about the effects of air pollutants on health is complex. Where air quality is poor there is generally a mixture of pollutants in the air and it is hard to work out the effects of the separate pollutants. The WHO has reviewed all the evidence of the health effects of three pollutants: nitrogen dioxide, ozone and particulates. The conclusion is that current levels of air pollution in Europe cause respiratory diseases and reduce life expectancy.

One classic study of the effects of particulates on health was the 'six cities' report published in 1993 by scientists from Harvard University. The study began in the 1970s. Groups of young people from six cities were selected for comparison. Their health records were monitored for the next 20 years. At the same time the concentrations of air pollutants in the cities were measured. Figure 3.19 shows some of the results.

The results in Figure 3.19 support the growing concern about the effects of particulates on health. The results of this epidemiological study suggest that the levels of particulates typically found in European cities reduce the average life span by 2–3 years.

Questions

24 a) Suggest three reasons why it is impossible to be sure that a single measurement of the concentration of nitrogen dioxide in a busy street gives a true value of the level of NO_2 pollution in the air.
b) What steps can scientists take to increase their confidence that the value they quote for the concentration of nitrogen dioxide in the air is close to the true value?

25 Which would you expect to introduce the bigger errors into the measurement of air pollution: the methods of sampling or the methods of analysis? Give your reasons.

26 The accuracy of estimates of PM_{10} emissions are improving but the level of uncertainty is still high. Estimates of PM_{10} emissions cover particles from diesel engines, from power stations, from industrial processes and from mining, quarrying and construction. Which of these estimates would you expect to be most reliable? Which would you expect to be least reliable? Give your reasons.

Figure 3.19

Death rates in six US cities plotted against (a) the PM_{10} concentration and (b) the $PM_{2.5}$ concentration. $PM_{2.5}$ particles are very fine, being less than 2.5 µm (micrometres) across. The 'rate ratio' is a measure of the death rate of the groups studied compared with a control group and allowing for factors such as age, body weight, income and smoking.

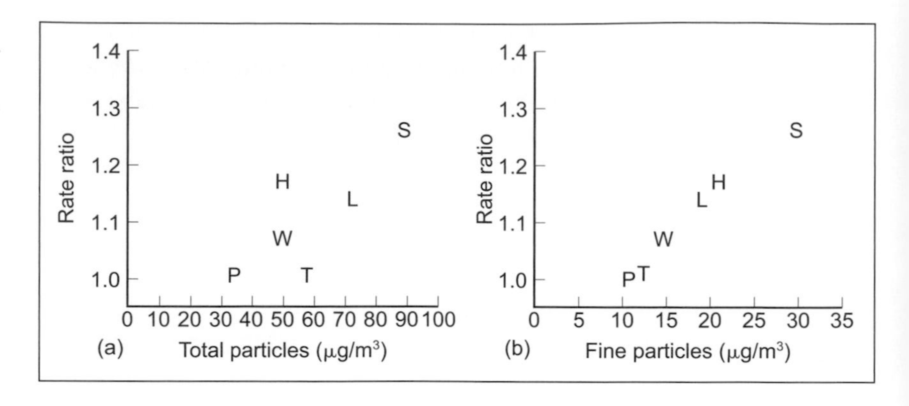

Questions

27 Study the two graphs in Figure 3.19.
a) Which graph shows a stronger correlation between the relative death rate and the concentration of particulates?
b) What conclusion can you draw from the two graphs?

28 Suggest reasons why the findings of the 'six cities' study were strongly attacked by scientists working for some energy and transport industries.

29 Why is an epidemiological study such as the 'six cities' study hard to replicate?

30 a) What would you say to someone who suggested that the small amount of pollution from car engines is not a problem because it quickly disappears into the atmosphere?
b) Does the pollution released into the air eventually disappear?
c) How does the weather affect the level of pollution?

The findings of the 'six cities' study have been fiercely debated since they were published. Some scientists challenged the interpretation of the data. There was a demand for replication and a search for supporting evidence from other lines of scientific enquiry.

In the years since 1993, more epidemiological studies have addressed the issue of long-term effects of air pollution on survival. One of these was carried out by the American Cancer Society. A report from this study in 1995 confirmed a significant association between the level of $PM_{2.5}$s and survival for people living in more than 100 different cities all over the USA. Since then, an extended follow-up of the study has confirmed the original findings and extended them by identifying a significant correlation between the levels of $PM_{2.5}$s and deaths from lung cancer mortality.

The main damage caused by the fine particles is probably caused by the chemicals they carry with them, which include unburnt hydrocarbons and carcinogenic compounds. A number of epidemiological studies have correlated the level of particulate pollution with a variety of effects on health such as asthma and heart disease. Generally the means by which fine particles cause these effects is not known. Much remains uncertain. Research continues to reduce this uncertainty and to improve our knowledge of the ways in which particles affect human health.

How can technology improve air quality?

Our town planning laws have been based on motor transport for over 50 years and huge sums have been spent building roads and setting up out-of-town shopping centres. As a result, our way of life now requires many people to travel a long way to work by road. This means that it would be economically very difficult to make the choice for unpolluted air rather than unlimited motor transport. What is appropriate in towns may be inappropriate in rural areas. So there are many difficult choices to be made if lowering the levels of pollutants in air is to be given priority.

To date, the main approach to improving air quality has been to find technical ways of changing the design of cars and the composition of fuels. More recently there has been growing interest in ways to cut down on traffic in towns and cities by means of a congestion charge and also by replacing petrol and diesel vehicles with electric cars, trams and buses.

Catalytic converters

As the number of vehicles on UK roads doubled between 1970 and the mid-1990s, the main method of cutting air pollution was to bring in catalytic converters. An EU regulation has required that all cars sold in Europe be fitted with catalytic converters since 1993 (Figure 3.20).

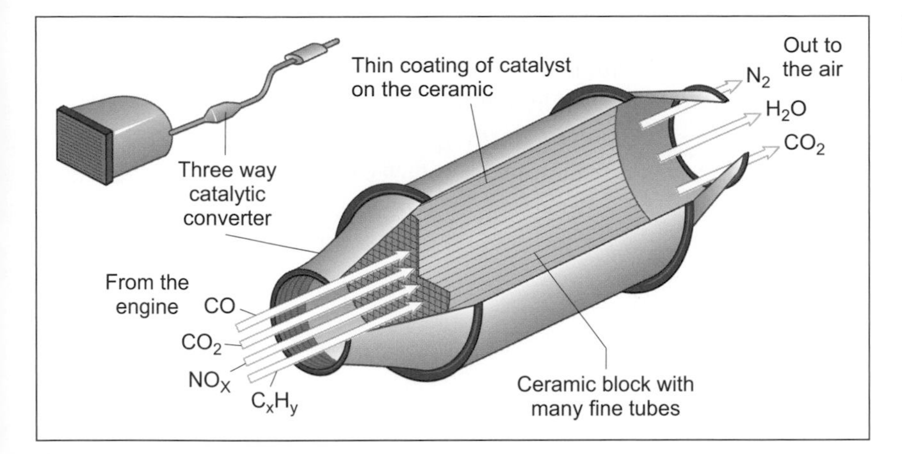

Figure 3.20

A catalytic converter.

The catalyst is an alloy of platinum and rhodium. These are both very expensive metals, but only very small amounts are needed. They are spread very thinly all over the surface of a ceramic block pierced with a vast network of fine holes. This gives a very large surface area of catalyst to speed up a whole series of chemical reactions which convert harmful gases (oxides of nitrogen, unburnt hydrocarbons and carbon monoxide) into gases which are harmless to health (carbon dioxide, nitrogen and water).

The catalyst is only effective when hot. On a short journey the catalyst may not get hot enough to start working. Also, converters only work effectively if the engine is properly tuned so that the cylinders are supplied with the right mixture of air and fuel.

Catalytic converters do nothing to reduce carbon dioxide emissions. In fact they may make matters worse by slightly increasing fuel consumption.

Questions

31 Show how a catalytic converter illustrates the idea that during a chemical reaction the atoms involved do not change but just recombine to form new groupings.

32 Suggest reasons for avoiding the use of cars for short journeys which could be made on foot or by bicycle.

Electric vehicles

In towns and cities, electric vehicles have the great advantage that they do not emit air pollutants. Electric trains and trams are powered by electricity directly from power stations. Electric cars carry batteries which have to be regularly recharged. To date this has limited the use of these vehicles because the driving range between charges is limited compared to vehicles running on petrol or diesel. Also it takes much longer to charge a battery than to fill up with fuel.

In the short term, the popular solution to poor battery life has been to develop hybrid cars which have both an electric motor and a petrol engine. In the UK these cars have a very low rating for road tax and are exempt from the congestion charge in London. When running on their electric motor at low speeds these cars do not emit air pollutants, but they are not as fuel efficient overall as the most economical conventional vehicles. The main benefit is the reduction in air pollution in slow-moving traffic.

New research is developing lithium-ion batteries that will power a vehicle to travel as far as a car will go on a tank of petrol. The aim is

Figure 3.21

The G-Wiz was produced for city use in the mid-2000s. The original model runs on eight lead–acid rechargeable batteries and has a range of about 50 miles per charge. The charging time is about six hours.

to provide batteries that are safe and can be recharged in minutes instead of hours.

Second generation hybrid electric vehicles have a small petrol engine and alternator that does not drive the car but just charges the battery. The battery pack can provide the energy needed for short local journeys and can be recharged from the mains overnight. On longer journeys, the engine runs to keep the battery charged.

An alternative to batteries is to use fuel cells. Transport for London ran a trial of buses powered by hydrogen fuel cells, which ended after nearly three years in January 2007 (Figure 3.22).

Like other electric vehicles, fuel cell buses do not emit pollutants that are harmful to health. However, if the hydrogen is made from natural gas, they still depend on fossil fuels whose processing releases carbon dioxide into the atmosphere.

Figure 3.22

A fuel cell is like a battery, but instead of containing a fixed amount of chemicals, it is supplied with a flow of hydrogen gas and air. The hydrogen combines with oxygen to make water and produce electricity. The only gas leaving a fuel cell is water vapour.

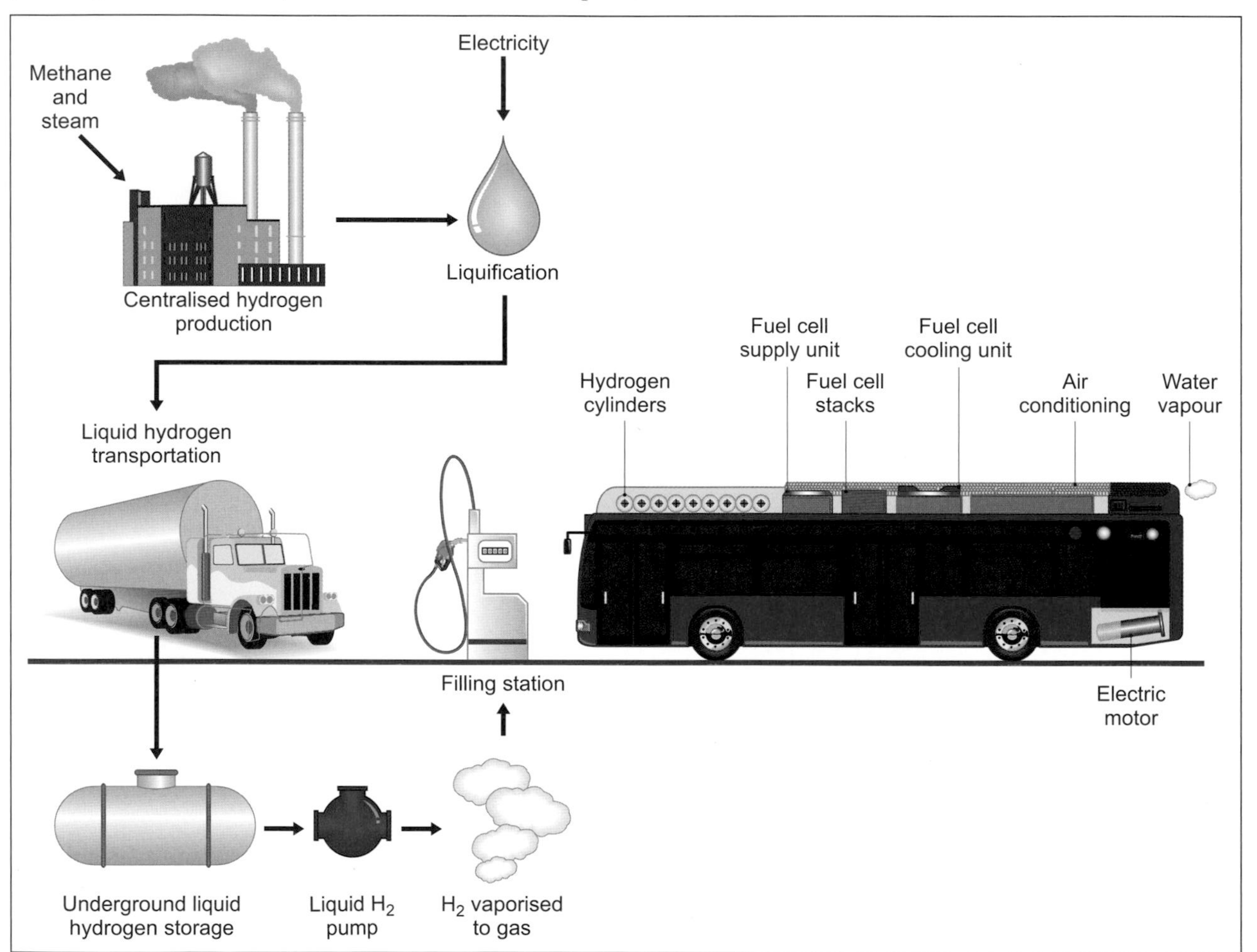

Fuel-efficient transport

Technology has made some contribution towards reducing the amount of fuel used for transport. More efficient engines give more kilometres per litre, and also reduce the amount of polluting gases that are released into the atmosphere. But really significant reductions in the amount of fossil fuels burned for transport cannot be achieved by science and technology alone. They require people to change their habits, both in the amount of travelling they do and in the way they do it.

Despite the fact that public transport is more fuel-efficient than using a car, many people prefer to use their cars. There was a huge increase in car travel during the final quarter of the twentieth century. This is predicted to continue during the first quarter of the twenty-first century. It has important implications not only for energy consumption but also for the quality of people's lives. More cars on the road increases congestion and results in slower and more frustrating journeys. Many people spend several hours each day travelling relatively short distances to and from work. The pollution caused by vehicles also reduces the quality of life for many city dwellers much of the time.

Nowadays, the average distance travelled each day by a person in the UK is about 20 miles (32 km) compared to just 5 miles (8 km) in 1950. Figure 3.23 shows the changes between 1995 and 2005 in how people travel in the UK.

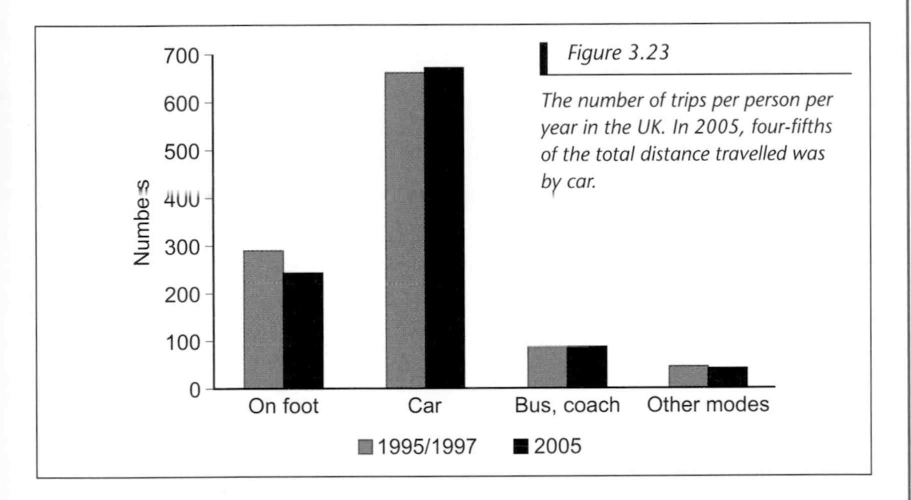

Figure 3.23

The number of trips per person per year in the UK. In 2005, four-fifths of the total distance travelled was by car.

The most obvious way to reduce the amount of fossil fuel used for transport is for people to reduce the amount of travelling that they do. But some travel is necessary, and so people need to be encouraged to travel by the most fuel-efficient means. One way to compare the fuel efficiency of different modes of transport is to look at the number of kilometres a vehicle can travel per litre of fuel. But it is more useful to compare them in terms of the number of passenger-kilometres, or tonne-kilometres, per litre of fuel (Figure 3.24).

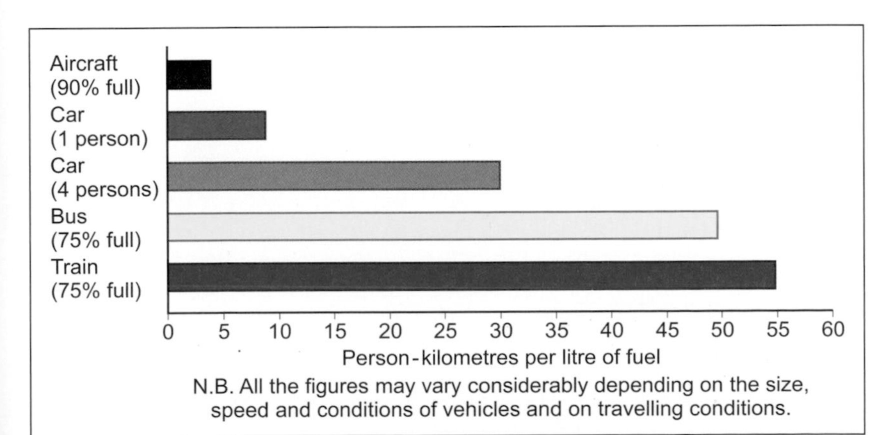

Figure 3.24

Fuel efficiency of different forms of transport.

Questions

33 Summarise in a table the advantages and disadvantages of these types of power supply for cars: petrol engine; battery and electric motor; hybrid engine and battery; fuel cell.

34 The amount of pollution created by electric vehicles depends mostly on the source of the electricity used to charge them. Explain why this is so.

35 Suggest reasons to account for these statements reported on the National Statistics website. From 1995 to 2005:
a) the number of walking trips fell by 16%
b) the average number of bicycle trips per person per year fell from 18 to 14.

36 a) Which are the two most fuel-efficient ways of travelling (other than on foot or by cycle)?
b) Which is the least fuel-efficient way of travelling?
c) Under what circumstances is a car a reasonably fuel-efficient way of travelling?

37 The number of kilometres a car travels on a litre of fuel is reduced when the car is full of passengers, but not by very much. Suggest a reason why.

38 Trains, buses and aircraft often last for 20–30 years compared to an average of about ten years for a car. They also travel many more miles each year than an average car. Explain how this helps to make them more energy efficient.

39 To what extent are these techniques practicable as ways to cut air pollution in towns:
- designing cars to be smaller and lighter?
- developing electric cars which can be recharged from the electricity mains?

False economy

Modern cars have more efficient engines than older cars, so some countries have tax incentives to encourage the scrapping of older 'gas guzzlers'. However, manufacturing a new car means that you have to produce the materials from which the car is made, and this consumes a very large amount of energy. Researchers in the Netherlands have calculated that scrapping cars three years earlier than normal results in 4% more energy being consumed overall.

Question

40 Identify examples from the news today where industry, commerce or the general public is resisting controls on transport which might otherwise improve health or the quality of life. What reasons are put forward in arguments intended to persuade the authorities not to bring in the controls?

How can regulations improve air quality?

Regulation to control pollution from vehicles

Since 1970 there has been a succession of directives from the EU to deal with air pollution. These have brought in tighter and tighter controls over the emissions from motor vehicles and industry. In 1991 the UK government introduced the Road Vehicles Regulations which set limits for emissions of carbon dioxide and hydrocarbons to be included in the MOT (Ministry of Transport) test for vehicles with petrol engines.

Particularly significant was the 1995 Environment Act, which created for the first time a statutory framework for local management of air quality and required the government to set air quality standards. Two years later saw the publication of the final version of the *National Air Quality Strategy*, which has subsequently been revised and updated. An important aim of the *National Air Quality Strategy* is to cut the number of times that air pollution is so serious that it is labelled as an 'episode'. The strategy sets targets for cutting the emissions of eight pollutants which have serious effects on health. Local authorities are responsible for gathering the data they need to assess air quality in their area and make sure that the targets are met.

Review Questions

41 Give an example to show that the properties of a chemical compound are very different from the properties of the elements it is made from.

42 The gases from vehicle exhausts are usually invisible and yet they are both changing the world's climate and causing local air pollution. Outline, in note form, the arguments you would use to try to persuade someone to avoid driving when possible so as to cut down on pollution.

43 Energy transfers in a car can happen in various ways. Give examples of these types of energy transfer:
- mechanically by forces acting
- thermally by heat flowing from high to low temperatures
- electrically
- by radiation.

44 For the same journey, diesel vehicles are more efficient than petrol vehicles so they give out less carbon dioxide. Their emissions of carbon monoxide and unburnt hydrocarbons are lower but their emissions of particulates and nitrogen oxides are higher when compared to petrol cars fitted with a catalytic converter. What information would you need to decide whether a petrol or a diesel car is 'greener'?

45 Is it true that most people benefit from widespread use of motor vehicles while only a minority suffer the ill effects? What are the implications of this in a society in which the many can outvote the few?

46 a) What are the scientific and technical uncertainties which make it hard to know what to do about air pollution?
b) What are the social, political and economic difficulties which make it hard to tackle air pollution problems effectively?

47 'Cutting down pollution from cars is easy, any driver can do it. All you have to do is drive less'.
a) Is this a realistic way of cutting air pollution?
b) What restrictions on the use of private cars do you expect to see coming in during your lifetime?

48 'The polluter should pay'. How might this principle apply to motor vehicles? Who should pay:
- drivers of private cars?
- passengers on public transport?

49 a) Suggest some of the factors that would have been considered when setting upper limits to the concentrations of pollutants in the air.
b) Why are the values for the upper limits not set at zero?

50 Transport policy has been based on the assumption that we have to preserve people's freedom to travel, whatever the costs. Politicians hope that there will be technical solutions to pollution from transport. If you were an adviser to the Minister of Transport, what strategies would you recommend to reduce fuel use for transport in the UK?

Chapter 4

Medicines to treat disease

Figure 4.1

Dispensing homeopathic pills. Homeopathic remedies such as these have not generally undergone rigorous scientific testing to show that they work, but they have been licensed for quality and safety. A quality medicine contains what it says it does, is free of contaminants and is made by a production process that is rigorous enough to guarantee consistent standards.

The issues

The production of new medicines can be controversial, because new treatments can make a big difference to people's lives but are very expensive to develop. The benefits of new medical drugs have to be weighed against the risks and the costs of development. In the UK the National Health Service has to make complex judgements about the treatments it prescribes, taking into account the evidence from clinical trials, the costs and the priorities of the public. There is also a lively debate about the effectiveness of complementary medicines.

The science behind the issues

Medical innovations based on science can help to improve the quality of life of many people, but at the same time there may be unintended and undesirable side effects. Clinical trials are a crucial method of collecting valid evidence that a new approach works, or is an improvement on existing treatments. These trials are carefully controlled, staged studies which involve testing on cells and tissues, then on animals and finally on human volunteers.

Clinical trials are carried out with a representative sample of the population. Carefully planned sampling methods are used to ensure the sample is selected without bias, and that the sample size is large enough for the effects of random variation to be insignificant.

What this tells us about science and society

Medicines are assessed for safety, quality and how well they work. Economic and ethical issues have to be considered when animal and clinical trials are carried out. Ethics and economics also influence national and international regulation affecting licensing and marketing of medicines on the international market.

Many complementary and alternative medicines are licensed to ensure they are safe enough to use in treatments. Scientists are uneasy that these medicines are often not tested with repeatable measurements and observations to see if they work (Figure 4.1).

Health and medicine

Science-based medicine generally emphasises cures targeted at particular symptoms or causes. This has been so since the first successes of medicines based on the germ theory of disease, such as Pasteur's dramatic success with vaccines for rabies (see Chapter 1) and Fleming's discovery of penicillin (see Chapter 2).

There are, however, many conditions causing chronic, long-term discomfort, where conventional medicines may help symptoms but cannot provide a cure. Back pain is the single biggest cause of missed time at work in the UK. It is sometimes stress-related and is often difficult to treat. Other conditions, such as eczema, asthma, allergies, general fatigue and headaches, do not always respond well to treatment either. Increasingly, people with these types of problems are turning to alternative therapies in search of a treatment that works. Others choose to rely on complementary medicines for a variety of reasons to do with personal preferences and beliefs.

Figure 4.2

*The Madagascar periwinkle (*Catharanthus roseus*) is a source of anticancer medicines such as vinblastine and vincristine. The discovery of these medicines illustrates the importance of listening to and understanding the ways that people practise their traditional medicine. Of the 250 000 flowering plants in the world, only 5000 have been tested for useful drugs in laboratories.*

Around 25% of modern medicines are based on chemicals originally found in plants. The Madagascar periwinkle, for example, is a small plant with a long history (Figure 4.2). In the 1950s its reputation as a folk remedy for diabetes led Canadian scientists to investigate the plant for substances which could be used in place of insulin. Instead they discovered that extracts of the plant had a powerful effect on cell division, reducing the number of white blood cells in mice suffering from leukaemia. Research continued, and in 1958 chemists successfully isolated the active chemical which reduced the number of leukaemia cells. The drug, called vinblastine, was first tested in 1960 on a patient with a large cancerous tumour. The tumour rapidly disappeared. For more than two years the patient was free of cancer, but then the symptoms reappeared. Fortunately, another drug isolated from Madagascar periwinkle, vincristine, brought the disease under control once more. In 2002 a further chemical from Madagascar periwinkle, vinorelbine, became available to the NHS for use in advanced breast cancer and some types of lung cancer.

Key terms

Drugs are chemicals which may be natural or synthetic. They are the active ingredients of medicines for the treatment, relief or prevention of disease. People also take drugs for pleasure, stimulation and relaxation.

Medicines are formulations of drugs mixed with inert materials. The mixture is made up to help get the drugs into the body. Medicines may be in the form of capsules to swallow, ointments to rub into the skin, solutions to inject or drops for the eyes and ears.

Question

1 Why is it so important to isolate the pure drug when researching plant extracts for a new treatment?

Key term

A drug is **efficacious** if it is effective at the dose tested and against the illness for which it is prescribed.

Developing new medicines

The first step in the development of a new medicine (Figure 4.3) is to identify a disease for which the existing treatments are inadequate or not available. Detailed research into the disease helps to pinpoint the processes in the body which could be affected by the medicine.

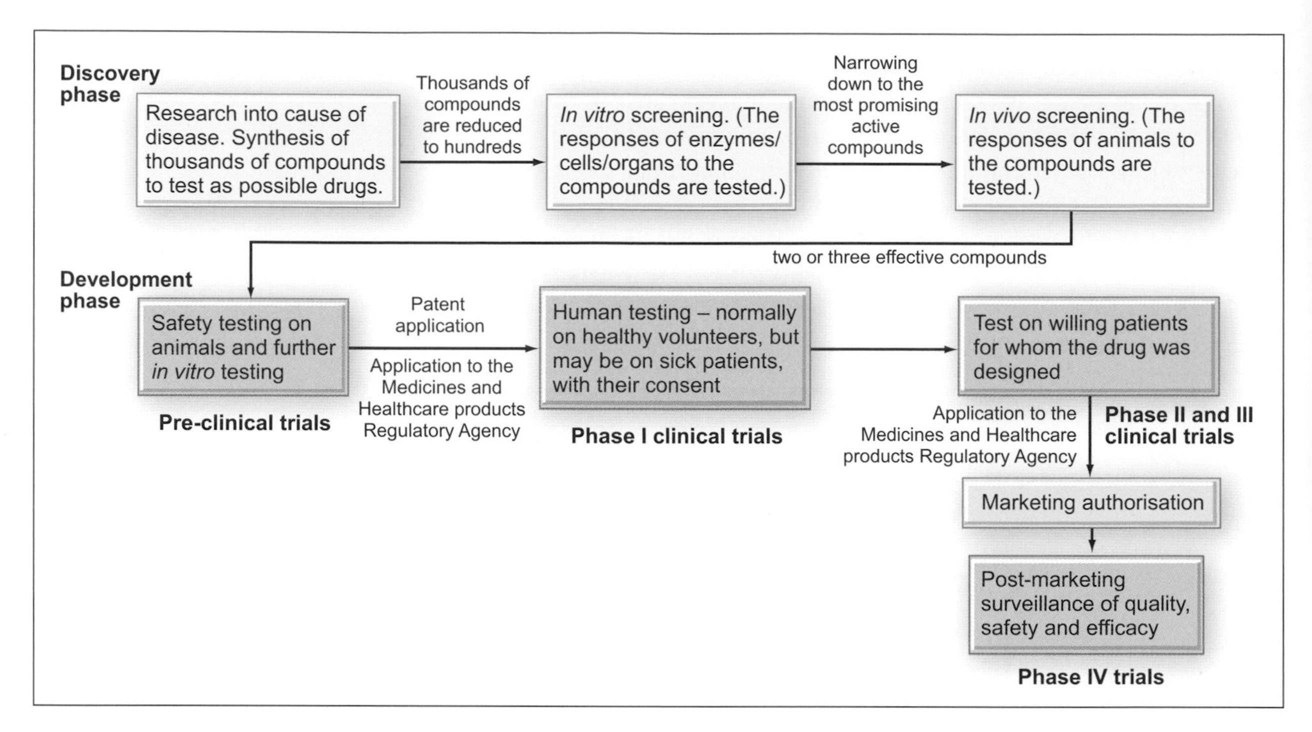

Figure 4.3

Stages in the development of a new medicine.

The action of a medicine is closely linked with the structure of its molecules. Nowadays a research team looking for alternatives for an existing treatment might start with the structure of an established drug and explore ways of changing it to produce a better product. Much of the early planning can be done by computer modelling (Figure 4.4).

For a research compound to become a marketed product it has to cross four critical barriers: it has to be safe; it has to meet a definite need for treatment; it has to be technically possible to develop and manufacture the product; and there has to be a big enough market to ensure that the development work will be commercially viable.

Figure 4.4

Computer design speeds up the discovery of new medicines. Even very small changes in the shape of a potential drug molecule can improve its performance, and computers help scientists to screen huge numbers of variations, a few of which might make a big difference in the effectiveness of treatment. Drug structures on screen are matched with likely sites in the body where they might be effective.

Testing new medicines

All claims made for the medical benefits of a new medicine and for its proposed use in treatments have to be supported by scientific evidence. Medicines must normally satisfy the requirements for quality and safety. It is also important that they work. In other words, that they are efficacious. Manufacturers are therefore obliged to carry out a rigorous testing programme before bringing a new medicine to the market.

In vitro testing

Testing a new medical drug begins with screening of thousands of chemicals to find some which may be suitable. Testing of a few chemicals of interest then takes place on preparations of cells, tissues and even whole organs. This is called *in vitro* testing. The preparations are placed in plastic containers, each with a solution of all of the substances needed to keep the preparation alive. The compound under test is added and its effects on the preparation observed.

If a compound has the desired effect *in vitro*, it begins to look interesting. The decision to proceed further with a promising compound is a crucial one, because the investment in time and money now becomes substantial.

Animal testing

Cells *in vitro* do not have the complex interactions between cells and tissues which exist in living bodies. *In vivo* testing on new medicines is needed, and this begins in animals. Several years are spent looking at every aspect of the new medicine (Figure 4.5).

Scientists use animal models of the disease in initial studies to see how well a new medicine works. If the drug progresses to later stages of development, studies may be performed in humans to test the medicine in double-blind, randomised controlled trials.

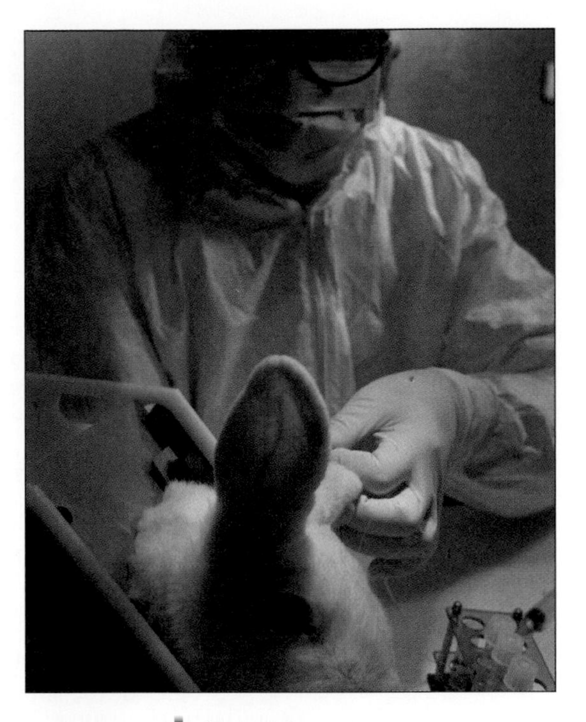

Figure 4.5

A researcher taking a blood sample from the ear of a white rabbit in a research project exploring antiviral medicines and gene therapy.

The stages of animal testing in development of medicines

There are international and national regulations to ensure standardised methods are available for use at each stage of testing.

Animal testing explores whether any action that looked promising *in vitro* is active in living organisms. Testing on animals often gives only limited information about how well a medicine will work in humans, but this stage of testing is necessary to see if the medicine causes damage to tissues and organs. Part of the purpose of *in vitro* and *in vivo* trials is to ensure that later testing in humans is as safe as possible.

Animal trials also help researchers to find out what happens to a medicine in the body. For example, the effects of a medicine on muscle may depend on how it is absorbed in the gut, and how efficiently it is broken down by the liver and excreted by the kidneys.

Toxicology testing is used to estimate the dose needed to produce signs of acute poisoning. The effects of lower doses over a longer period of time are also studied. Acute toxicity testing is one of the more controversial animal tests, because it may kill the animals.

The purpose of toxicity studies is to identify the dose that leads to seriously harmful effects in a short time. Scientists use the results to calculate the minimum dose that kills. The information from the tests is then used to choose the dose for a larger group of animals that is given this dose several times at regular intervals. Later the results of all these tests can help the scientists estimate the safe doses for human trials.

It is very important to check that a medicine which is safe for adults does not harm human reproduction. Safety tests are carried out in animals to find out whether new drugs are toxic to embryos, fetuses or babies. This includes tests to see if medicines pass through the placenta to the fetus or pass into breast milk.

Questions

2 It is important to doctors and patients that medicines are tested to make sure they are efficacious and adequately safe.
a) Explain why all medicines will have side effects.
b) What do you think 'adequately safe' means?

3 The drug thalidomide was prescribed to pregnant women to treat morning sickness. It was very effective. Sadly it caused birth defects in more than 10 000 children in the early 1960s. Animal trials were limited, by modern standards, before the drug was released for use with people. Which of the tests described here do you think may have been introduced as result of the thalidomide tragedy?

4 Explain why, in the testing of the effects of new medicines:
a) computer modelling cannot replace all animal testing
b) animal testing cannot replace trials in humans.

Key terms

Inferences are conclusions based on data from a study or experiment.

Systematic sampling is a technique to ensure no one group in a population is over or under represented in the study sample. A representative proportion of each group is allocated to the study samples.

Animal models

Scientists use, as models in their research, animals with a disease that is either the same as, or very similar to, a disease in humans. Animal models are used to test new treatments before they are given to humans. Animal models can be used to test medicines because many of the chemical processes carried out in cells are the same in all living organisms.

Sometimes bacteria and invertebrates such as worms and fruit flies can be used as models for early stages of testing, depending on the particular process being affected by the medicine. Where there is strong scientific justification, scientists are allowed to test some types of medicines on monkeys and apes. These primates are closely related to humans.

Clinical trials

Sampling

Sampling methods are important in science and social sciences. Correct sampling methods allow reliable inferences to be made about a wider population using data collected from a relatively small sample.

The method of sampling for large-scale studies depends on the context of the study. The larger the sample size, the more confident it is possible to be about any inferences drawn from the sample. Large samples mean that scientists can be more certain that any effects being observed are not just due to random variation normally found in populations. However, there are limits, because larger samples increase the cost and complexity of any study.

Random sampling is used in situations where valid conclusions rely on being able to show that no one individual is more likely to be selected to take part in a study than any other. In a trial to discover the effect of a new medicine, random sampling is appropriate because it is important to be sure that there is no bias involved in the selection of the treatment and control groups. Any bias can mean that the people selected are not representative of the wider population. Selecting an unrepresentative proportion of particular groups invalidates the experiment. The sample has to be representative of social status, age, race, lifestyle and other relevant factors.

A refinement of random sampling is stratified, or systematic sampling. This sets out to make sure that a sample includes the correct proportion of different sub-groups in the population as a whole. Individuals are still allocated randomly to any treatment or control group.

Figure 4.6

Technicians counting tablets into plastic bottles in preparation for a clinical drug trial.

Testing in humans

Testing a new medicine in humans starts with phase I trials. These are short-term studies in a small number of volunteers to test how well humans tolerate the medicine. Phase I volunteers are normally healthy, but there are circumstances where medicines are tested first in sick patients. Phase I trials in sick patients might be carried out, for example, when testing a medicine to treat late cancers which has unpleasant side effects.

Several hundred patients are involved when phase II trials begin. These are short-term studies to identify the relationship between dose and response. Phase III normally involves large studies in several thousand patients to confirm that the drug works and is acceptably safe.

Once a medicine has been approved for marketing, tests continue, to extend the evidence about the medicine's safety or how well it works. This is phase IV of the clinical trials, and is often also funded by the pharmaceutical company. During phase IV, data can be collected from much larger samples of patients – nationally and internationally. This allows more valid inferences to be made from the data. The whole clinical trial programme may take as long as 8–10 years to complete, with phase IV continuing even longer.

The 'gold standard' for clinical trials

The most reliable clinical trials involve randomised and controlled double blind studies. People are only included in the trial if they agree to take part, and they understand any potential risks. The patients who are treated with the new medicine are compared with a control group (Figure 4.7). People in the control group receive either a placebo or a standard treatment, which is the best currently available treatment for the disease. Obviously the new medicine and the placebo or standard treatment must look similar to doctors and patients.

The patients who have been accepted onto the trial are assigned randomly either to the group to receive the new medicine or to the control group. Randomisation and numbering is done by computer. The computer programme takes note of personal and medical details to make sure that treatment and control groups are similar.

Each patient has a number and receives a numbered box containing their medicine. Neither the patient nor the doctor caring for the patient knows which medicine the patient receives. In this sense, both are *blind* – hence the name 'double-blind' studies. Double-blindness effectively prevents bias creeping into the results, which would otherwise happen if doctors chose which patients were to receive the medicine. Figure 4.7 summarises the sequence of events.

Key terms

In a **randomised controlled study** there are two groups: one group receives the treatment being tested, while the control group receives either no treatment or some standard treatment. Patients are randomly assigned to each group.

A **double blind study** is one in which neither the patient nor the doctor knows whether the patient is receiving the new medicine or the control treatment.

A **placebo** is a 'dummy pill' given as a control during a clinical trial. Symptoms can be alleviated when a placebo is administered, partly due to patients' expectations rather than any property of the medicine. This is called the **placebo effect**.

Figure 4.7

Randomisation and 'blindness' are essential components of clinical trials.

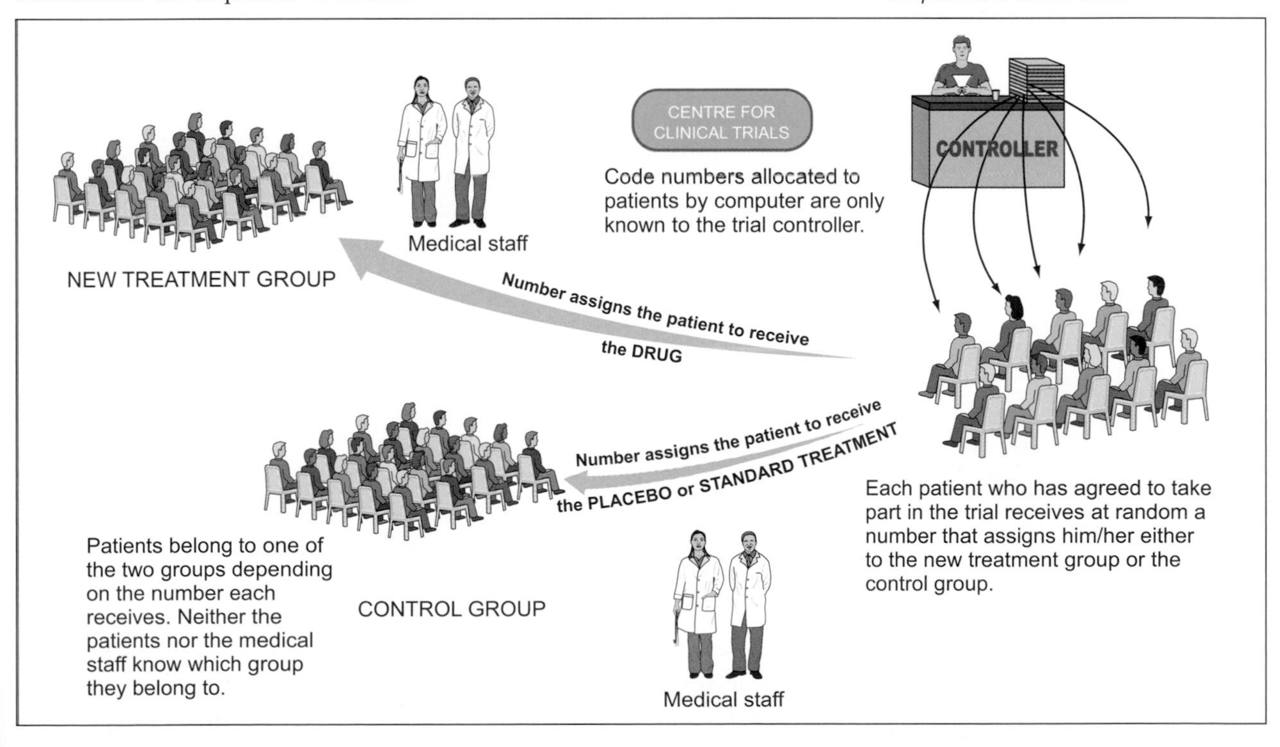

Questions

5 During clinical trials the control group might be given a placebo or the standard treatment.
a) Why might it be ethically acceptable to give the control group a placebo in a phase II trial?
b) Why is it not normally ethically acceptable to give the control group a placebo during a phase III trial?
c) Suggest two circumstances under which it would be acceptable to give the control group a placebo in a phase III trial.

6 It is important to select a representative sample of the population for testing a new medicine during phase III of clinical trials.
a) What group of people make up 'the population' in this context?
b) Suggest some difficulties that might arise in making sure that the selection is really representative.

Scientists do not intend to be biased, but bias can be introduced without a research team realising it. For example, doctors running a trial might subconsciously avoid putting sicker patients into a treatment group if they were to know that the medicine being tested has unpleasant side effects. This would mean that as the trial went on, the control group would have more and more of the sicker patients in it. The less sick people in the new medicine group might then do better than the control group, making it look as if the new medicine is better than it actually is, compared with the standard treatment.

The scientists analyse the data from randomised double-blind controlled clinical trials to see if there is evidence that the new drug works. A treatment that has 'worked' is one where any difference between a treatment and placebo group is big enough not to be attributable to normal variation, including genetic differences affecting how people react to a new medicine.

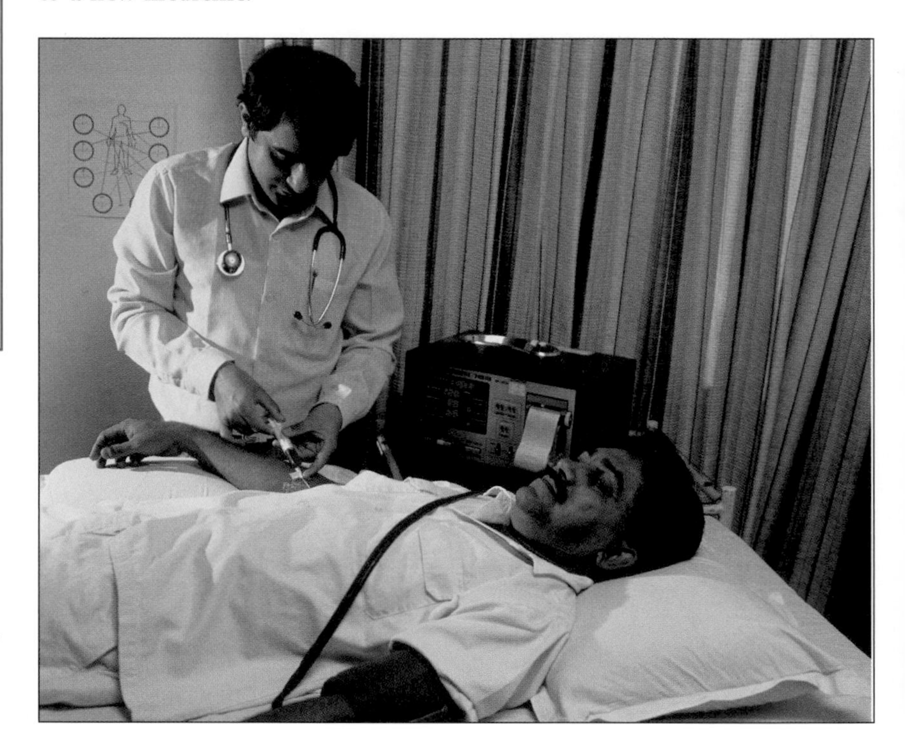

Figure 4.8

A doctor taking a blood sample from a volunteer during the clinical trials of a new drug.

Regulation of medicines

The regulators

There are two ways that medicines can be licensed for use in the UK:

- through the European Medicines Evaluation Agency (EMEA), covering all EU countries
- through the Medicines and Healthcare products Regulatory Agency (MHRA), which grants licences for the UK only.

The MHRA acts on behalf of the government to make sure that the benefits of new medicines outweigh the risks of harmful side effects. The MHRA is involved in the development phase of a new drug because it authorises clinical trials in humans once the results of tests with animals are known (see Figure 4.3, page 52). In addition, an independent ethics committee has to approve clinical trials before they start (see Chapter 5).

A series of independent advisory bodies support the MHRA's work. As well as experts, these bodies also include some members of the public with no particular relevant expertise. The MHRA works within the EU to negotiate and then implement and enforce relevant aspects of European law, and to transpose EU directives into UK law.

Two further organisations involved in overseeing the production and use of medicines in the UK are the National Institute for Biological Standards and Control (NIBSC), and the National Institute for Health and Clinical Excellence (NICE).

NICE is an independent organisation responsible for providing national guidance on promoting good health and preventing and treating ill health. NICE appraises the cost effectiveness of new medicines, and makes recommendations to the NHS. If NICE approves a new medicine or treatment, then the NHS in England and Wales is required to provide the funding to make it available to patients.

Evaluating clinical trials

In the UK, staff at the MHRA review the data from clinical trials. Once the MHRA concludes that the clinical trials offer evidence that a new medicine works and is safe, it grants a licence which allows the manufacturer to market the medicine (Figure 4.3). In the USA permission to manufacture and market a new medicine is given by the Food and Drug Administration (FDA). Other countries also have their own regulatory bodies.

Outcomes of clinical trials which are published in peer-reviewed journals are taken more seriously within the scientific community than media reports of trials (Figure 4.9) for example. The editors of peer-reviewed journals send all articles to expert reviewers before they are published. They choose reviewers with a good reputation in the field of research. This means that research published in these journals has credibility. Other scientists know that someone who understands the research has critically evaluated the process, data and inferences from the trials (Figure 4.10).

Figure 4.10

Research scientists spend a lot of their time communicating their research to the scientific community. This involves publishing in peer-reviewed journals and attending scientific conferences where research is presented and discussed.

Question

7 Some clinical trials involve three groups: an active group that receives the medicine being tested, a placebo group that receives a placebo, and a third group who receive no treatment of any kind. This third 'natural history' group's condition is allowed to run its course. Explain how the practice of adding a natural history group to a clinical trial might help separate any real benefits of the active medicine from placebo effects.

THE LANCET

"These two trials encourage cautious optimism that interleukin-6-receptor blockade might be a new treatment option in rheumatoid arthritis and juvenile idiopathic arthritis."

Figure 4.9

Articles published in The Lancet *have high status. This is a peer-reviewed journal. Scientists compete for their work to be published in this journal.*

Key term

Peer review is a process used for checking the work carried out by scientists. Scientists working in a similar field, but from outside the immediate research team, assess the quality of the research.

Questions

8 a) Explain, with the help of examples, why society exercises control on the development of new medicines.
b) What factors do regulatory bodies have to take into account when deciding whether to license or authorise the use of a new medicine?
c) Explain the difference between the roles of the MHRA and NICE in managing the availability of medicines in the UK.

9 Why is it important to have procedures for people to report side effects of new medicines?

Licences for medicines

Licences for new medicines are often very specific. A medicine may be licensed for one type of cancer, for example, but not for another. Alternatively a medicine may only be licensed for a particular stage of a disease.

Doctors in general practice are permitted to prescribe whichever medicines they like, even if they are not licensed. Sometimes doctors may prescribe 'off list' if they consider this may benefit individual patients.

An important task of the MHRA is to monitor licensed medicines. The organisation collects reports of adverse reactions to drugs from doctors, dentists, pharmacists, patients and manufacturers.

Some doctors are reluctant to prescribe as soon as a medicine is licensed, preferring to wait for more information about long-term effects and efficacy. Some NHS hospitals or trusts won't let their doctors prescribe new medicines which they know are under review by NICE until they have been recommended for use in England and Wales, or recommended by the Scottish Medicines Consortium (SMC) in Scotland. NICE assesses medicines on the basis of their cost-effectiveness. There can be a backlog of new medicines waiting to be assessed, so sometimes new treatments take a while to become available on the NHS.

Regulation of animal experiments

Manufacturers and research scientists have legal obligations to the animals used during trials of new medicines. Britain made history in 1876 when it became the first country to pass a law which controlled research with laboratory animals. The law was updated in 1986, and the Animals (Scientific Procedures) Act established three main areas of legal control over animal research.

- The people who carry out the research must be competent to do so (Figure 4.11). Competence to carry out *particular* types of experiment on *named* species of animal in the course of an *approved* project is established by the issue of a personal licence.
- The facilities for looking after animals before, during and after experiments should be suitable (Figure 4.12). A certificate of designation is only granted to scientific institutions where the places for keeping animals meet the necessary high standards.
- The likely benefits of the research should outweigh any possible distress to the animals. Only when the balance in favour of human welfare is established will a project licence be issued for the work to go ahead. Even then the law aims to protect animals from unnecessary distress. Scientists are required to prevent or relieve pain wherever possible.

Figure 4.11

Laboratory mice are used for safety tests on medicines. The size and genetic make-up of mice makes them suitable for medical and toxicity tests. Licences are issued to the individuals who are responsible for animal experiments.

The Animal Procedures Committee was appointed under the terms of the 1986 Animals Act. Its job is to advise the Home Secretary about scientific procedures involving animals. The committee examines applications for animal experimentation, licence them and advises on reported infringements.

Membership of the committee reflects a wide range of views on animal use. The Act stipulates that at least two-thirds of the members have full registration as medical practitioners or vets, or that they have relevant biological qualifications. At least one member is a lawyer, and no more than half the membership can have held an animal-testing licence during the last six years. There is normally an academic philosopher on the committee.

Complementary medicine

Complementary medicine is a term which covers an enormous range of different approaches to healthcare and healing. Complementary treatments may be used alongside orthodox therapies or in place of them. Most complementary therapies involve long consultations with the patient, but some can take place without the patient even being present. Many of these therapies have been around for hundreds if not thousands of years, but some are relatively new.

People turn to complementary therapies for a variety of reasons. If conventional medicine has failed to cure an illness then people may look for alternatives (Figure 4.13). Some people do not like the idea of taking conventional medicines for a long time, and are looking for an alternative way to manage or cure their condition. Yet others simply want to use a system that they perceive to be more 'natural'.

Explanations for the way some complementary therapies work do not fit with accepted scientific explanations for the way the body works. Many scientists and doctors cannot accept that these alternative treatments work because of the lack of evidence for their efficacy.

Some scientists support research into alternatives where there seems to be evidence of efficacy. Examples include the treatment of pain or nausea by acupuncture, and treatment of allergies by homeopathy. Finding a mechanism for how the treatments work could be part of the research. Others think that a system of medicine which has no plausible scientific explanation should not be investigated further.

People continue to debate whether or not complementary medicine should be subject to the same degree of scrutiny as conventional medicine. There are examples where lack of regulation has caused serious harm, such as the use of a Chinese medicine slimming pill which led to a number of cases of kidney failure and cancer in Belgian women in the 1990s.

A choice of therapies

People using conventional medicine know exactly where to go if they are unwell – they visit their doctor. But when people use alternative therapies, they have to decide which therapy and which practitioner to choose. Different therapies seem to work better in different conditions, but to add to the confusion some conditions can be helped by a variety of different therapies – it all comes down to personal choice.

One clear example is the case of allergies. Allergies have become increasingly common over the last 20 to 30 years, affecting people of all ages. They are caused by an overreaction of the immune system to materials in the environment. Examples of illnesses caused by allergies are hay fever, asthma, perennial rhinitis (a constantly runny or stuffed

Figure 4.12

Animal welfare has to be taken into account when institutions are licensed to carry out experiments on animals. Laboratory mice are the most commonly used animals for biological and medical experiments.

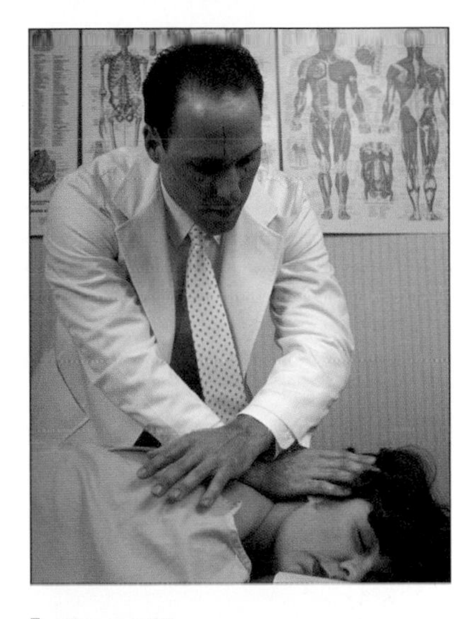

Figure 4.13

A chiropractor at work with a patient. Chiropractic involves manipulation of the joints between the vertebra of the spine.

Key terms

Alternative medicine describes treatments that differ from orthodox, drug-based medicine. Examples include homeopathy, Chinese medicine and acupuncture.

Complementary medicine includes alternative diagnostic and therapeutic approaches which are used alongside conventional medicine.

up nose), and the type of eczema which gives sufferers an inflamed, itchy and flaky skin (Figure 4.14).

The body may be sensitive to a whole range of things, which can include grass pollen, fungal spores, animal hairs, dust mites and certain types of food or medicines. The allergic response to some foods, such as peanut allergy, can be very violent, with swelling of the lips, mouth and tongue, nettle-rash and vomiting. In really severe cases, people collapse and die because swelling in the mouth blocks the airway and the heart fails.

However, the majority of allergy sufferers have much milder symptoms. Often conventional medicine can only offer medicines to treat the symptoms, which patients have to take for a long time, even years. Increasing numbers are looking for alternative ways to treat their allergies (Figure 4.15) – and there is a wide choice open to them.

There can be hazardous side effects from some alternative treatments too. Some herbal remedies contain powerful active ingredients which have to be used with care. Now that herbal remedies are regulated, the public are encouraged not to buy unlicensed remedies where the levels of active ingredients can vary from one preparation to another.

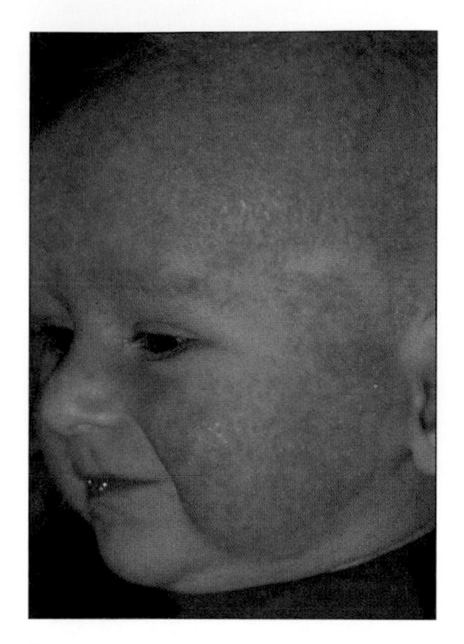

Figure 4.14

An eczema rash covering most of the skin on a baby's face. When the immune system becomes over-sensitive the body itself comes under attack, which can lead to distressing symptoms like this eczema – and much worse.

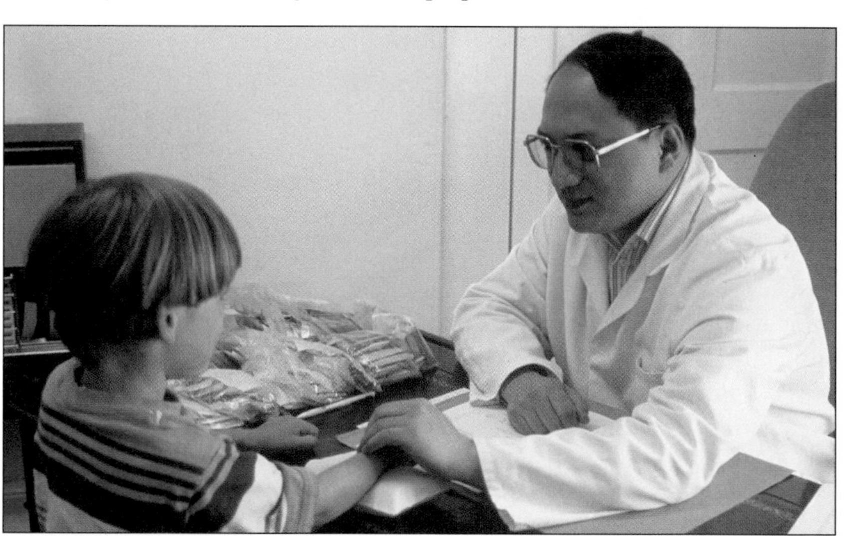

Figure 4.15

A Chinese herbal doctor taking the pulse of a young boy. On the desk are various herbal remedies. Chiropractic, Chinese herbal medicine, acupuncture, homeopathy and clinical ecology all claim to help reduce both asthma and eczema.

Conventional doctors are licensed, but complementary practitioners don't have to be. Unqualified practitioners, or people self-medicating, may not be aware of potential dangers such as interactions between complementary and conventional medicines. For example, new evidence suggests that St John's Wort can interact with a range of medicines including oral contraceptives, some anti-convulsant medicines and medicines taken by patients with HIV infection.

Does complementary medicine work?

Nearly 40% percent of US adults aged 18 years and over use some form of complementary therapy. In the UK about 10% of the adult population consults a complementary practitioner every year, and 90% of this contact happens outside the NHS.

Many people clearly believe that complementary medicine works. This is shown by the growing number of people who are turning to alternatives to conventional treatment, either for a cure for their complaints or to help them maintain a sense of wellbeing. More than 50%

Questions

10 What experience do you and your friends, acquaintances and families have of complementary medicine? Which are the popular forms of therapy? What sorts of people look for alternatives to conventional medicine? How do they respond to treatment?

11 Do you consider voluntary self-regulation, recommended for some unregulated complementary treatments, is satisfactory when there are mandatory regulations for others? Argue your case.

Figure 4.16

A homeopath selecting a treatment from a drugs container. Homeopathy is based on the principles of 'like cures like' and 'less is more'. Medicines containing successive dilutions of a chemical which produces the same symptoms as an illness are given to stimulate the body's defence systems. The assumption is that the remedies used encourage the body's ability to heal itself. Treatments are based on very dilute solutions of plant extracts, minerals and other chemicals from natural sources.

of GP practices now offer some form of complementary therapy to their patients. Two-thirds of Scottish general practitioners prescribe herbal or homeopathic medicines (Figure 4.16).

One of the main problems with assessing complementary treatments is that it is difficult to find hard evidence for their effectiveness. The therapies tend to work on a number of very subtle levels, and often the close and confidential relationship with the therapist is important. This makes double-blinding impossible. A therapy may make a patient feel better almost immediately, even if physically measurable symptoms remain. Does this matter if the patient is now pain free and feeling healthy? Because of difficulties such as these, complementary therapists can be loathe to subject their work to scientific evaluation. They and their patients believe that it works, and people continue to come back – and pay – for continued treatment (Figure 4.17).

Questions

12 Why do you think people suffering from allergies turn to complementary therapies? Do you think they are wise to do so?

13 What type of evidence do scientists need before they accept a new theory or treatment?

14 Explain the difference between the terms 'complementary medicine' and 'alternative medicine'.

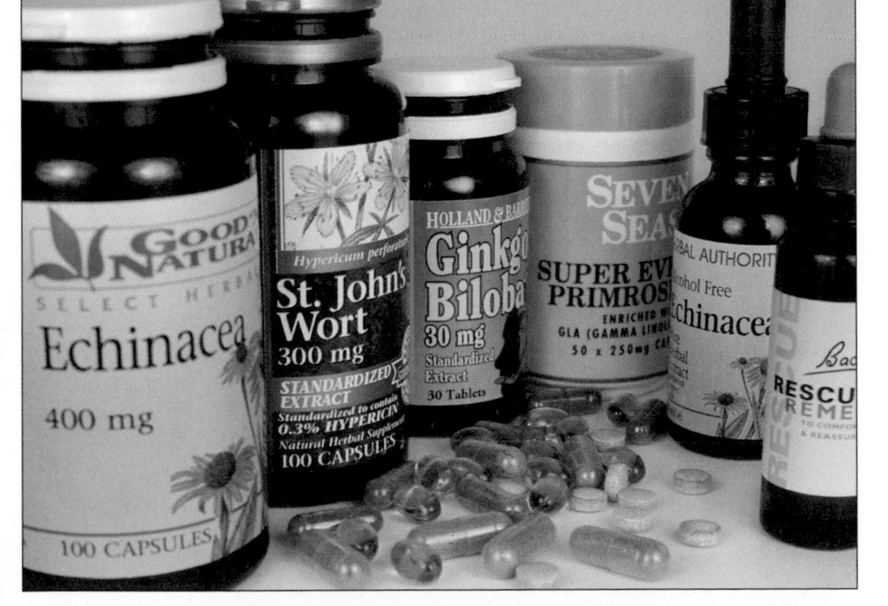

Figure 4.17

A range of alternative medicines. Many of these are available over the counter in health food shops, supermarkets and pharmacies.

Key term

A **cohort study** is a study in which patients who are receiving a particular treatment (or have a particular condition) are followed over time and compared with another group who are not being treated (or are not affected by the condition).

Long-term studies of effects

To investigate whether a factor increases the likelihood of a particular outcome, scientists compare a sample exposed to the factor with a control sample that is not. Cohort studies follow a group of people with different interventions (for example, use of alternative therapies). The health of the cohort is studied over time, and compared with the control group. Causal links between the intervention and any outcome can be inferred because the participants are allocated to treatment or control groups randomly, before the outcome is known.

Cohort studies are expensive and time-consuming, and have many logistical problems. Assumptions made in cohort studies include that those participating are no different from the rest of the population. For this reason everyone selected for a study must participate; those not participating may be different from those who do, causing another type of bias.

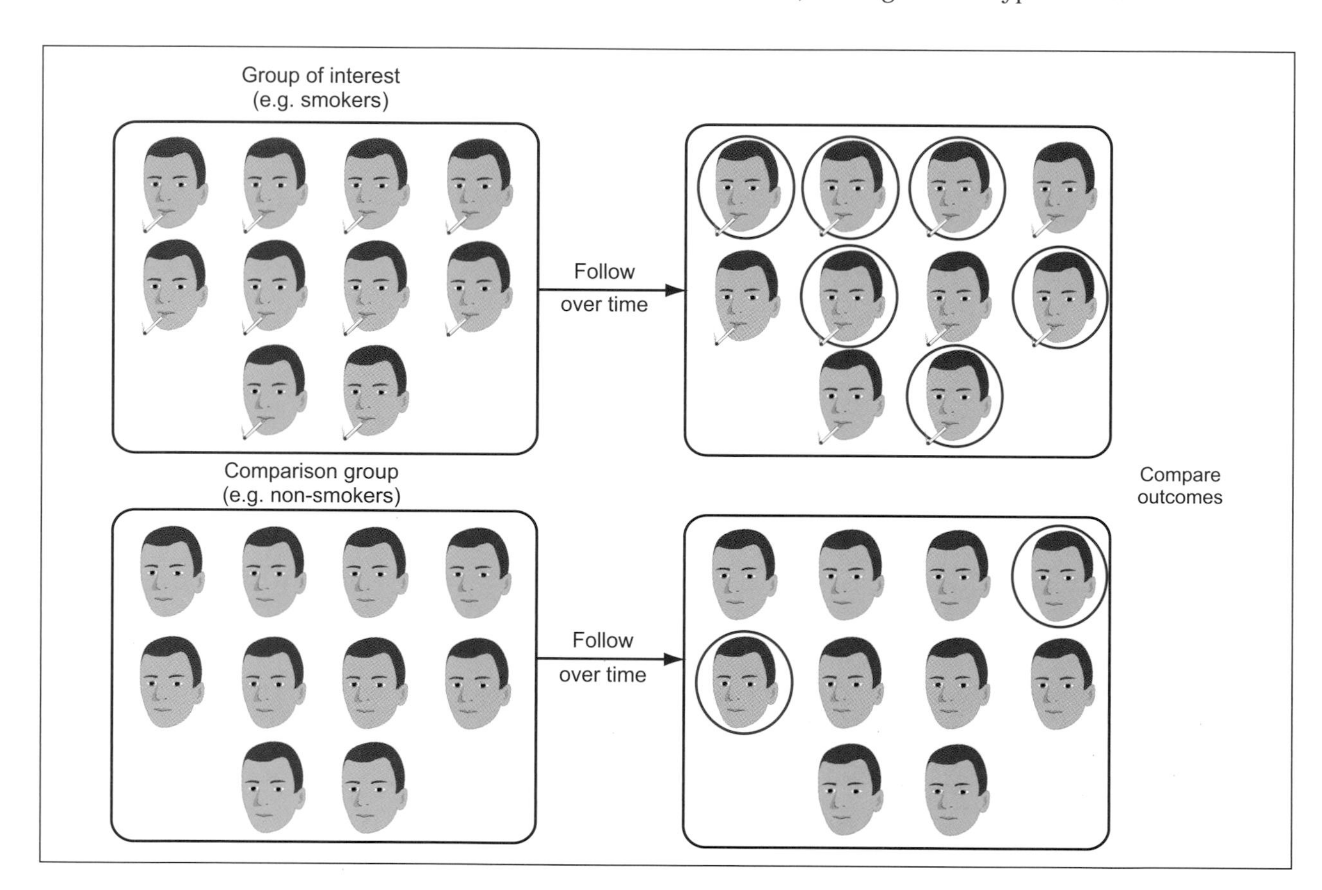

Figure 4.18

A cohort study following smokers and non-smokers over time might compare the percentage of people in each group affected with an associated disease such as heart disease. In the diagram, individuals who were affected with heart disease during the study are circled in red.

Questions

15 Why is it often impossible to carry out a double-blind controlled trial with placebos for alternative medicines?

16 Applying different regulations to complementary and conventional medicine could be described as 'double standards'. Using your knowledge of the purpose of regulation, do you think 'believing that a therapy works' is a good enough reason to allow its use?

17 Explain how bias could be introduced to a cohort study of the health benefits of regular massage, if some of the participants refuse to take part, or drop out early in the study.

Review Questions

18 Use the example of toxicity testing to illustrate why it is important to have systematic methods for measuring and observing scientific procedures.

19 All animals are composed of cells which work in similar ways, yet drugs tested on non-human animals may have different effects when tested on humans. Suggests reasons why this is so.

20 Explain the importance of the following in the design of trials to test new medicines:
a) systematic sampling
b) randomisation
c) sample size
d) control samples.

21 When you come across an account of a new medical breakthrough on a website, in a newspaper or on TV, what are the questions you should bear in mind when assessing the information you are given about the research and the findings?

22 a) Explain, with the help of examples, why society exercises control on the development of new medicines.
b) What factors do regulatory bodies have to take into account when deciding whether to license or authorise the use of a new medicine?

Chapter 5

Ethical issues in medicine

The issues

Scientific advances in the past 25 years mean that much more can now be done to help infertile couples and people suffering from serious diseases. Some of these scientific advances raise serious ethical issues. This makes it more important than ever that scientists engage in public debate with citizens, politicians and policymakers.

Scientists have to take into account people's legitimate concerns, their priorities, and the questions they would like answered. Personal beliefs and values affect how policymakers interpret and apply scientific evidence, and how society responds to new scientific processes and techniques. Decisions based on values and beliefs are part of a scientist's work, but they are not science. Even so, scientists still have a responsibility to ensure their practice is ethical.

The science behind the issues

Stem cells from human embryos can offer treatments for serious illnesses such as Parkinson's disease and cancer. Embryonic stem cells have the potential to grow into many different types of specialised cell, and to replace tissues damaged by injury or disease.

As scientists practice and develop techniques, the procedures become more refined and safer. More research has the potential to reduce the number of animals needed for testing medicines, and to cut the number of human eggs needed to produce stem cells. Deciding whether or not to carry out this research involves discussions based on economics, safety and values.

What this tells us about science and society

New biotechnologies raise ethical issues about the use of animals in research and testing as well as the use of human embryos. Testing new treatments faces doctors with decisions about the people to include in, or exclude from, trials where the outcome may be beneficial or have harmful side effects. Yet without rigorous testing, they cannot try out new therapies which may save lives in the future.

Society asserts values which apply to scientific practice in the form of guidelines, laws and regulations. These controls are based on scientific evidence, but also on principles such as respect for living things and fairness. Individuals may have values or religious beliefs that oppose regulations made by decision makers. Decisions involving society have to balance the rights and interests of individuals and groups in society.

Figure 5.1

These people are taking part in an animal rights demonstration.

What is ethics?

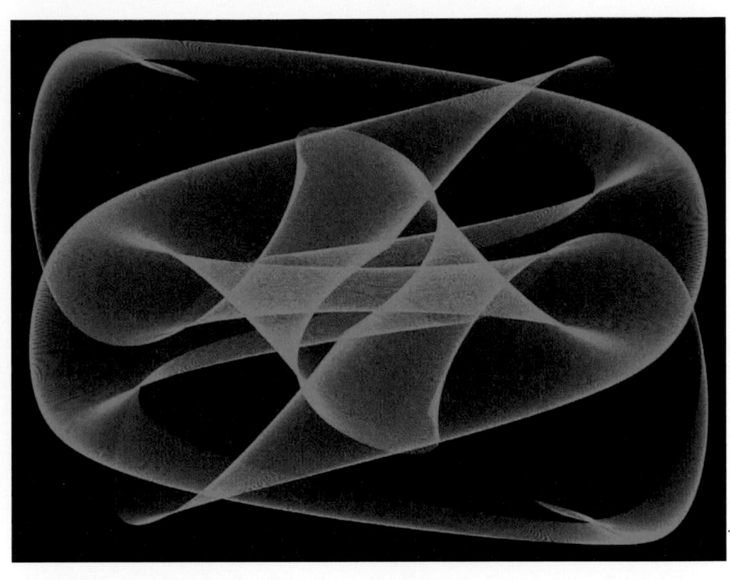

Figure 5.2

Mathematical model based on a sine wave. Ethical conclusions cannot be unambiguously proved in the way that mathematical theorems can. However, this does not mean that all ethical conclusions are equally correct, any more than all scientific or historical theories are equally true.

Ethics is the branch of philosophy concerned with how we should decide what is morally wrong and what is morally right. An example of an ethical question is: 'Is it right that embryonic stem cells are used to develop new therapies?'.

Answering this type of question is rarely simple; we all have moral views. For example, you might believe that lying and abortion are always wrong while telling the truth is always right. But in order to maintain that something is ethically acceptable or unacceptable, you must be able to provide a reasonable explanation as to why this is the case. Science can provide a factual basis for answering ethical questions, but is not enough on its own (Figure 5.2).

One way to decide if something in science or technology would be ethically acceptable is to look at the likely consequences. For example, to decide whether or not non-human animals such as mice or macaque monkeys should be used in medical experiments, think of all the consequences:

- of using such animals
- of not using such animals.

In this particular case, obvious major consequences include: the likely medical benefits to humans of using animals in medical experiments and the likely harmful effects on the animals from their being used. It is still difficult to work out these consequences in any detail. Even if a careful review of the historical evidence shows how useful medical experiments on animals have been for human health in the past, this cannot accurately predict how useful medical experiments on animals will be in the future.

Decisions and opinions on ethical issues are based on values. An individual's view on an issue may be based upon a religious or moral position.

There is no one universally accepted way of deciding whether something is ethically acceptable or not. Instead, there are a number of ethical frameworks, each of which allows you to work out whether a particular action would be right or wrong if you accept the principles on which the framework is based. Often you arrive at the same answer no matter which framework you adopt.

So how do people carry out ethical thinking, or in other words, 'do ethics'?

Ethical frameworks for ethical decisions

There are four widely used ethical frameworks. These can be used to help decide whether you feel something is ethically acceptable or not when considering issues such as the use of animals for medical research and development of medicines, and the use of embryonic stem cells. Both of these issues are raised in this chapter.

Figure 5.3

The Magna Carta: this was the first document setting out commitments by a king to respect legal rights of the people.

1 Balancing rights

Where do rights come from? Some people with a religious faith find them in the teachings of their religion. For example, the ten commandments in the Jewish scriptures talk about not stealing, not murdering, telling the truth and so on. Many rights are social conventions built up over thousands of years. If you want to live in a society, you are expected to generally abide by its conventions (Figure 5.3).

Most of us tend to feel that there are certain human rights that should always be respected. For example, we talk about the right to life, the right to a fair trial and the right to freedom of speech. There are international conventions about these rights, which signal their international recognition.

Some decisions involve balancing the rights of certain individuals and groups against those of others, regardless of the consequences.

Consider whether scientists should use non-human animals in medical experiments. Some people believe that this is unacceptable, partly because the animals may suffer, but more fundamentally because the animals used in such research have not given their consent. After all, scientists do not use children to see whether medicines are toxic, so why is it acceptable to use rats?

Someone arguing in this way might maintain that non-human animals have certain 'rights', just as humans do. On the other hand, there are those who argue that it is acceptable for us to use non-human animals for such tests precisely because humans have a right to use animals for their own purposes, provided their right to good care and welfare is respected.

2 Maximising the benefits

A utilitarian approach looks at the potential consequences of an action when deciding if that action is ethically acceptable. Utilitarianism argues that the right decision or choice is the one that leads to the greatest good for the largest number of people – maximising the benefits. It can also be argued, however, that some actions are wrong, even if they lead to good outcomes.

No one believes that we can ignore the consequences of an action before deciding whether or not it is right. The deeper question is not whether people need to take consequences into account when making ethical decisions, but whether that is all that they need to do. Are there certain actions that are morally required – such as telling the truth – whatever their consequences? Are there other actions – such as experimenting on people without their consent – that are wrong whatever their consequences?

Utilitarians do not believe any particular action is 'right' or 'wrong' – this depends on how much benefit the action brings about. Instead, it would be necessary for a utilitarian to look in some detail at particular cases, such as whether doctors should tell the truth to their patients, in order to decide the correct course of action. In this case, a utilitarian would consider whether telling the truth would lead to a greater increase in benefit compared with not telling the truth.

3 Making decisions for yourself

Autonomy is concerned with the respect due to individuals. People act autonomously if they are able to make their own informed decisions and then put them into effect. There was a time when, for example, doctors told their patients what was best for them. Now though, there is a move towards helping patients to act autonomously. The principle of autonomy is the reason why people should be provided with access to relevant information, for example, before consenting to a medical procedure or taking part in a medical trial.

4 Ensuring justice

Justice is about equality, fair treatment and the fair distribution of resources or opportunities. Considerable disagreement exists, though, as to what precisely counts as 'fair treatment' and a 'fair' distribution of resources. For example, private medical care could be seen as making superior resources available only to those who can pay; alternatively, it could be seen as providing 'choice'.

The ethics of researching and testing medicines with animals and humans

People vary greatly in their views about whether or not it is acceptable to use animals in medical research. Most people agree that it is wrong to make animals suffer. Most also agree, however, that there are some circumstances when the interests of animals may be outweighed by the interests of human beings. One example is the use of animals to develop and test new medicines. Opinion polls consistently show that around 70% of people find this acceptable so long as everything possible is done to minimise pain and distress (Figure 5.4).

Figure 5.4

The use of animals in medical research is strictly regulated in the UK and many other countries, but is still considered controversial by some people. These mice are being used for antibody production.

Others, however, think that all use of animals for medical research and testing is wrong, no matter how great the benefits. In the UK only a minority of people want an absolute ban on the testing of medicines with animals. Some members of this minority hold to their views passionately and are prepared to fight to win the argument.

Animal rights

Animal rights is an ethical framework which guides some people's moral position about humans' use of animals. Giving laboratory animals rights at one level means that they are entitled to such things as food, water, exercise and veterinary treatment. Accepting that animals have rights on a level with human rights would make it difficult to justify using them in medical research for the benefit of humans.

Animal welfare

A much more widespread position than believing that animals have rights, is the belief that humans should treat animals well so far as is possible. Here the emphasis is on animal welfare. This is pretty much the position in European law. In Europe there are strict guidelines to show which animals can be used for research, what they can be used for and the way they should be treated.

The view of organisations which defend the use of animals in medical research:	The view of an animal welfare organisation:	The view of anti-vivisectionist organisations:
'Without this research, the drugs we now take for granted would not exist, and the chances of finding new or improved treatments in the future would be extremely limited.'	'As long as animals continue to be used in experiments they must be given the maximum protection from pain and suffering, whatever the purposes for which they are used.'	'Animal experiments are both unjust and unnecessary and cannot be justified, for any reason.'

Figure 5.5

Contrasting views on the use of animals in medical research.

Both the animal rights approach and the animal welfare approach assume that animals can suffer pain and experience pleasure. A utilitarian who holds the belief that the right course of action is one that maximises the amount of overall happiness or pleasure in the world would allow certain animals to be used in medical experiments, provided the overall expected benefits are greater than the overall expected harm. Suppose, for example, to oversimplify greatly, that it takes the lives of 250 000 mice used in medical experiments to find a cure for breast cancer and that 50 000 of these mice are in pain for half their lives. There could still be a utilitarian argument for using the mice.

Obligations of pharmaceutical companies

Pharmaceutical companies have responsibilities towards the fair pricing of medicines, the availability of medicines, research and development to produce new medicines, and medicine safety and quality. They also have a responsibility to act ethically in the treatment of animals and humans in research. Ethical issues are raised and discussed in local research ethics committees, appointed to the centres where the research is to be carried out.

Medical ethics committees

The task of an ethics committee is to advise pharmaceutical companies on whether the potential benefit of a clinical trial is worth the risk to the volunteers. For this reason the committee should be independent of the company producing the medicine and the research team carrying out the trial. The top priority is the health and safety of the volunteers.

An ethics committee for clinical research typically consists of two doctors (one from general practice), two people from professions allied to medicine (such as nursing or pharmacy) and an equal number of lay people (ideally including a lawyer and someone from a religious organisation). The committee should mirror the make-up of the local community too in terms of the gender and ethnic background of its members.

Ethics committees advise on the recruitment of suitable volunteers for phase I trials so that they are not sought from groups which seem too vulnerable. The ideal volunteer is someone who would like to earn some money for taking part in the trial but is not desperate for it; the money is

Questions

1. Look at the statements in Figure 5.5. Which is closest to your view?
2. Which ethical framework are you using to justify your view?
3. Why include lay people on ethics committees? Why not leave the decisions to experts?
4. Is it right that healthy people are paid to take part in the early phases of clinical trials? Who would you expect to be willing to volunteer to take part in such trials?
5. Many clinical trials take place in developing countries where controls and regulations are less strict or well enforced. How can the rights of the participants be protected in these situations?

intended to cover expenses rather than act as an inducement to take part. An important aspect of the work of the committees is to ensure that volunteers understand what is involved so that they can give informed consent.

Patents

One way that pharmaceutical companies balance their needs and obligations is through patenting medicines early in their development.

Scientists publish new methods, processes and products as patents. Freedom to publish scientific findings may be restricted in some circumstances, for example, where the research is being carried out by a company to develop a new product, or where the results have security implications.

Some processes and products may be declared non-patentable if their use would be 'contrary to the public order or morality'. British law, for example, bans the patenting of any techniques to clone human embryos, modify human sex cells or use human embryos for industrial and commercial purposes. (Such bans have an ancient history. Queen Elizabeth I refused to grant a patent in 1596 to Sir John Harrington on his design for a water closet, on the grounds of propriety.)

> **Key term**
>
> A **patent** is a licence granted to an inventor, which gives the inventor the legal right to stop anyone else from making, using or selling the invention without his or her permission. Patents generally cover technological advances – new products and processes. A patent normally lasts 20 years, after which anyone can use the invention.

Factors affecting the availability and use of medicines

Patents are territorial rights. The patents issued by the UK Patent Office only grant protection within the UK. However, the UK has also signed up to the European Patent Convention, so inventions patented in other European countries are also protected here, and vice versa. Under the 1995 Trade-Related Intellectual Property Rights (TRIPS) agreement, similar patent rules are supposed to be in place across almost the entire world.

The process of research and development is long and slow, so it is very expensive to bring a new medicine to market (Figure 5.6). Companies take out patents which apply in many countries. This makes sure that only the company which developed the medicine is allowed to sell it. As patents are taken out before clinical trials start, there may only be

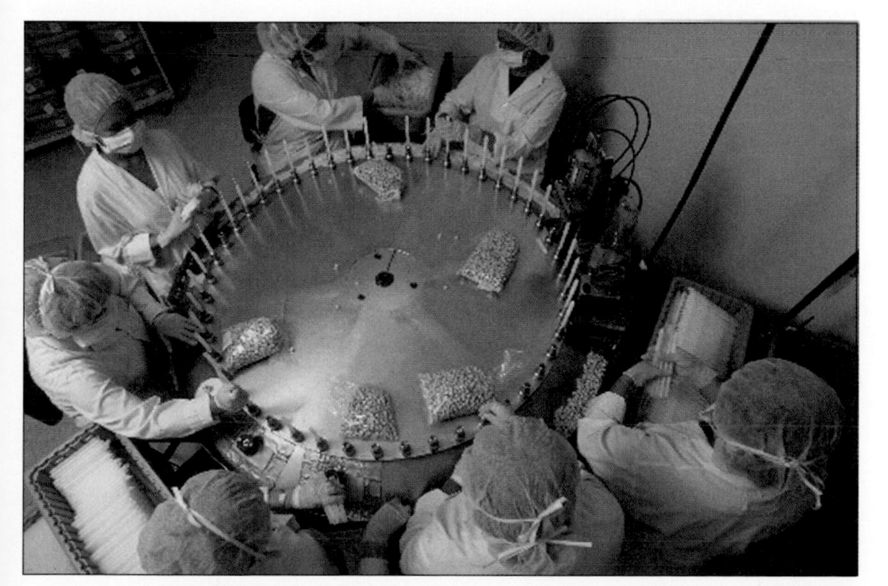

Figure 5.6

Pharmaceutical technicians involved in the production of medicines. The time needed for research and development, the expertise of staff and the specialist equipment needed means medicine production is an extremely expensive process.

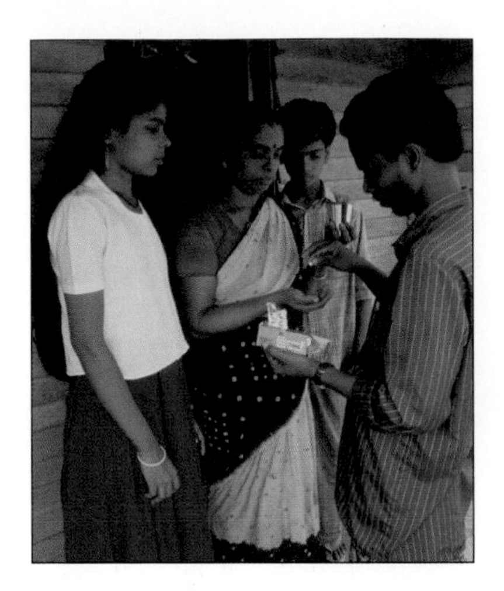

Figure 5.7

A health worker distributing a drug used to cure and prevent the spread of filariasis, a group of diseases caused by parasitic worms which invade the lymphatic system, the skin or the eye. The larvae of the worms are spread by blood-sucking insects.

Questions

6 In the context of the development of new medicines to tackle diseases in developing countries:
a) explain why the current approach is not adequate
b) suggest some actions that might be taken if there were an international political commitment to tackling the problems.

7 Compare the issues arising from the need to develop orphan drugs with the issues related to the provision of drugs for developing countries. How are these issues:
a) similar
b) different?

8 More medicines are becoming available in developing countries. However, a significant proportion of these are sub-standard or fake drugs. In some countries in Asia half the anti-malarial drugs are fake or sub-standard.
a) Why do these present a very serious threat to health in these countries?
b) Suggest some measures that a government could take to protect its citizens from the dangers of these fake drugs.

a few years left to run once the medicine is licensed. The limited time during which companies can recoup their costs pushes up the price of medicines. Companies need to make a profit and finance new research and development.

However, some experts in health policy argue that the case for high prices is overstated. They cite two main pieces of evidence: about 40% of investment by the pharmaceutical industry is spent on marketing, not on researching new drugs; many of the new drugs being developed are 'me-too' drugs, very similar to those already on the market but patented by a rival company.

Sales of medicines which treat very rare but life-threatening diseases are low. The 'orphan' drugs in these medicines are not normally profitable enough for pharmaceutical companies to invest in them. For this reason, the EU and the US governments provide incentives to encourage companies to develop orphan drugs. The incentives include lower licensing fees, a longer period when the company can market the drug exclusively, and money towards research and development.

Drugs for developing countries

Generally, multinational drug companies protect their patents in order to maximise profits. This does not meet the needs of developing countries, which cannot afford the cost of essential medicines let alone more expensive drugs. There is little financial incentive for research and development into treatments needed in the developing world, where the need is great but the profits are low.

Individual countries can negotiate with the pharmaceutical companies to buy medicines at lower cost. Or they can obtain a licence to manufacture cheaper, unbranded 'generic' versions of drugs. In some cases governments are able to override patents to produce their own generic drugs (Figure 5.7). However, if such discounted or generic drugs are smuggled out of poor countries and are sold in richer countries at discounted prices this may make pharmaceutical companies reluctant to offer discounted prices.

Stem cells and cloning

Stem cells are the mother cells of all the cells in the human body. They can divide to produce copies of themselves or to form many other types of cell. They are involved at every stage of development from embryo to adult. Stem cells taken from embryos that are just a few days old can turn into any of over 200 different types of cell that make up the adult body.

Stem cells are found in the early embryo, the fetus, the placenta and umbilical cord and many tissues of the body. Scientists have focused their research on stem cells in developing embryos and in adult tissues such as skin.

Embryonic stem cells

Embryonic stem cells can be taken for research from spare human embryos left over from fertility treatments, or from cloned human embryos developed in the laboratory. One method for making cloned embryos is called nuclear transfer (Figure 5.10).

Early embryonic development

A human egg when fertilised then starts to divide. After three stages of cell division, it consists of eight identical cells. Each of these cells could develop into a complete, healthy human being. If two or three of the cells develop separately, the result is identical twins or triplets. Twinning can occur up to 14 days after fertilisation.

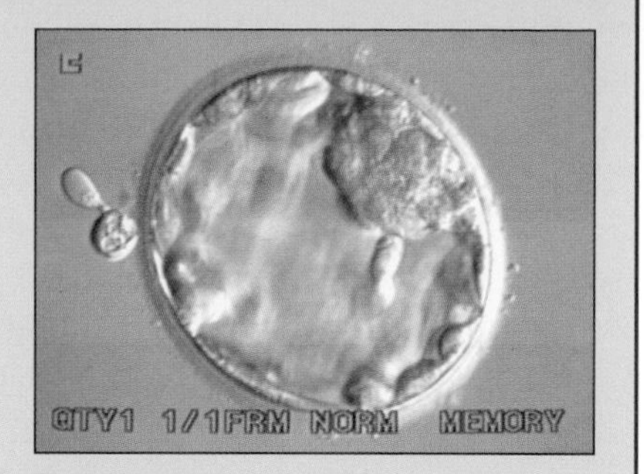

Figure 5.8

Five days or so after conception, the embryo is a hollow ball of cells. Between 6 and 12 days after fertilisation the embryo implants in the wall of the uterus.

Some of the cells from the early embryo go on to form the placenta. The inner cell mass goes on to form the tissues of the developing embryo. These inner 50 cells have already started to specialise. They can give rise to many but not all cell types (Figure 5.9).

Figure 5.9

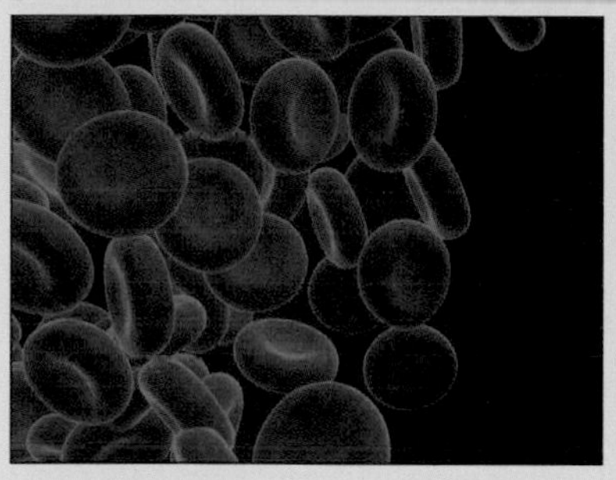

(a) Red blood cells

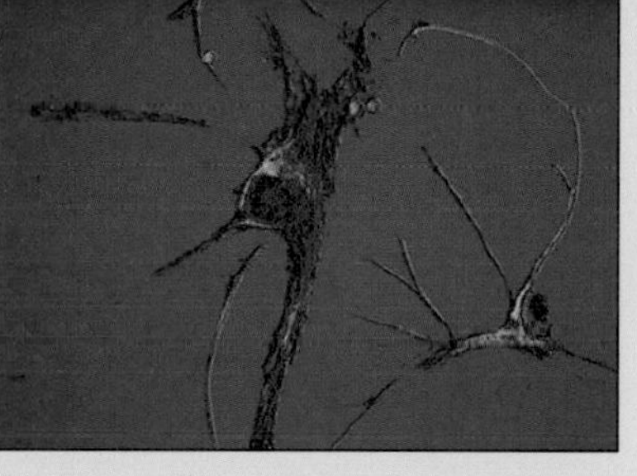

(b) Nerve cells

As the embryo develops, the cells which form a new human become increasingly specialised into different tissues such as blood or nerves. Most cells in specialised tissues are no longer able to develop into a wide range of cells.

Key term

Stem cells are cells that are not differentiated themselves but can either divide again and again to make more stem cells, or differentiate to form specialised cells.

Figure 5.10

Embryos can be cloned by nuclear transfer. The nucleus from a donor cell Is Inserted Into an empty egg cell. After the resulting hybrid cell has been 'activated' (normally using an electrical pulse), it begins to divide, creating new cells. The embryonic cells produced can be used for therapeutic cloning. The cloned embryo could be used for reproductive cloning. This is illegal for human embryos.

Producing a cloned embryo

Nucleus from a body cell (e.g. a skin cell) from the donor. This contains the donor's genetic material.

An egg cell which has also been donated. The nucleus is removed so there is very little or no genetic material left from the egg donor.

The nucleus is transferred into the empty egg cell.

Electrical pulse

The egg cell divides to produce an embryo.

REPRODUCTIVE CLONING

The embryo is implanted into a uterus where it can develop normally.

THERAPEUTIC CLONING

Cells can be cultured in a lab. They can be stimulated to develop into specialised tissues, which could have wide-ranging therapeutic potential.

Key terms

Cloned cells are all genetically identical to the parent cell or organism. A clone is a group of cells or an organism which is produced asexually.

Reproductive cloning uses nuclear transfer to create an embryo which is then implanted in the womb of an animal.

Therapeutic cloning uses nuclear transfer to create an embryo, and then taking the stem cells from that embryo for research purposes.

Question

9 Explain why embryos produced by nuclear transfer are clones of the person who donated the body cell, not of the person who donated the egg.

Tissue stem cells

Stem cells can be derived from various tissues in adults. Scientists have found stem cells in bone marrow, blood, skin, muscle, liver, brain, the cornea and retina of the eye, the lining of the gastrointestinal tract, and pancreas. These stem cells help to maintain, and in some cases repair, the tissue in which they are found. For example, stem cells in the skin give rise to new skin cells to replace old or damaged skin cells. Adult stem cells are relatively unspecialised. They can usually only give rise to specific cell types. Stem cells found in muscle, for example, normally only give rise to muscle cells. This means that tissue stem cells can only turn into a limited range of cell types.

Uses of stem cells

Stem cells (Figure 5.11) have four properties that distinguish them from other types of cells in the body and make them interesting to scientists:

1 Stem cells are unspecialised

A stem cell does not have any specialised physiological properties or functions. Examples of specialised cells are red blood cells, which carry oxygen in the blood, and muscle cells, which contract to produce movement.

2 Stem cells can divide and produce copies of themselves

Stem cells can divide and produce identical copies of themselves, over and over again. In contrast, specialised cells such as blood and nerve cells do not normally replicate themselves, which means that when they are seriously damaged by disease or injury, they cannot replace themselves.

3 Stem cells have the potential to produce other cell types in the body

A stem cell is an unspecialised cell type. When it divides it can either produce identical daughter cells, or it can produce more specialised cell types. A central aim of stem cell research is to understand what determines this choice between self-renewal and differentiation.

4 Stem cells provide an ideal model for studying the development of organisms

Stem cells may help scientists understand how a complex organism develops from a fertilised egg. Identifying the factors that determine whether a stem cell chooses to carry on replicating or differentiates into a specialised cell type, will help scientists to understand what controls normal cell development.

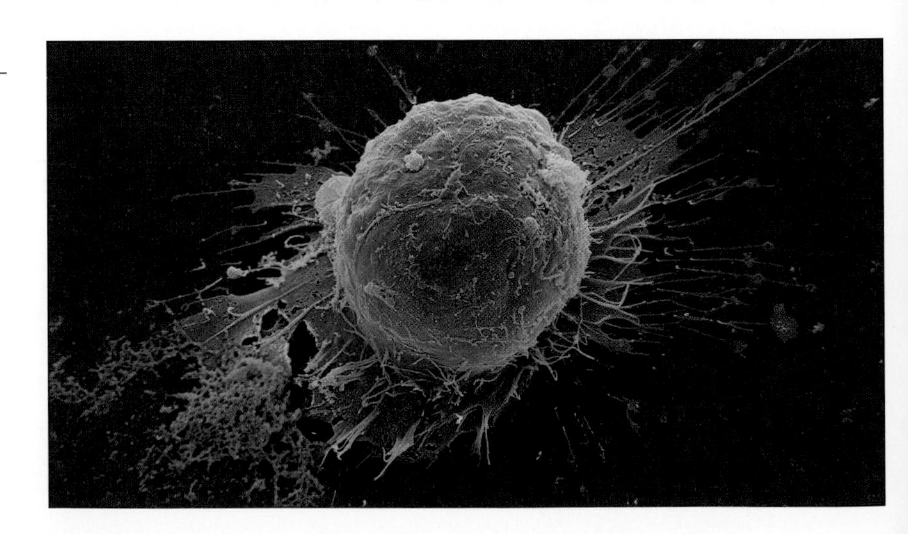

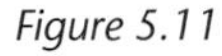

Figure 5.11

A coloured scanning electron micrograph (SEM) of a human embryonic stem cell. These cells are able to differentiate into any of the 200 or more cell types in the human body. The type of cell they mature into depends upon the biochemical signals received by the immature cells. This makes stem cells a potential source of cells to repair damaged tissue in diseases such as Parkinson's and insulin-dependent diabetes. An embryonic stem cell has a typical diameter of around 10 μm

Regenerative medicine

Stem cells may hold the key to replacing cells lost in diseases which doctors cannot now cure. One of the reasons for the great interest in stem cell research is that it might lead to cures for diseases such as Parkinson's, heart disease and diabetes.

Scientists are studying what happens when they transplant stem cells into the damaged or diseased tissue of animals and humans. The hope is that the transplanted stem cells can regenerate missing cells.

Bone marrow transplants and skin grafting are established examples of regenerative medicine. Normally, bone marrow continuously makes new blood cells to replace old ones. During a bone marrow transplant, stem cells are removed from healthy bone marrow and transplanted into a patient affected by a disease, such as leukaemia, which stops the bone marrow making blood cells normally.

It is hoped that researchers will find ways of treating other disease by directing stem cells in the laboratory to produce the specialised cell types needed for transplantation. A lot of current research is therefore focused on identifying which factors consistently induce stem cells to become certain cell types.

If scientists find ways to grow cells that are all the same, then pharmaceutical companies could use them to test drugs. The early stages of testing the effect and toxicity of medicines on cells could be investigated in these cells, and the amount of testing done in animals reduced further.

Stem cells and bone marrow

Leukaemia patient Stephen Knox, aged 31, was given just months to live. In July 2002, he became the first adult in the UK to undergo a bone marrow transplant using blood from a baby's umbilical cord; after the treatment, his cancer went into remission.

The procedure was developed by Professor Stephen Proctor of Newcastle University (Figure 5.12), who injected Knox with blood stem cells from discarded placentas and umbilical cords (Figure 5.13). The stem cells grew new bone marrow and, to doctors' astonishment, Knox's white blood cell count was up to an adequate level within five weeks.

Following successful animal research, stem cell transplants are now routine in the treatment of several types of cancer. They enable patients to undergo high doses of chemotherapy despite the toxic effects of the drugs on bone marrow. After treatment, the patient can be given their own or a donor's bone marrow stem cells, which graft back into the bone.

A further development, pioneered in animals and now entering clinical practice, is to use donor stem cells to treat certain serious and often fatal inherited blood disorders such as Fanconi anaemia or severe combined immune deficiency (SCID). The patient's own bone marrow, which produces diseased blood cells, is destroyed using high-dose chemotherapy and replaced by donor stem cells. In time it may also be possible to treat sickle cell anaemia and thalassaemia in this way.

Figure 5.12

Professor Stephen Proctor of Newcastle University carries out research into leukaemia in adults.

Questions

10 Why do scientists think that stem cells may help to cure diseases such as Parkinson's disease and diabetes?

11 Explain why research on embryonic stem cells may reduce the need for embryo research in the future.

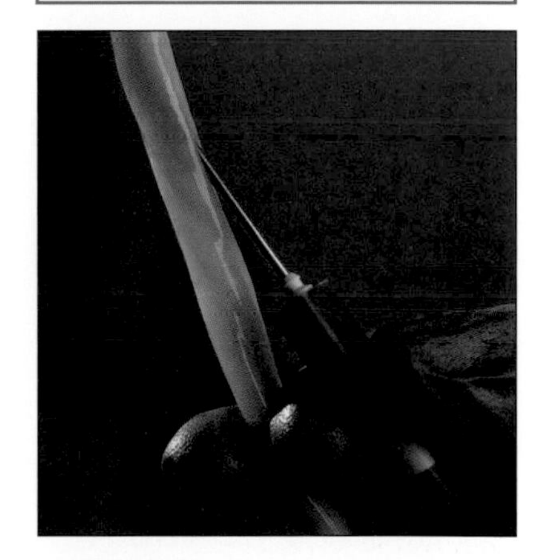

Figure 5.13

Collecting blood from a baby's umbilical cord using a needle. The donated blood will be frozen and stored to provide a source of blood stem cells. These cells can be used to treat diseases such as leukaemia, a cancer that causes the overproduction of abnormal white blood cells. Blood stem cells can become specialised to form red blood cells, white blood cells and platelets. Photographed at the London Cord Blood Bank, part of the National Blood Service, England.

Peer review

Muscle cells die during heart attacks, when they are deprived of oxygen. In 2001, research in rats and mice suggested that when bone marrow stem cells were injected into the heart, new heart muscle cells grew. This technique was then tried in humans, and doctors thought that heart muscle was regenerated. The patients' hearts seemed stronger and it was presumed that this was due to growth of new muscle.

When findings such as these are published by an individual scientist or research group, their data, and their interpretations of their data, are carefully checked by the scientific community. This has to happen before new ideas are accepted as reliable scientific knowledge. This process of 'peer review' helps to detect invalid claims in reports of scientific work and adds weight to valid ones (see Chapter 4). Scientists value observations and measurements that are replicable. They are generally sceptical of findings that cannot be repeated. Scientists publish their work in technical journals, in conference papers and on the Internet. This allows other scientists to see and comment on the data collected, the methods used and the interpretations made.

In April 2004, two papers in the science journal *Nature* challenged the idea that stem cells could help to repair heart muscle. Using state-of-the-art genetic tools, they discovered that bone marrow stem cells in the heart did not turn into muscle. Instead, they turned into blood cells. Any improvement in patients, they said, seems to have been caused by growth of new blood vessels. This was an unexpected side effect of treatment. Reports of treatments with bone marrow stem cells are now undergoing rigorous scrutiny as a result of these findings.

Ethical concerns about the use of stem cells and cloning

Stem cell research raises many ethical and societal issues relating to the source, uses and storage of stem cells. Other issues arise when the research is taken up by commercial companies and patented. Clinical use of stem cells raises concerns about fair access to treatment and the risks to patients.

Embryonic stem cells

The ethical issues which arise from research with embryonic stem cells depend on the moral status of a human embryo. Some people believe that an embryo should have full human rights from its conception. From this point of view, any activity which does not directly benefit the embryo is 'wrong'. Other people think that the embryo's rights increase as it develops. The potential benefits from research should be weighed against the respect given to the embryo at each stage of its development. Utilitarians consider the seriousness of the disease and the potential number of people that research could help when deciding whether or not the use of embryos can be justified.

The source of embryonic stem cells raises other issues. Scientists take stem cells from early embryos which would otherwise be discarded (Figure 5.14). These include embryos from abortions, embryos from fertility treatments and embryos produced specially for the research.

Questions

12 In the early stages of research into stem cell therapy, why is it more likely that trials take place on patients suffering from serious diseases, such as cancer and Parkinson's, rather than patients with sports injuries who might need repair or replacement of tissues?

13 When deciding whether or not to use a new therapy or medicine, why do doctors take more notice of articles in peer-reviewed journals than the brochures of pharmaceutical companies?

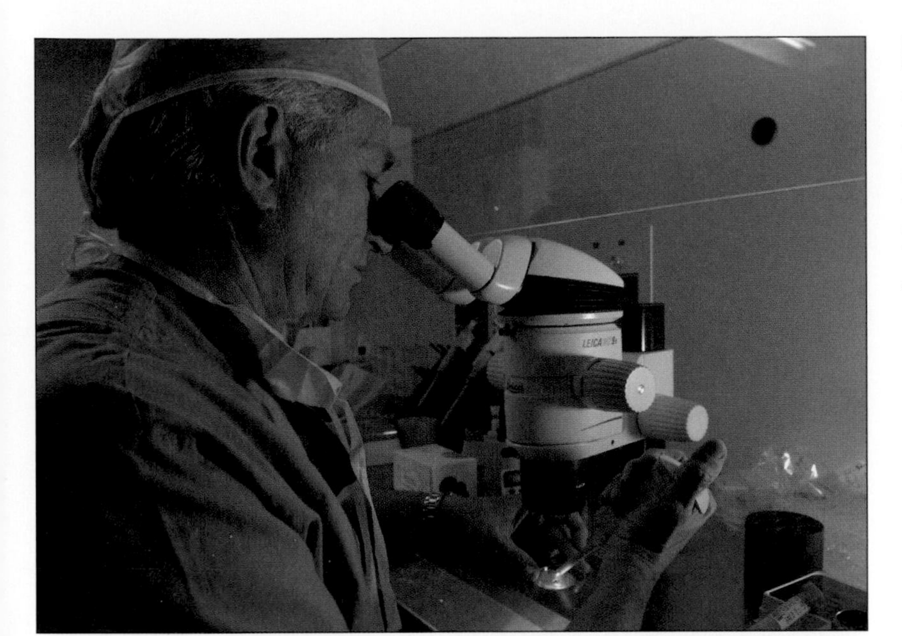

Figure 5.14

Embryo selection: a technician at an IVF clinic examining embryos under a microscope. The embryos have been produced by fertilising harvested eggs with sperm cells in a laboratory. The technician selects the most healthy-looking embryos, which are then either transferred to the patient's womb (uterus) or frozen for future use. Unwanted embryos can be used in research.

The ethical acceptability of using fetal tissue from abortions is closely tied to views on abortion. The potential of embryonic cells to create valuable cells for transplantation raises the possibility that women could be offered incentives for aborting fetuses for this purpose.

An embryo produced by IVF has been created with a view to implantation in the uterus. If it is not implanted, it no longer has a future and, in the normal course of events, it will be discarded. Removal and cultivation of cells from these unwanted embryos need not indicate lack of respect for the embryo, and could be regarded as being a similar process to tissue donation.

Issues of autonomy arise in this type of research. Informed consent from donors is complex, as the future use of the embryonic cells may be unclear. For example, cells reproduced from embryonic cells have the potential to continue to divide and grow for ever. These cells could in time be used in a wide range of treatments. This means that informed consent for specific types of use is difficult.

Fully informed consent would involve the researchers disclosing any potential commercial benefits and sources of funding of embryo research. It is worth noting that obtaining informed consent from embryo donors assumes the parent 'owns' the embryo, and the embryo doesn't have rights of its own.

How does society decide what is right and what is wrong?

Scientists have to ensure that their practice is ethical. In this they may be guided by a professional ethical code. At the same time they have to abide by national or international regulations, which may also have an ethical basis.

EU member states have different positions regarding the regulation of human embryonic stem cell research. Opinions on the legitimacy of experiments using human embryos are divided according to the different ethical, philosophical and religious beliefs throughout Europe. These differences are reflected in the laws of each country.

Questions

14 Currently the law allows embryos to be created for the purposes of research only in the case that there are insufficient embryos from IVF available. Is it more morally acceptable to use embryos that would otherwise be discarded, as opposed to embryos created for the purpose of providing stem cells? Set out the arguments using ethical frameworks.

15 Nuclear transfer techniques which produce cloned human embryos for research can now be licensed for some types of research. Explain what ethical issues this raises, apart from those of the embryo's rights.

Decision makers are influenced by the mass media, by special interest groups and by public opinion as well as by expert evidence. Decisions about science and technology may be affected by prior beliefs or vested interests, which can affect how people interpret and evaluate evidence when making decisions.

Most people agree that there are no ethical objections to using stem cells from adults, so long as the donors give their consent. For example, adults may provide bone marrow for transplants. The problem is that most scientists believe that these stem cells are likely to be less valuable for research and in developing new treatments than early embryonic stem cells. Even if adult stem cells are used for new research, they may be of little use unless a better understanding is gained of how they specialise. This understanding may only come from research on embryonic stem cells or animal stem cells.

In the UK, the Human Fertilisation and Embryology Authority (HFEA) regulates research on human embryos under the 1990 Human Fertilisation and Embryology Act. Until 2001, UK law only allowed the use of human embryos where the HFEA considered their use to be necessary or desirable:

- to promote advances in the treatment of infertility
- to increase knowledge about the causes of congenital disease
- to increase knowledge about the causes of miscarriage
- to develop more effective methods of contraception
- to develop methods for detecting gene or chromosome abnormalities in embryos before implantation.

In the UK, scientists are only allowed to use human embryos for this restricted range of types of research. To conduct research on human embryonic stem cells, scientists need a licence from HFEA. Research on human embryos is only allowed up to a limit of 14 days following fertilisation (Figure 5.15), and provided that no embryo which is subjected to research procedures is re-implanted in the uterus.

Questions

16 Describe the size and appearance of a 14-day embryo.

17 Comment on why 14 days may have been chosen as the cut-off for research.

18 What are the ethical issues involved in storing umbilical blood in stem cell banks?

Figure 5.15

This photo (a) shows a cluster of cells which is a human embryo, seven days after conception. The dot in (b) shows its actual size. After 14 days, cells begin to line up to form what will eventually be the start of the spinal cord. Before this time, no nervous system is present.

(a)

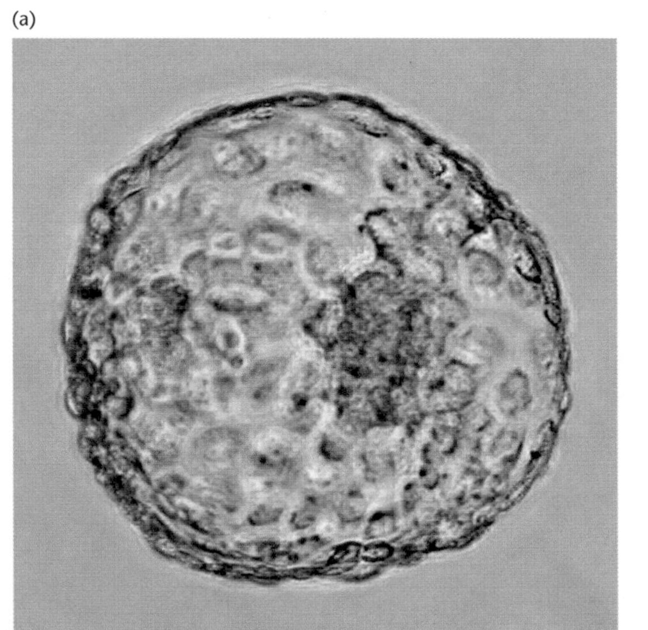

(b)

•

In December 2000, MPs in the House of Commons voted by 366 votes to 174 to change the rules governing research on human embryos. This change in the law was confirmed in January 2001, when the House of Lords voted by 212 votes to 92 to support the same proposal. As a result, so-called 'spare' embryos from IVF treatment can now also be used as a source of embryonic stem cells for research into serious disease. In February 2002, parliament also approved fundamental research with stem cells necessary to understand how cells can become reprogrammed to produce different types of specialised cells.

Stem cell research is coordinated by the UK Stem Bank. This is Europe's first centralised stem cell resource, providing scientists with shared access to research on a variety of different human stem cells (Figure 5.16). The bank was set up with full support of the UK government and with funding from the Medical Research Council and the Biotechnology and Biological Sciences Research Council. Its aim is to make possible the huge amount of research that is needed to understand stem cells.

Figure 5.16

A technician placing metal boxes of umbilical cord blood into a cryostorage tank at the Northern Cord Blood Bank, Newcastle England. Umbilical blood containing embryonic stem cells can be stored in liquid nitrogen.

Is human cloning legal?

On 11th August 2004, the HFEA announced they had granted a licence to researchers at the International Centre for Life in Newcastle to perform therapeutic cloning for the first time. Stem cells created under this licence are for research purposes only.

Human reproductive cloning is illegal in the UK and many other countries because it has not been proven to be medically safe, scientifically sound, or socially and ethically acceptable. Embryos can be created for research purposes only, and kept only up to 14 days. It is a criminal offence to implant them in a woman, under the Human Reproductive Cloning Act 2001.

Reproductive cloning as a technique is difficult and unpredictable. It took scientists 277 attempts before they produced one viable lamb, Dolly, at the Roslin Institute in Scotland. In addition, many cloned animal offspring are born abnormally large and with other severe abnormalities. This raises the possibility that any surviving human clone could face serious health problems, and that the surrogate mother would also be at risk during pregnancy.

It is not difficult to understand how individuals who want to replace a lost child or who are unable to bear children through conventional methods might be swayed by the idea of human cloning. However, even if it were possible, no human clone would ever be truly identical to its genetic parent. Whilst they might look similar, they would grow up in a different time and environment, with different social and cultural influences. In addition, the knowledge that you are genetically identical to your 'parent' could add an immense psychological pressure (Figure 5.17). On the other hand, twins are 'natural' clones.

Figure 5.17

Human clones can, theoretically, be produced through nuclear transfer techniques. Science determines what is possible, but society must decide what should be allowed.

Review Questions

19 Explain, with the help of examples, the difference between an ethical and a scientific approach to an issue such as stem cell research.

20 a) Describe a situation in which a doctor might be torn between telling the truth to a patient and not telling it.

b) In your opinion, should doctors always tell the truth?

21 One way to reduce the number of human eggs needed for research is to use eggs from other animals for nuclear transfer. Cow eggs can receive a nucleus from a human body cell, and develop into an embryo for the provision of embryonic stem cells. Early media reports about this technique suggested that scientists were creating animal-human hybrids such as half man half cow. Explain why embryos produced by this technique would not produce such hybrids, even if they were allowed to develop.

Chapter 6

Reproductive choices

The issues

For many hundreds of years people have known that some diseases run in families. These genetic diseases are not infectious, but they can be passed on from one generation to the next.

Today it is possible to test people to see if they are carriers of some genetic diseases. It is also possible to test to see if a fetus in the womb is likely to be affected by a disease inherited from its parents. The results of these tests can face people with agonising decisions.

At the moment there is no cure for any genetic disease, only treatments which can alleviate the symptoms. The new techniques of genetic engineering offer some hope that cures may be found, but the prospect of gene therapy raises new dilemmas.

The science behind the issues

The 'instructions' for development are made up of many pairs of genes. One gene in each pair comes from the mother and the other from the father. Each egg and each sperm cell contains a randomly selected half of each parent's genes. The single cell that they form at conception contains a full set of genetic information (Figure 6.1).

Some inherited diseases are the result of a faulty form of a single gene. In most cases this makes it easier to predict the chance that a baby will suffer from the disease. Most diseases are a result of the influence of several genes *and* environmental factors.

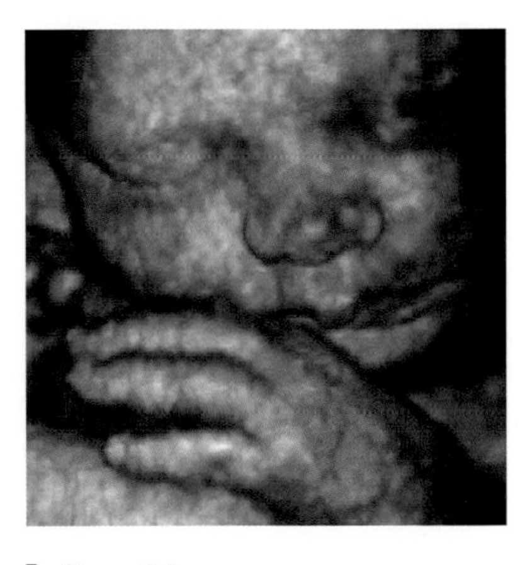

Figure 6.1

A coloured, three-dimensional ultrasound scan of the face and hands of a fetus. The genes for diseases are present from conception. As methods of investigating the secret world of the unborn fetus get better and better, it is becoming possible to identify problems before birth, but the options for action are very limited.

What this tells us about science and society

Like all scientific measurements, genetic tests are not perfect. There is always a chance that the results may be wrong. It is important to be able to estimate the reliability of a test.

New techniques of genetic engineering may make possible treatments which people find unacceptable. Just because something can be done does not mean that it ought to be done. Science cannot provide all the answers to these issues. They involve social, economic and ethical decisions.

There can be tensions between the rights of individuals and the interests of society. Some groups have supported women's rights to have genetic testing and possibly a termination. Other groups support the rights of the unborn child, even if affected by an inherited disease.

What are genetic diseases?

The primary source of Jewish law, the Talmud, specifies that if a baby boy dies of bleeding after circumcision, his younger brothers should not undergo the ritual and his male cousins on his mother's side are also exempt. This shows that, many centuries ago, there was a recognition that blood which does not clot runs in families, and is inherited from the mother rather than the father. This condition is now known as haemophilia and it is an example of a genetic disease.

Some genetic diseases are the result of problems with whole chromosomes. One of the most common problems of this type is Down's syndrome. People born with Down's syndrome have an extra copy of one chromosome (chromosome 21). This causes slow and limited mental development and many physical problems. Chromosome abnormalities

Key terms

Every individual carries two copies of each **gene**, one inherited from the mother and the other from the father. A gene is a length of DNA on a chromosome that carries the genetic instructions for a particular feature such as eye colour, skin pigmentation or height.

In the nucleus of a cell, DNA is packaged into long, coiled strands called **chromosomes**. The nuclei of human body cells contain 46 chromosomes.

Deoxyribonucleic acid, **DNA**, is the chemical which carries a code making up the genetic information in a cell. DNA is a long-chain molecule made up of separate units joined together. Its complex double-helix structure was discovered in the 1950s by James Watson and Francis Crick.

At **fertilisation**, the nuclei of an egg and a sperm cell fuse, producing, in humans, a cell with 23 pairs of chromosomes. Each chromosome pair contains one chromosome from the mother and one from the father.

Gene expression is the process by which a gene's coded information is translated into an inherited characteristic.

Sex cells and fertilisation

Figure 6.2

When sex cells are formed, the number of chromosomes is halved so each human egg or sperm contains only 23 chromosomes. The genes carried on the chromosomes are exchanged and combined in different ways so each new sex cell has an unique mixture of genes.

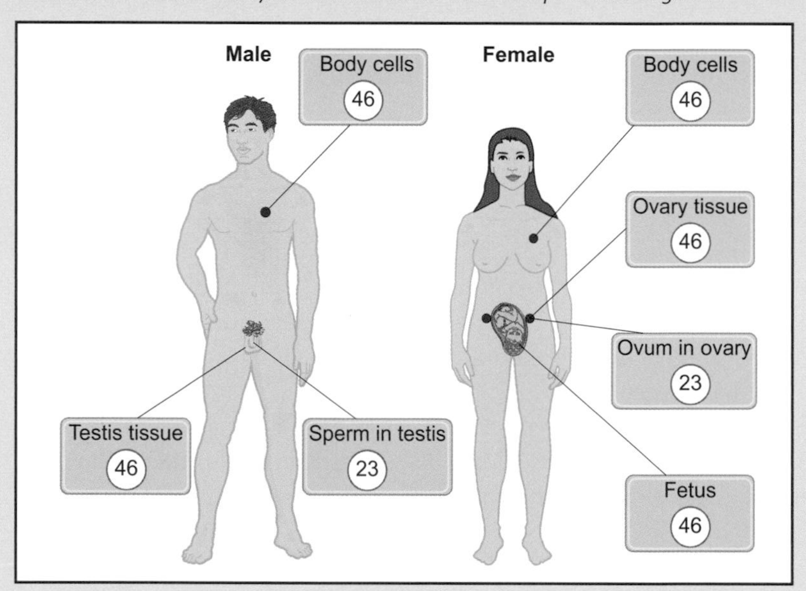

In the nucleus of every cell is the genetic material, made up of DNA. DNA is packaged into chromosomes, which carry the 'instructions' for the development of a new individual in the form of many pairs of genes. All the cells of an individual have the same genetic information in their DNA but the information that is acted on in a particular cell depends on its local environment in the body. There is a control system which decides that in a specialised cell particular genes are 'expressed' but others are not.

Normal human body cells contain 46 chromosomes in 23 pairs, red blood cells being an exception. In sexual reproduction (Figure 6.2), eggs and sperm contain one of each chromosome pair only, so at fertilisation the full complement of 46 chromosomes is restored.

In 22 of the 23 pairs of human chromosomes, the two members of the pair are the same size and shape and carry genes for the same features. The one exception is the sex chromosomes (Figure 6.3).

Question

1 Why is it important that sex cells have only half as many chromosomes as body cells?

Sex chromosomes

Figure 6.3

Females have two X chromosomes, males have an X and a Y. The Y chromosome is much smaller than the X and carries only a few genes – most of which control male sexual development. This means that women have two copies of any genes on the X chromosome, where men have only one. The X chromosome carries many genes that are important for growth and development. A faulty gene on one X chromosome can be made up for by the second gene in women, but men could suffer from a disorder as they have no second gene to compensate. Genetic diseases associated with genes on the X chromosome are known as sex-linked diseases. Haemophilia is an example of a sex-linked genetic disease. Red-green colour blindness is also an inherited condition which is sex linked.

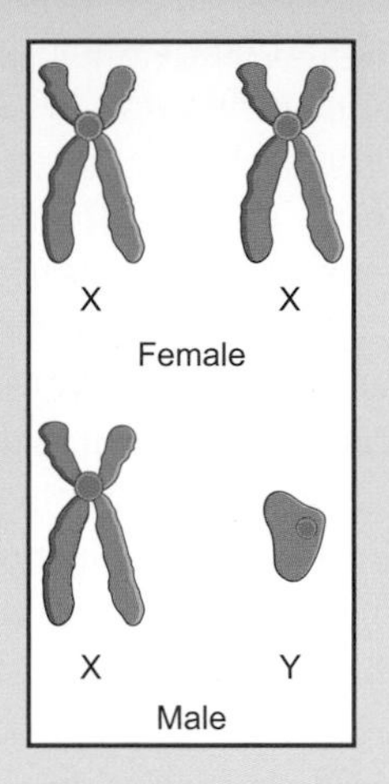

Key terms

Alleles are different forms of a gene. Some alleles are **dominant**. Any dominant allele present determines the characteristic controlled by that gene. A **recessive** allele only determines the characteristic controlled by that gene if the dominant allele is not present.

Mutations are changes in the DNA which may affect the way genes work. Mutations can be caused by environmental factors such as ionising radiation and smoking, or they can be the result of mistakes made when DNA is copied during cell division.

are not usually passed from one generation to the next as the people affected are often sterile, and often do not live long enough to reproduce.

Cystic fibrosis

Some genetic diseases are the result of just one faulty gene. One person in every 25 in the UK carries the faulty gene for cystic fibrosis, and about 1 in every 2500 babies born to white Europeans has the disease. It causes severe problems in the breathing and digestive systems linked to thick, sticky mucus which the body cannot shift. It shortens the life of sufferers, who may need physiotherapy several times a day, and take many drugs to keep their bodies functioning.

The cystic fibrosis gene occurs in many alternative forms or alleles. There is one allele which is the normal form. This is the dominant allele. There is also a large number of mutated alleles which are faulty forms of the gene. The mutated alleles are all recessive. As long as a person inherits at least one normal form, he or she will be healthy. A person has to inherit faulty alleles from both parents for the symptoms to show up.

Both parents must be carriers of the disease if their child suffers from cystic fibrosis (Figure 6.4). Carriers have both a faulty allele and a normal allele for a particular gene. They may not be aware they are carrying the faulty version if its effect is masked by the expression of a dominant normal allele. If a child inherits a faulty gene from both parents they will have no working copy of that gene, and so will be affected by cystic fibrosis.

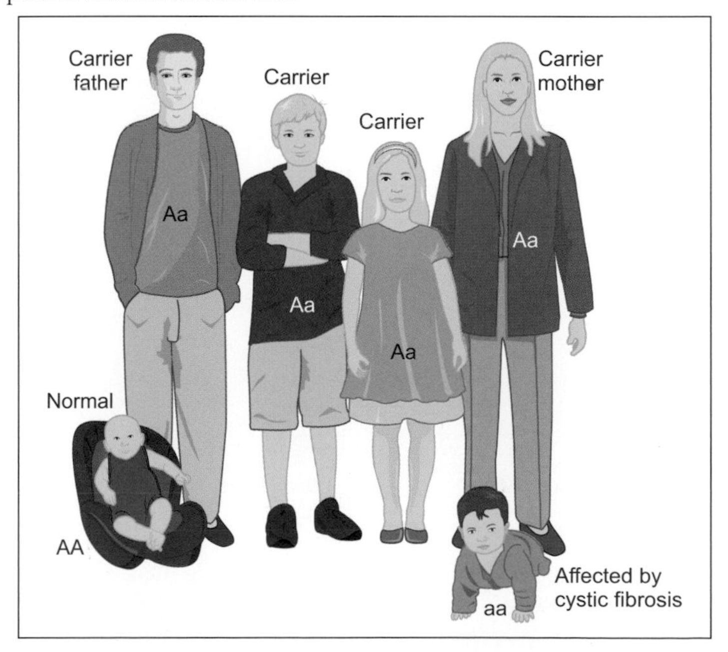

Figure 6.4

Inheritance of cystic fibrosis. Both parents are healthy. They each have one normal gene (A) and one faulty gene (a). Each child has a one in four chance of inheriting two faulty genes and showing the symptoms of the disease.

Questions

2 How would you explain to a couple, only one of whom is a carrier of the cystic fibrosis gene, why they are at no risk of having a child with the disease?

3 Explain why only 1 in 2500 babies born to white Europeans have cystic fibrosis when 1 in 25 people carry the faulty gene for the disease.

Key term

Genetic tests look for mutated genes known to cause a particular disease.

Questions

4 What is the chance that a baby will be affected by Huntington's disease if one parent carries the faulty form of the gene but the other parent has two normal forms of the gene?

5 Many genetic diseases show themselves within the early years of a child's life, but Huntington's disease does not show its symptoms until middle age. Suggest why families may not want to take part in genetic tests which show whether or not they have inherited the disease.

Huntington's disease

Huntington's disease is caused by a single dominant gene (Figure 6.5).

Until very recently, the victims of Huntington's disease only knew they were affected when symptoms of the disease began to show. This usually happens between the ages of 30 and 50, so many people have already had children before they discover their own disease. It is very rare, affecting about 1 in 20 000 people of Western European descent, and 1 in 1 000 000 people of Asian or African descent.

Huntington's disease involves the gradual destruction of nerve cells in the central nervous system. The person begins to have involuntary jerking or writhing movement of their arms and legs, and strange facial grimaces. Changes in personality occur, including laughing and crying at the wrong time, inappropriate anger, memory loss and bizarre behaviour. The pattern of symptoms varies a lot, but the patient usually loses the ability to communicate several years before their inevitable death.

In 1983 Dr James Gusella, working in Boston, Massachusetts, decided to compare the DNA of family members affected by Huntington's disease with those who did not have the disease. He discovered a DNA sequence associated with the Huntington's gene.

This discovery was the basis for a genetic test which has been developed for use with family members of Huntington's sufferers. In the past people faced many years of uncertainty before knowing if they had inherited the disease. The test opens the door for people who wish to face and plan their own future, and also to plan whether to have children. For example, if one partner has the Huntington's gene, a couple may choose

Figure 6.5

Anyone who inherits the faulty gene for Huntington's will get the disease, and there is a 50% chance of any children they may have inheriting the disease too.

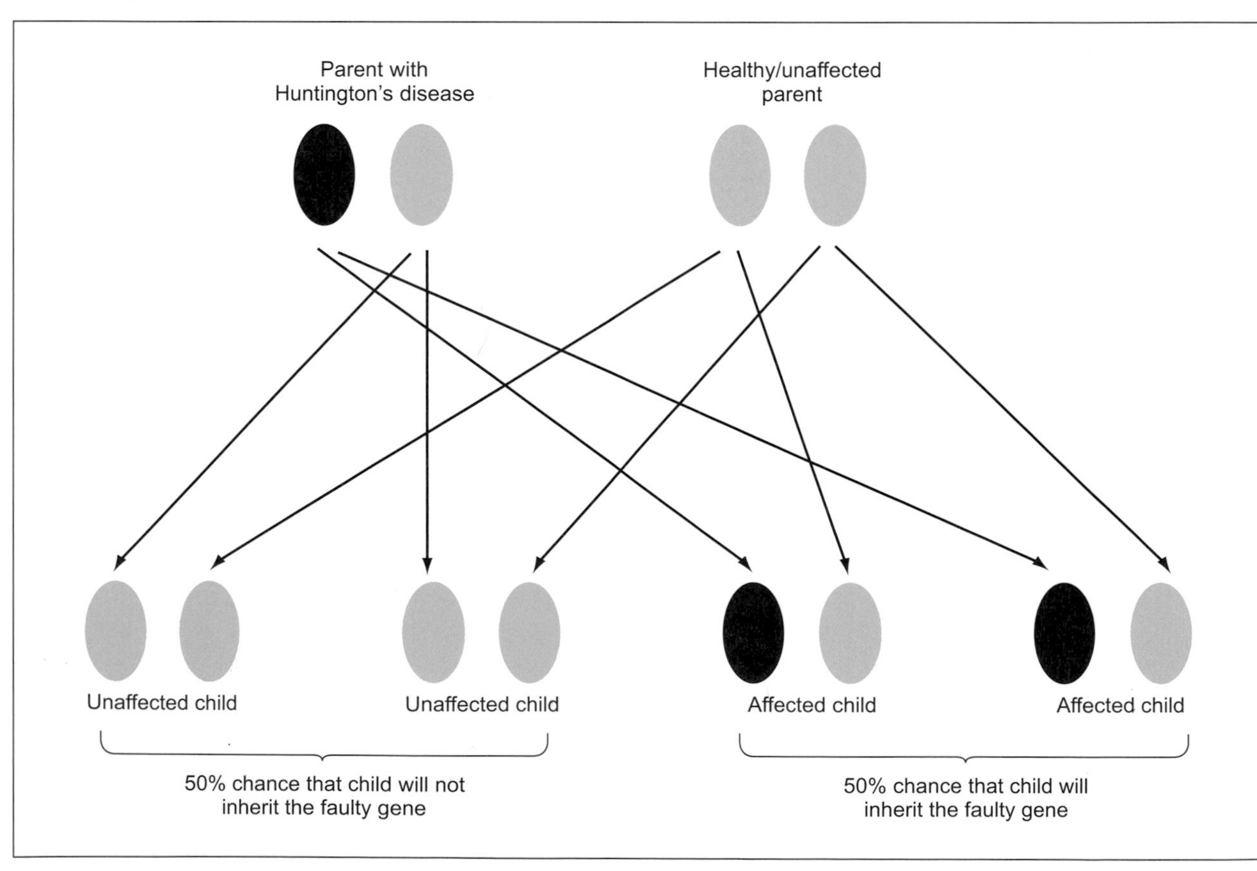

not to have children and so remove the risk of passing on the gene. On the other hand, they may decide to have children, but test the fetuses, ending the pregnancy if the Huntington's gene is present.

At first sight such knowledge appears invaluable, but some people might find it unbearable to learn that they have inherited a completely untreatable fatal disease. In America, where genetic testing for Huntington's disease is available, most family members choose to refuse screening. An uncertain future with a degree of hope may be preferable to one filled with the foreknowledge of their own decline and death. Figure 6.6 shows how Huntington's disease was inherited in one American family.

Figure 6.6

Because the symptoms of Huntington's disease do not appear until early middle age, and because the genetic nature of the disease was only recognised relatively recently, the disease has cut a path through families for generations. The individuals who inherited Huntingdon's disease can be seen clearly in this American family tree.

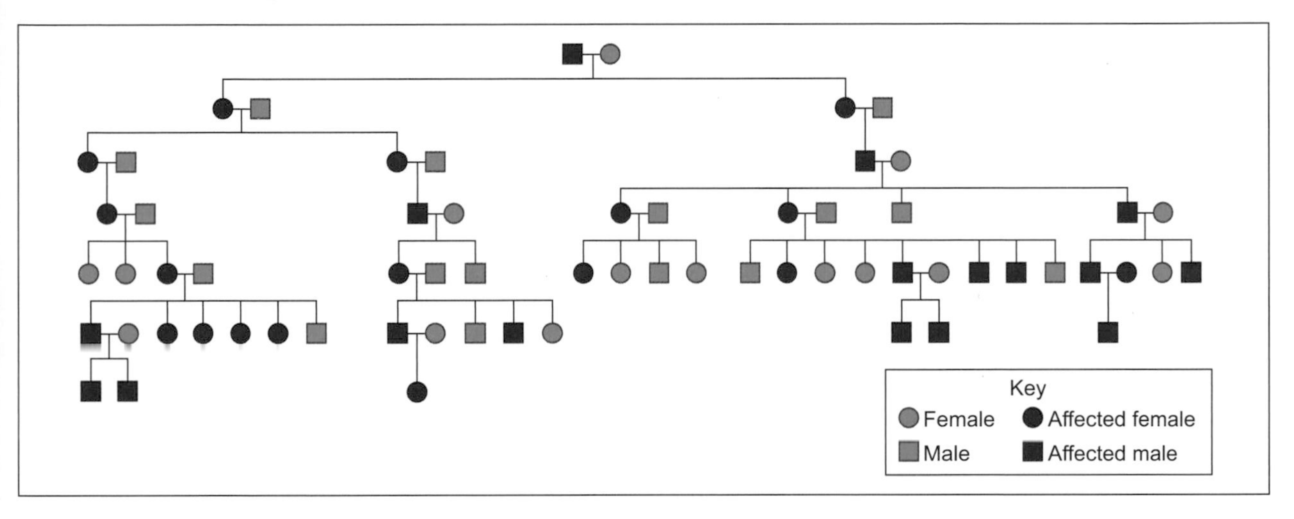

Multifactorial inheritance

The genetic diseases considered so far in this chapter are the direct result of the mutation of one gene. It is becoming clear that most inherited characteristics and medical conditions are influenced by multiple genes as well as multiple environmental factors. Many of the common diseases that affect the West are called multifactorial disorders because they have some

Multifactorial diseases

Some inherited diseases are the result of a single malfunctioning gene. Cystic fibrosis, sickle cell anaemia and Tay Sachs disease are examples of single gene disorders. Patients with the same disorder may have different symptoms because, even in single gene disorders, other genes and the environment affect the way the gene is expressed. A disease may be caused by one of many possible faults in the gene responsible; symptoms of cystic fibrosis can be caused by over 1000 different malfunctions of the same gene, each with slightly different effects.

More genetically complex diseases include asthma, diabetes, epilepsy, hypertension, coronary heart disease, manic depression and schizophrenia. These diseases have a low tendency to be inherited within families compared with single gene disorders. For example, only 2–6% of the close relatives of diabetics also suffer from diabetes. Coronary heart disease shows around the same level of heritability. This is much lower than would be the case for a single gene disorder like cystic fibrosis.

Low heritability indicates that no single genetic factor is responsible for the disease. Many genes may contribute both positive and negative effects, as do environmental and lifestyle factors such as diet, exposure to radiation or toxic chemicals.

The incidence of any complex disease depends on a balance of genetic and environmental risks. Too many negative factors tip the balance towards disease. Some links between genetic and environmental or 'lifestyle' factors have been strongly established, such as the relationship between smoking and lung cancer. Others are more controversial, often because the exact relationship between genes and lifestyle is so complex and difficult to prove (Figure 6.7) (see also Chapter 8).

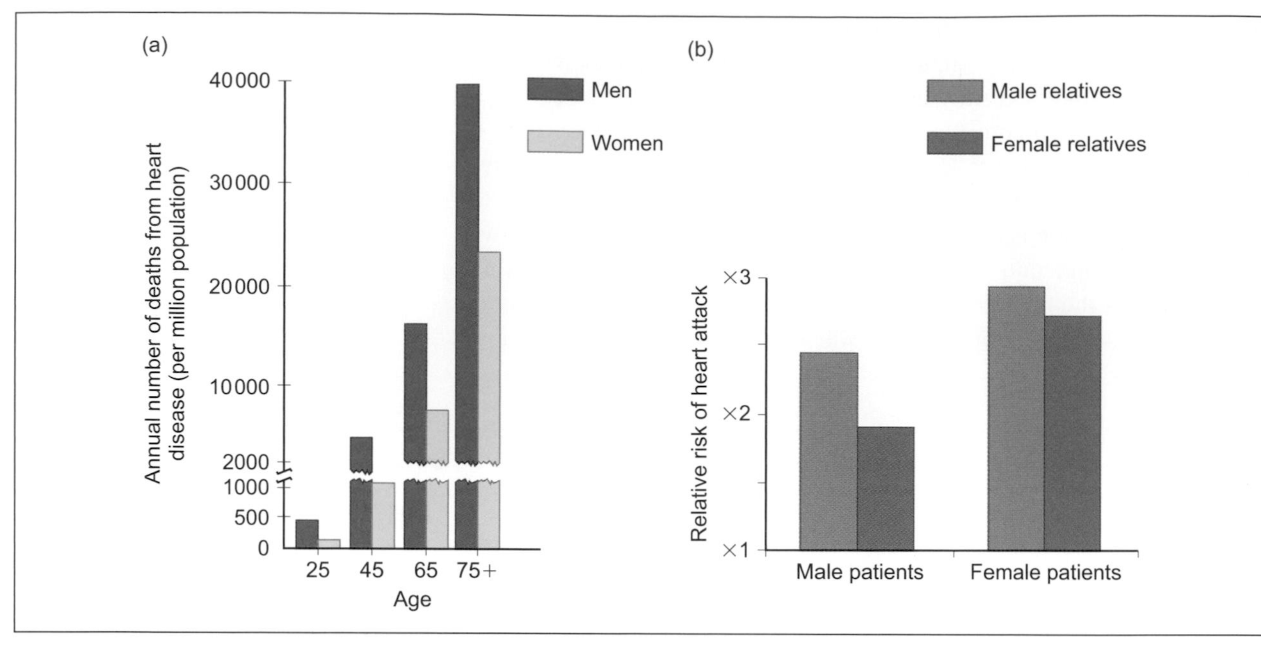

Figure 6.7

Unavoidable risk factors. (a) Deaths from heart disease according to age and sex, for England and Wales. (b) Relatives of people who have a heart attack are more likely to suffer one themselves. Genetic risks for multifactorial diseases have to be considered along with other risk factors. For example, some families have a genetic tendency to have problems in the way they metabolise fats, leading to furring up of the arteries and possible heart attacks. However, this genetic tendency may never show itself if the members of the family all eat a very low fat diet and take lots of exercise throughout their lives.

inherited genetic component but are also influenced by the environment. These diseases, such as cancer and heart disease, present a real challenge for medical genetics due to their complex nature. Some of the genes that make individuals susceptible to cancer and heart disease have been identified, and the presence of mutations can now be tested for. For example, different genes that influence breast cancer susceptibility have been found on chromosomes 6, 11, 13, 14, 15, 17 and 22.

Case-control studies

Large-scale studies can help to identify genes that contribute to complex diseases since individuals with alleles that predispose them towards a disease are statistically more likely to be affected. Similarly, those with alleles that protect them from a disease are statistically less likely to be affected.

Case-control studies select particular groups for study, for example people with a particular level of exposure to environmental factors, or with a particular health condition. If a disease is rare, case-control studies allow large numbers of sufferers to be studied, so this method is commonly used for studies of cancers and other rare diseases. This method is quicker, cheaper and easier to carry out than the cohort studies described in Chapter 4.

The participants are normally compared with a healthy control group to see if the differences account for why some people become ill and others don't. Statistical analysis is used to identify any difference between cases and the controls. Results are expressed as a relative risk

Key term

A **risk factor** is anything that increases a person's chance of developing a disease such as cancer or heart disease.

Question

6 a) In Figure 6.7 are the relatives of female patients or male patients more at risk?
b) Is the inheritance of risk from heart attack sex-linked?

Key term

Case-control studies are studies in which patients who already have a certain condition are compared with people who do not. For example, a study in which lung cancer patients are asked how much they smoked in the past, and the answers compared with a sample of the general population would be a case-control study.

for the disease (Figure 6.7). A relative risk of 1 means no difference in risk between the two groups. The probability that the relative risk differed from 1 purely by chance will also be reported in studies. If the probability of a relative risk being different from 1 purely by chance is less than 5%, that association is usually described as 'statistically significant'.

By collecting DNA with medical and lifestyle data from hundreds of thousands of people, and following their long-term health, researchers can work out why some people develop a particular disease while others do not. 'Prospective' biobanks carry out this type of study; they assess and take samples from participants at the start of the study, and then follow their health over subsequent years, even decades (see Chapter 8).

Does cancer have a genetic cause?

In a healthy adult, the rate at which cells divide to produce new cells is kept under very strict control. The rate of formation of new cells matches the requirements for growth and for replacing cells that have been lost through wear and tear or injury. When a person has cancer, some cells are somehow able to escape from this control and multiply to produce a growth, or tumour.

Not only do the cells in the tumour make no contribution to the body, they also take up space and get in the way of the activity of normal cells in the organs affected, eventually causing death. The ability of cancer cells to spread around the body and invade other tissues (metastasis) is one of the characteristics of the disease.

All cancer is genetic because it is triggered by altered genes. Even though all cancer is genetic, just a small portion (perhaps 5 or 10%) is inherited.

Questions

7 When doctors talk about cancer they may use these words: tumour, benign, malignant. What do these terms mean?

8 Identify and list some of the differences between normal body cells and cancer cells.

Cell division in cancer

Figure 6.8

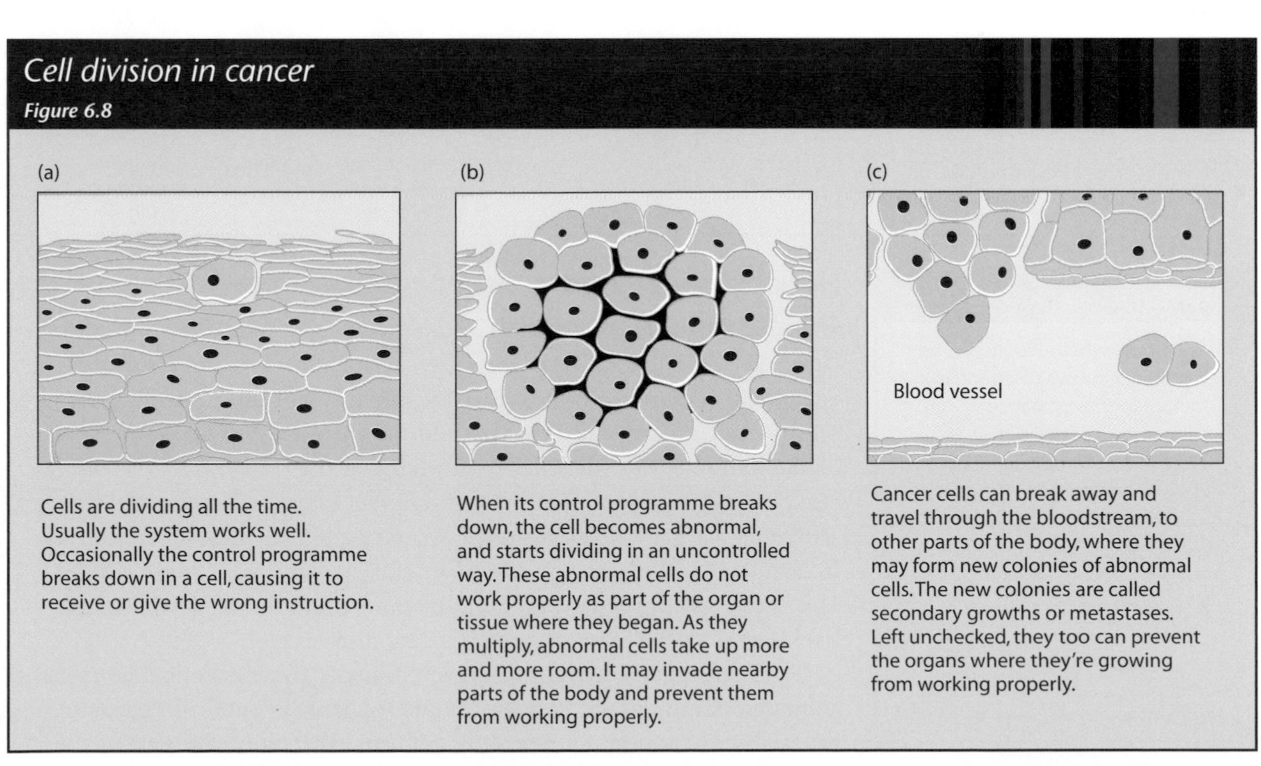

(a) Cells are dividing all the time. Usually the system works well. Occasionally the control programme breaks down in a cell, causing it to receive or give the wrong instruction.

(b) When its control programme breaks down, the cell becomes abnormal, and starts dividing in an uncontrolled way. These abnormal cells do not work properly as part of the organ or tissue where they began. As they multiply, abnormal cells take up more and more room. It may invade nearby parts of the body and prevent them from working properly.

(c) Cancer cells can break away and travel through the bloodstream, to other parts of the body, where they may form new colonies of abnormal cells. The new colonies are called secondary growths or metastases. Left unchecked, they too can prevent the organs where they're growing from working properly.

Key term

Mortality is a measure of the rate of death from a disease within a given population.

Most cancers come from random mutations that develop in body cells during one's lifetime, either as a mistake when cells are going through cell division or in response to injuries from environmental agents such as ionising radiation (see Chapter 7) or certain chemicals. When damaged cells reproduce, the new cells inherit these mutations. Mutations in body cells may lead to cancer. The risk of cancer rises as we get older because the number of mutations in our body cells increases with age.

A minority of mutations affect genes in our sex cells, and these mutations are potentially serious as they can lead to the inheritance of a genetic disease.

Research carried out by Cancer Research UK suggests that, although our genes affect our susceptibility to certain types of cancer, our environment and lifestyle also play an important role. Bowel cancer is the third most common cancer in men in the UK, and the third most common cause of cancer death. Less than 1 in 10 cases of cancer of the bowel are due to an inherited genetic defect. In most cases, a combination of factors such as diet, exercise, age, previous medical history and genetic factors contribute to the likelihood of a person developing the disease. As a result, there are only a few diseases where genetic tests can predict the future outcome with certainty.

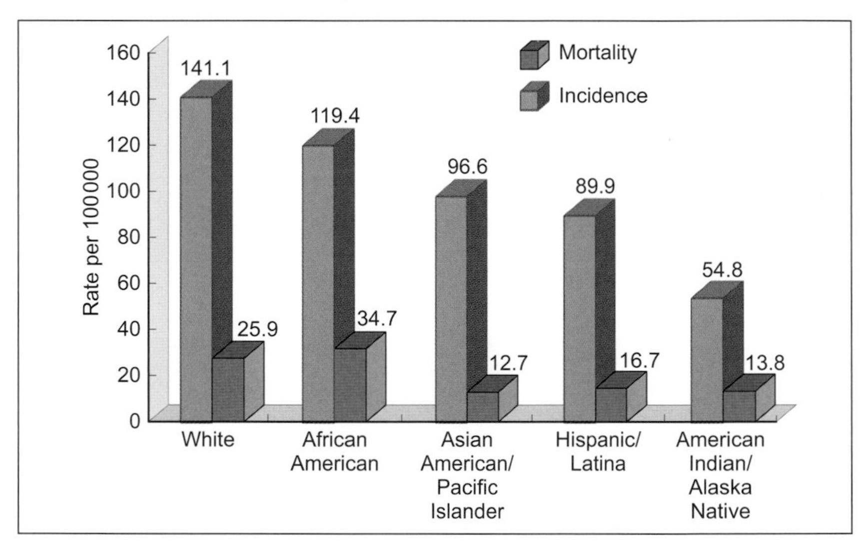

Figure 6.9

This graph compares the breast cancer rates of some different ethnic groups across the United States, 1998–2002.

Question

9 a) Discuss what evidence, if any, Figure 6.9 shows of ethnic risk factors for breast cancer.
b) Suggest an alternative explanation for the differences shown in Figure 6.9.

Heart disease and genes

Research into heart disease

Heart disease is the UK's largest killer, accounting for 105 000 deaths every year (Figure 6.10). In July 2007, the results from research carried out by the University of Leicester and the University of Leeds, in collaboration with colleagues in Germany at the Universities of Lubeck and Regensburg, were published in the *New England Journal of Medicine*. This research has begun to show how our genetic make-up can lead us to develop heart disease, and how we can predict who is most at risk. The study confirmed six new combinations of genes that increase the likelihood of developing coronary artery disease.

The first important clues to the identities of these genetic factors came from an analysis of all the genes in almost 2000 people with coronary artery disease and in 3000 healthy controls. The study was part of the

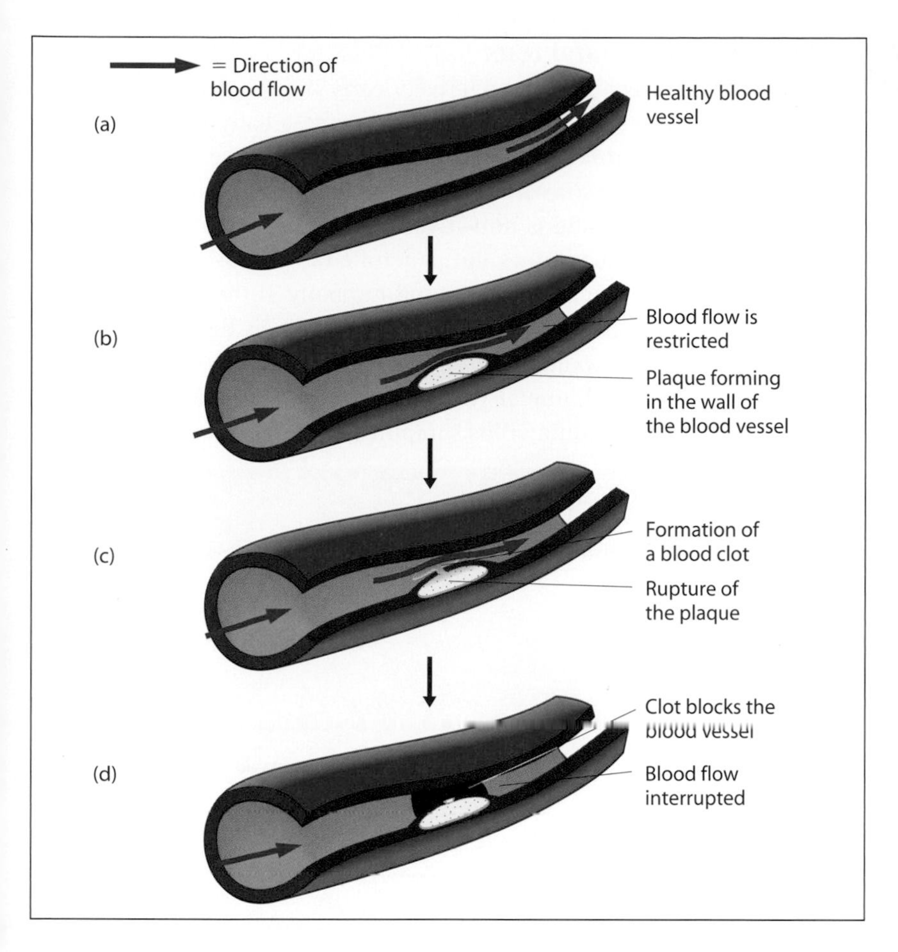

Figure 6.10

Coronary heart disease. A person suffers a heart attack when the reduction in blood supply through the coronary artery interrupts the blood flow to the heart muscle. Obstruction is caused when the smooth lining of the blood vessels becomes roughened and thickened by fatty plaque deposits. The plaque causes the blood vessel to rupture, and blood clots form, blocking the flow of blood. This figure shows: (a) section through a healthy artery; (b) fatty material building up inside an artery; (c) and (d) formation of a blood clot or 'thrombosis'.

Wellcome Trust Case Control Consortium (WTCCC), the largest ever study of the genetics of common disease. The findings were then compared with the German MI Family Study, with almost 900 additional cases and 1600 additional controls.

The researchers found that changes in the DNA on chromosomes 1, 2, 6, 10 and 15 were associated with an increased risk of developing coronary artery disease and heart attacks. The study also confirmed the importance of a gene on chromosome 9 associated with increased risk; this gene had been previously identified in an independent study.

Although the researchers know that some of the genes implicated are involved in cell growth and cell division, they are now working to understand exactly how this leads to heart disease. This understanding may enable the development of new preventative strategies and treatments.

'Is the baby all right?'

For most of human history, the only way to find out if a baby is 'all right' has been to wait and see, and for much of the world this is still the case. But the technology now exists to find out an amazing amount about a baby before it is born. Antenatal testing is becoming more and more important in the care of pregnant women. It is used not only to pick up genetic problems in the developing fetus but also to identify problems during development, where the fetus fails to form properly for a variety of reasons. An example of such a problem is spina bifida, when the backbone fails to develop properly.

Key terms

Screening is a public health service, where people are offered a test to identify those individuals at risk from a particular disease or condition. These people may then be offered more specific tests to identify a disease (such as genetic tests).

Genetic testing involves examining a person's DNA to see if they carry a particular allele or mutation that may cause disease. The aim of testing is to estimate the risk of further affected babies in a family with a history of the disease, or to identify the people who are carrying the faulty gene.

Antenatal tests

Screening tests relatively early in pregnancy pick up women who are at risk of having a baby with Down's syndrome and other genetic diseases. The simplest methods involve blood tests and ultrasound scanning. Ultrasound is non-invasive – meaning that the equipment does not penetrate the body – and it shows if the fetus is developing normally. If there are any concerns a sample of fetal cells allows the chromosomes to be examined for any abnormalities. To obtain the material, one of two tests is carried out – amniocentesis or chorionic villus sampling (CVS) (Figure 6.11).

Tests for genetic diseases

Figure 6.11

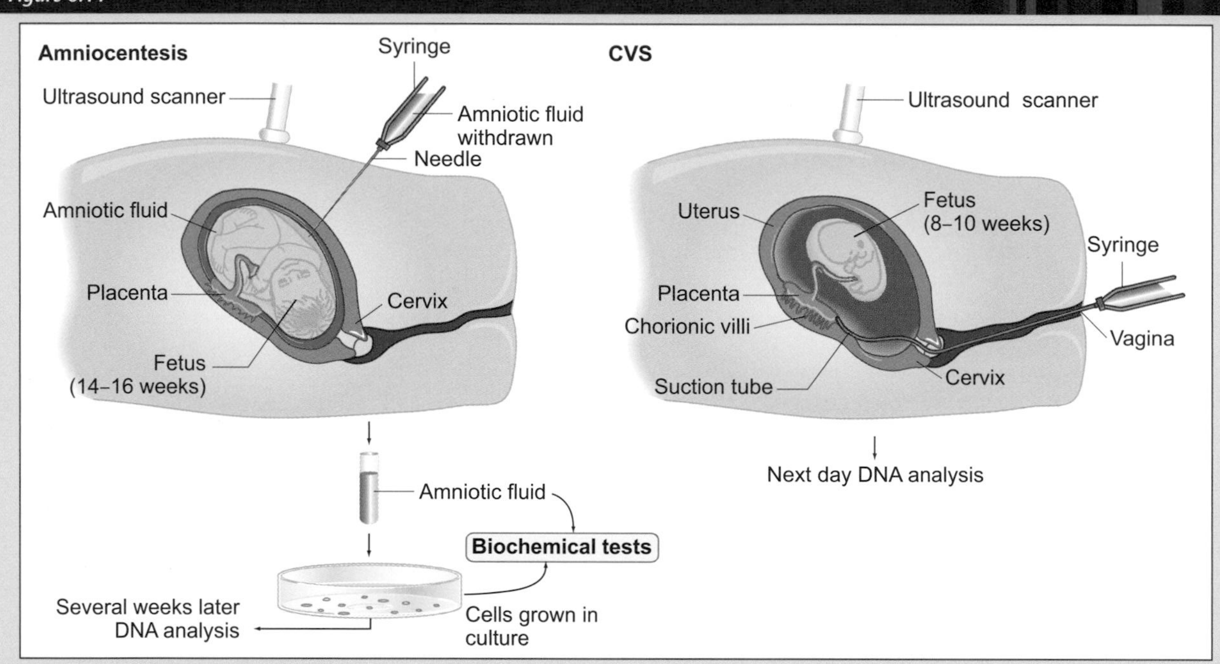

Amniocentesis and CVS are techniques used for genetic testing, to make an accurate diagnosis of Down's syndrome and other genetic disorders possible before birth.

Amniocentesis

Amniocentesis involves removing about $20\,cm^3$ of the amniotic fluid which surrounds the fetus using a needle and syringe. The test is carried out between 15 and 20 weeks of pregnancy, but ideally at around 16 weeks. Fetal cells can be recovered from the fluid, and, after the cells have been grown in a culture for a couple of weeks, the chromosomes in the cells are examined. A number of chromosomal abnormalities as well as the sex of the baby can be determined from examination of the chromosomes.

Disadvantages:

- The test can only be carried out relatively late in the pregnancy.
- The results are only available one to two weeks after the test.
- The test carries about a 0.5% risk of miscarriage after the procedure, whether or not there is something abnormal in the genetics of the fetus.

CVS

Used for DNA analysis, CVS involves taking a small sample of tissue from the developing placenta. Many more fetal cells are available for examination compared with amniocentesis, and the cells are tested for a wide range of genetic abnormalities. This diagnostic technique can be carried out earlier in the pregnancy (at 8–10 weeks) and it gives the information more rapidly (within 48 hours).

Disadvantages:

- There is a 1.5% risk of a miscarriage after the procedure, whether or not there is something abnormal in the genetics of the fetus.
- The test can't detect problems with the development of the fetus such as spina bifida.

The results

People tend to assume that by having a medical test they will be given a definite result which will make the situation clearer. Unfortunately, this is often not the case. Sometimes a test will give a false negative. For example, many pregnancy tests can be carried out as soon as the monthly period is missed. However, a negative result at this early stage does not necessarily mean a woman is not pregnant – it may simply be that her hormone levels have not built up sufficiently to be picked up, and she needs a more sensitive test, or to try again a week later.

Blood tests to measure hormone levels can indicate the risk of problems such as Down's syndrome in a fetus. These tests may also give a false negative, so later tests find the problem, or a baby with Down's syndrome is born.

False positive tests are also quite common. The blood tests which precede amniocentesis are for screening, used as a guide to select women most at risk of having a baby with a chromosomal disorder. There will always be a significant number of false positive and false negatives with this type of test. Amniocentesis and CVS are genetic tests, which have a high level of accuracy (approaching 100%).

Tests are rarely clear-cut – they have varying degrees of uncertainty. It is important to recognise this, and take account of it when such results are being used to decide whether a pregnancy should be terminated or continued to allow a baby to be born. It is still not possible to offer a genetic test for many conditions. Even for a common disease such as cystic fibrosis, there are hundreds of possible mutations, and only a few of these can be tested for. So, it is possible to have cystic fibrosis, but not have a test which will show this.

Result of test	*Disease present*	*Disease absent*
Positive	95	9990
Negative	5	89910
Total	100	99900

Figure 6.12

Results of genetic tests for a condition that affects 1 in 1000 babies

Difficult choices

Many pregnancies end in miscarriage when the fetus has a genetic disease. If the pregnancy continues, and antenatal testing shows that a fetus has a genetic disease, it can be very hard for parents. They have to come to terms with the fact that the child they are expecting will not be the 'normal' healthy baby they had hoped for. The choices facing them are often stark (Figure 6.13). At the moment there are no cures for many of the genetic diseases that can be identified by testing. The only choice is to end the pregnancy by an abortion or continue with it and, if the child survives birth, look after it and cope with the effects of the genetic disease.

Parents may choose to terminate a pregnancy rather than have a child with a serious disability or with the possibility of a prolonged and painful death. The parents have to make judgements about the quality

Questions

10 a) Suppose that you believe that the human fetus has a right to live. What duties does this mean that a pregnant women has to her fetus? Does anyone else have duties to her fetus? (See also Chapter 5.)
b) In some situations the right of the fetus may conflict with the woman's rights. How might these conflicting rights be resolved? Who should be involved in the discussions?

11 What is meant by a false positive and a false negative result to a medical test such as screening?

12 The rate of false positives or false negatives matters, as does the level of incidence of the genetic disease. Figure 6.12 represents the results of 100 000 tests for a condition that affects 1 in 1000 babies. The test has a false positive rate of 10% (number of false positive results as a percentage of the number of all true negatives) and a false negative rate of 5% (number of false negatives as a percentage of the number of all true positives).
a) How many false positive results are there when 100 000 babies are tested?
b) How many false negative results are there when 100 000 babies are tested?
c) Is the test worth using?
d) What harm is done by a false negative result?
e) What harm is done by a false positive result?
f) Is it more important to develop the technique to reduce the rate of false positives or to reduce the rate of false negatives?

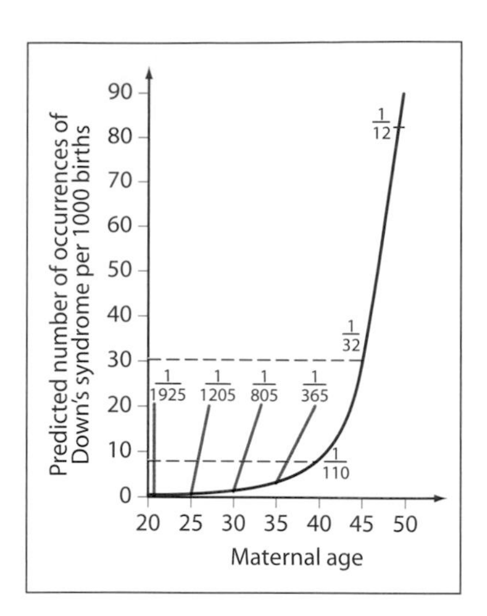

Figure 6.13

The likelihood of having a baby with Down's syndrome increases with the age of the mother. As many women in developed countries are delaying having their children until later, more and more will inevitably face difficult choices when problems arise.

Figure 6.14

A boy with Down's syndrome with his older sister. Some people think having a child with even a mild handicap would be devastating, while other people live enormously full lives in spite of serious disabilities. Parents also have to weigh in the balance the effect a sick or disabled child would have on their own lives and on the lives of any other children they might have – and many feel unable to shoulder such an unasked-for burden. Yet others find their lives enriched by the experience of having a family member with a disability.

of life of an individual who is not yet born (Figure 6.14). This is very difficult to do. Added into this are all the ethical issues associated with abortion.

Key terms

Abortion is the medical termination of a pregnancy. It is legal in the UK up to the 24th week of pregnancy. However, if there is a substantial risk to the woman's life or if there are fetal abnormalities there is no time limit.

Miscarriage is the term used for natural abortion when the fetus dies and is lost from the body. Miscarriages are very common in the first three months of pregnancy and in many of these cases the fetus has some sort of genetic problem.

Genetic counselling

For most people, finding out about genetic diseases in their family is very traumatic. Overnight they change from seeing themselves as normal, healthy people who will go on and produce a normal, healthy baby to realising that they carry a gene which could make them (or more probably their children) seriously ill or disabled. All of the issues discussed above are suddenly of immediate and personal relevance. Because for many couples the first realisation they have of a problem is when they are pregnant or shortly after their baby is born, the trauma is even harder to bear.

Genetic counsellors are trained to help people come to terms with the situation of carrying an abnormal gene which can cause genetic disease. A family pedigree (Figure 6.15) is worked out, which may be

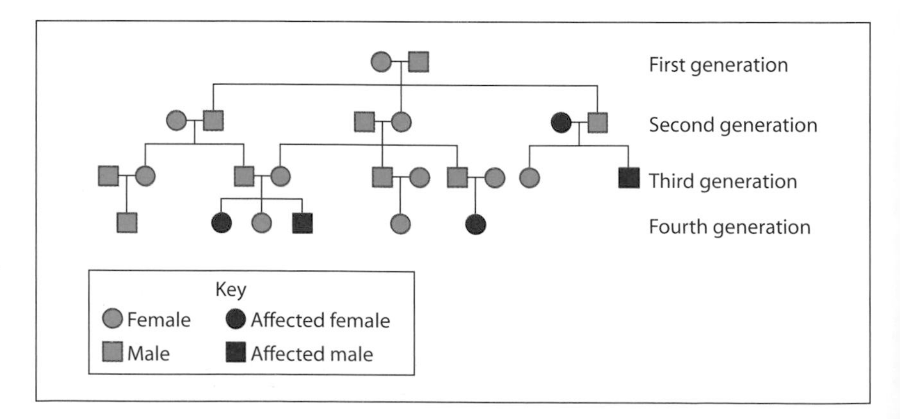

Figure 6.15

A family pedigree showing the inheritance of cystic fibrosis over four generations in one family. A genetic counsellor uses family pedigrees to work out the risk present for a particular pregnancy.

useful in confirming the diagnosis and can also be used to indicate any other individuals who might be at risk of having an affected child. This, however, raises another difficult issue. Should those other potential carriers, some of whom may be quite distant relations, be told or not? Genetic counsellors help people come to a decision about issues like this. They will also assess the statistical risk of a couple producing another child with the same defect. If the problem is caused by a single gene then the risk can easily be calculated. Even if it involves lots of genes or a whole chromosome abnormality, it may be possible to calculate the risk using accumulated medical data.

For most genetic defects, the options open to couples are:

- to avoid having children, or any more children
- to undergo prenatal testing and abortion of affected pregnancies
- artificial insemination with donor sperm to fertilise the mother's egg
- egg donation, followed by *in vitro* fertilisation (IVF) and implantation of the embryo in the mother's uterus
- adoption.

Again, the role of the genetic counsellor is to help couples recognise the options and work their way through to the alternative which is right for them, within their own framework of moral, family, religious and social beliefs and traditions. Information about treatment and management of the disorder is provided, so the couple can consider the full implications of having an affected child.

Possible solutions

One of the options open to couples who find they are at risk of producing children with genetic defects is IVF with preimplantation genetic diagnosis (PGD). Only those embryos (Figure 6.16) free of the problem

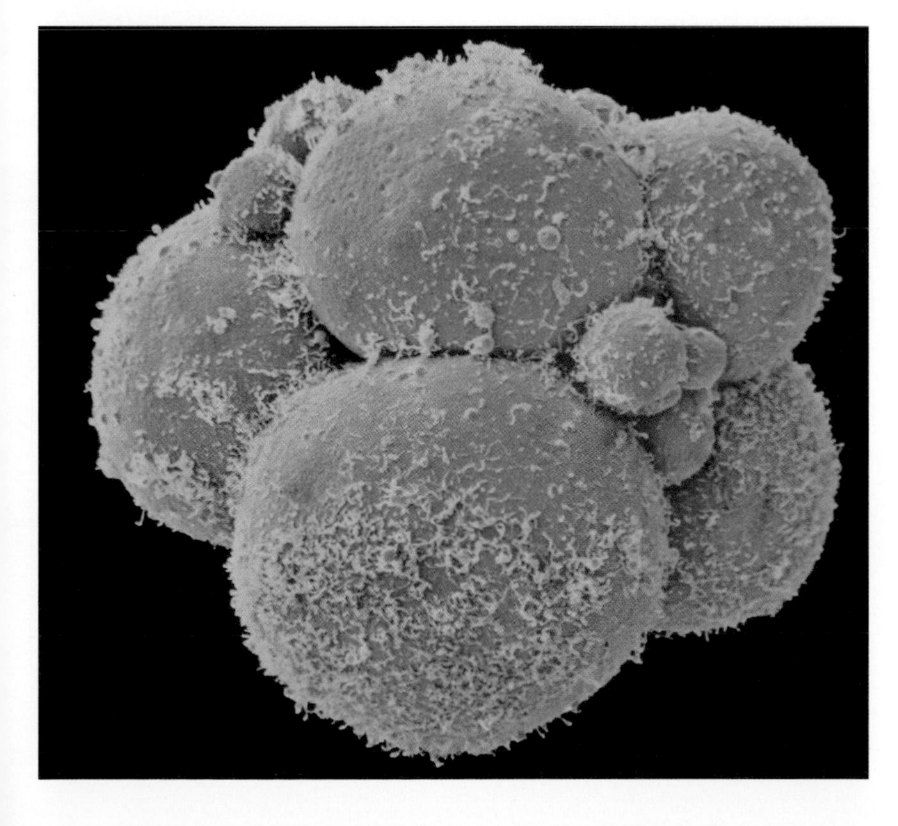

Figure 6.16

Coloured electron microscope picture of a human embryo at the eight-cell stage about three days after fertilisation. This is magnified about 900 times. When a human embryo is a small ball of cells, a single cell can be removed to check the genetic makeup without causing harm to the development of the embryo.

Questions

13 Why is it important to ask about the reliability of any antenatal test you might be offered?

14 Identify the ethical issues associated with taking a genetic test to see if a fetus is suffering from a potentially serious genetic disease. Outline the possible views that different people may take, using ethical frameworks to describe these views.

15 Explain why you would not normally see a recessive disease such as cystic fibrosis in three generations of a family, as shown in Figure 6.15.

16 What unusual feature of this family tree explains the presence of cystic fibrosis in three generations?

Key terms

***In vitro* fertilisation (IVF)** is a method of assisted reproduction where an egg is combined with sperm in a laboratory dish. The resulting embryo is transferred into the woman's uterus in the hope that it will implant in the uterine lining and start to develop. IVF normally requires hormone treatment that stimulate the ovaries to produce multiple eggs in order to increase the chances of successful fertilisation and implantation.

Preimplantation genetic diagnosis (PGD) involves removing a single cell from each early embryo produced by IVF and testing for the genetic disease. This does not harm the embryos, so healthy embryos can go on to be implanted in a uterus.

genes are placed in the mother's uterus to implant and grow (Figure 6.17). For example, in the case of sex-linked diseases such as haemophilia only female embryos would be replaced.

Another aspect of this complex picture is that with techniques of genetic engineering, the possibility of inserting a healthy gene into the cells of an individual with a genetic disorder is becoming a possibility. Trials are already going ahead in people with cystic fibrosis, which may allow their cells to produce normal, healthy mucus. The results of this gene therapy technique show some promise. Another disease which is being considered for this sort of treatment is haemophilia.

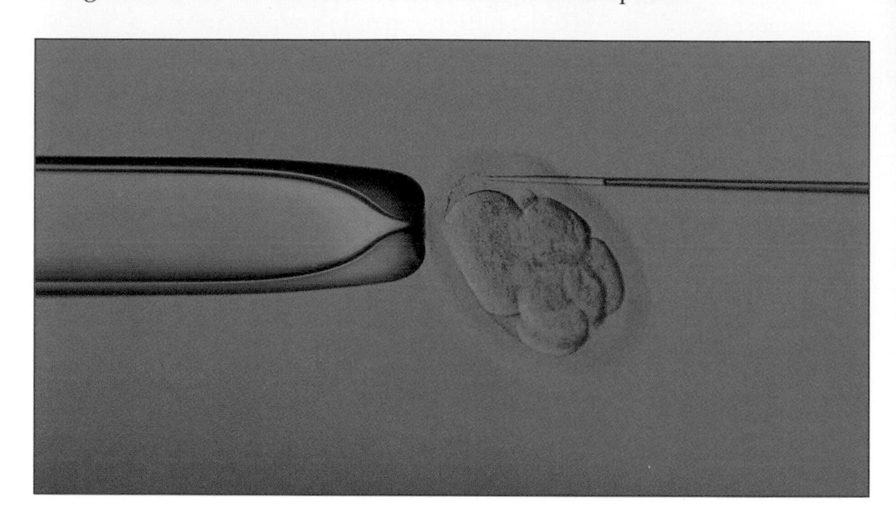

Figure 6.17

An IVF embryo. The eight-celled embryo produced by IVF in the laboratory is being manipulated in order to have one of its cells removed. A pipette (left) holds the embryo. A micro-needle (right) is first used to break through the embryo membrane, before a cell is removed with another pipette. The cell is genetically tested to check for disorders such as cystic fibrosis.

At present, genetic manipulation of the sex cells is forbidden in the UK and most other countries. This means an affected individual may be helped in their lifetime to be free from the effects of the genetic disorder they have inherited, but they will nevertheless still be at risk of passing it on to their offspring. Treatment has to continue through the generations, which is an expensive option.

It can only be a matter of time before it is suggested that manipulation of the sex cells as well as the body cells should be allowed – so as to treat not just the individual concerned but all their potential future descendants. While this sounds almost miraculous in theory, once it is permitted to manipulate the genes of the sex cells for medical reasons, there are those who fear that pressure would increase for changes to be made in the genes for intelligence, or beauty, or height or aggression. The days of eugenics, of manipulation of the human breeding stock and the creation of sets of 'designer humans', might then become not science fiction, but science fact. For now, this technique is still very risky; there may be undesirable results that could affect people for generations to come.

A question of balance

The benefits of gene testing have to be weighed up against the limitations. A negative result can create a tremendous sense of relief and may eliminate the need for frequent check-ups and tests that are routine in families with a high risk of cancer. Even a positive result can relieve uncertainty and allow a person to make informed decisions about the future. A positive result can also let a person take steps to reduce risk before disease has a chance to develop.

There are also limitations of gene testing. For example, some disorders that 'run in families' can be traced to shared environmental exposures rather than any inherited susceptibility. In addition, some mutations detected by a positive test may never lead to disease. Some disease-causing mutations may escape detection because existing tests look only for the more common mutations in a gene. Perhaps the most serious limitation of gene testing is that test information is not matched by therapies. To receive positive test results when there is no adequate treatment can be tragic.

Advantages and disadvantages of genetic screening

Genetic screening is part of a health policy to reduce disease. Governments, health authorities and whole communities decide whether or not to introduce large-scale screening programmes.

At first sight genetic screening seems an excellent idea – identify the people with faulty genes (and everyone has some) before they pass them on to any children they might have in the future. You can also warn people about their own health risks, and give them an opportunity to change their lifestyle. But what are the implications of this type of knowledge? What do individuals, doctors and the authorities actually do with it? And can genetic screening as it stands give us all the information we need? There are questions which need to be addressed, because the Human Genome Project has produced detailed maps of the whole of the human genetic material. It will eventually be possible to identify the genes responsible for all the single-gene disorders, and to provide screening tests for them. It should also make clearer the situation when genes and lifestyle both affect the health – for example, in conditions such as heart disease.

As discussed earlier in the context of Huntington's disease, knowledge of one's own genetic problems can be hard to bear. Another problem is who should have access to all of this information.

If an individual is screened and found to have a gene for a specific disease or to be a carrier of a problem gene, then one possibility is for the information to be confidential to them. Each person would then have to decide whether to tell partners, parents and friends, whether to have children and whether to tell their insurance company or employer (Figure 6.18).

Key term

Genetic screening involves testing a group of people to identify individuals at high risk of having or passing on a specific genetic disorder.

Questions

17 Describe the difference in the possible uses of gene therapy in body cells and gene therapy which involves the sex cells.

18 Is there an ethical difference between the two types of therapy, when both involve manipulation and changing the genetic material?

Figure 6.18

Genetic testing raises issues about who owns the information, and who has a right to know.

On the other hand, some people argue that society has a right to know. People should feel confident that their partner will not knowingly pass on abnormal genes. Financial institutions should not find themselves lending money or offering cheap life insurance to people who have not declared a genetic make-up suggesting a high risk of future disease.

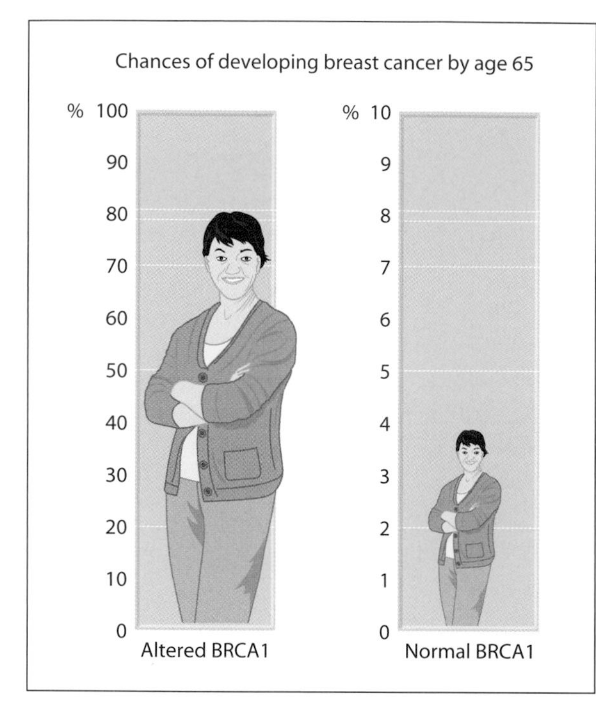

Figure 6.19

The presence of a gene mutation may indicate increased risk of a disease. This does not mean risks can't be reduced, for example by reducing lifestyle factors which may contribute.

Genetic tests for some multifactorial disorders are now available. Specific genetic mutations have been identified with links to several types of cancer. For some cancer types, this information has been converted into clinical tests. For example, scientists have identified gene mutations that are linked to an inherited tendency to develop colon or breast cancer. Tests for an inherited susceptibility to these cancers are commercially available.

In most of these cases, a positive test result does not automatically mean that you will succumb to the disease. It means that you have a particular mutation, and a much higher than average risk of developing the disease (Figure 6.19). Knowing this information allows people to make informed choices about their lifestyle. For example, healthy eating can help reduce the risk of developing certain types of cancer. Individuals can also choose to participate in screening programmes aimed at early detection.

Many people carry genes which mean that if they smoke, or eat too much, or make other lifestyle choices, they are far more likely to develop heart disease or cancer than others with exactly the same habits but different genes. It is possible that, in the future, an individual with genes leading to heart disease might be refused a driving licence after the age of 25 because of the increased likelihood of their dying at the wheel and causing a serious accident. If it is known that a young person is at high risk of serious disease later in their life, will they be disadvantaged when they look for college places, loans, jobs and relationships? Might they become fatalistic, taking the attitude that they are going to die young anyway, so they may as well smoke and drink to excess? Or, should we all try to have a healthy lifestyle, whether or not we are susceptible to heart disease? These sorts of questions and scenarios can sound a little far-fetched, but it is better for society to consider the possibilities now and set some guidelines in place than wait until the situation is happening.

Questions

19 a) Give two advantages of genetic screening and testing for a family.
b) Give two disadvantages of genetic screening and testing for a family.

20 a) What are the advantages of genetic screening and testing for society?
b) What are the disadvantages of genetic screening and testing for society?

How far should we go?

It can certainly be argued that it is unfair to have children affected with a serious genetic disease when the knowledge is there to prevent it. But knowledge does not solve the problems, it simply raises further dilemmas. Should all couples who plan to start a family or, indeed, who plan to have sex, be screened to see if their genetic weaknesses are compatible? Among the Ashkenazi Jews of North America, a devastating genetic disease called Tay-Sachs disease is very common. The disease causes damage to the nervous system, resulting in paralysis, blindness, dementia and early death, usually by the age of five years.

In an effort to reduce the incidence of Tay-Sachs disease, the rabbis have organised screening of young people in schools and colleges. Any couple considering entering into a serious relationship can contact the confidential screening service, Dor Yeshorim, to see if they are compatible. If it is found that they are both carriers of the Tay-Sachs gene the couple are counselled, and most choose not to continue the relationship. The human cost for those couples who find they are genetically incompatible must be great, yet the saving of human misery if Tay-Sachs can be reduced in or removed from the population would be enormous. Since Dor Yeshorim started, approximately 1 in 100 individuals tested prove incompatible with their chosen partner, and no Tay-Sachs babies have been born to people using the testing.

Screening all couples at risk of carrying a particular gene, however, can be enormously expensive. In a case like cystic fibrosis, screening would need to be done over the whole reproductive population of Europe, and even then only 70–75% of the possible cases of cystic fibrosis would be prevented. The economic value of screening programmes can be looked at to see if they are worth carrying out in economic terms.

A decision was taken to roll out routine cystic fibrosis screening for all newborns in England from 2007. Blood taken from a heel prick carried out on all newborn babies is used to test for a range of conditions, including cystic fibrosis. This newborn screening is to allow early diagnosis and treatment, known to result in longer and healthier lives. The costs of newborn cystic fibrosis screening is offset by the saving in treatment costs (Figure 6.20).

Questions

21 Explain what is meant by autonomy. Does the Tay-Sachs screening programme enhance or diminish the autonomy of a couple who are planning to marry? Suggest arguments for and against this particular screening programme.

22 Why is there no national screening programme for Huntington's disease?

23 Do you think that it is ethical to have screening tests available for a number of diseases but not to use them because of the expense?

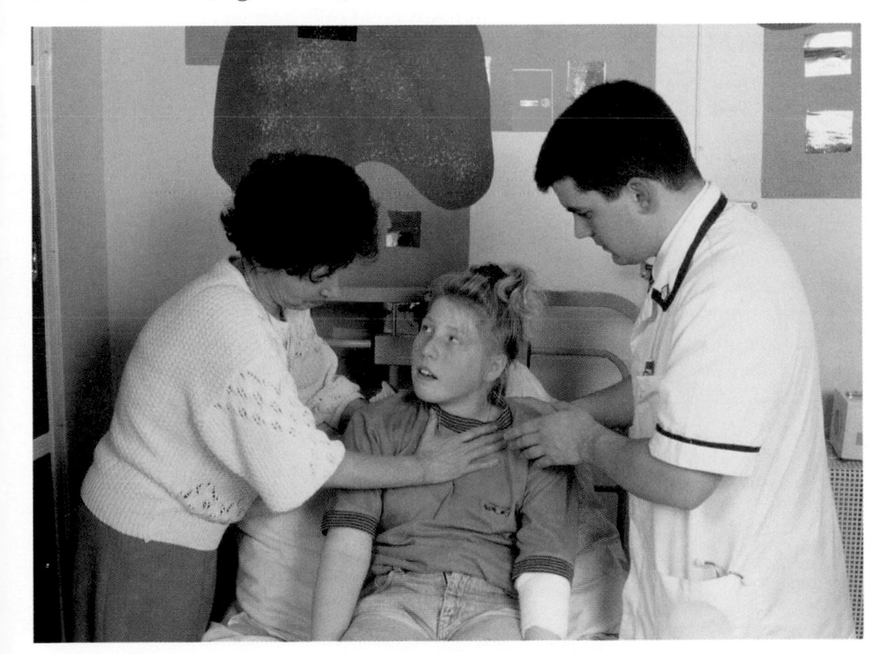

Figure 6.20

A physiotherapist teaching a mother of a young person with cystic fibrosis how to carry out a massage to relieve the build-up of mucus in the girl's lungs. Economic calculations look only at the financial costs of care. The human suffering of affected individuals and their families is not, and probably cannot be, quantified.

Review Questions

24 A family is told they have a one in four chance of having a child with cystic fibrosis. Does this mean that:

- if they have four children one will definitely inherit the disease?
- if they have three normal children already, their next baby will definitely have cystic fibrosis?
- if their first child inherits the disease, then their next child will not?
- there is a one in four chance of any child they conceive being affected, and this risk is the same for each pregnancy?
- if their first child does not have the condition, the chance of the next one having it is greater than one in four?

25 In the future it is quite possible that everyone will have access to information about all their genes. This raises a number of questions about who else should have access to this information. Consider each of the following:

a) You and your partner are planning to start a family. You know that you have a gene which could cause a genetic disease in your children if your partner carries a similar gene. So far you have not shown each other your genetic information. What do you think you should do? Is it the same as you suspect you would do in reality?

b) Some people have genes which mean that they are at higher risk of sudden death or illness. What sort of jobs might justify an employer having the right to know about an employee's genes?

c) A research post is advertised for someone to be part of a team carrying out long-term experiments into human ageing. One applicant (who is now aged 25) stands out as having the ideal qualifications for the job and being the right sort of personality to fit in with the rest of the team. However, his genetic make-up shows that he has an increased risk of suffering from mental health problems in his forties. Should this affect his chances of getting the job?

26 Screening tests for Down's syndrome now involve being able to measure the fluid-filled area at the back of the baby's neck during an ultrasound scan. This is known as the nuchal translucency measurement, and can take place when the fetus is 11–14 weeks old. The test gives a risk factor for Down's, as does a blood test of the mother's blood at the same time. What advantages do these tests offer, considering that if a high risk is indicated by these tests, most people will go on to have amniocentesis or CVS?

27 Justice is about fair treatment and the fair distribution of resources or opportunities. If testing for genetic diseases becomes widespread, do you think people should have to pay for the tests or should they be provided free by the National Health Service? Justify your views in terms of the concept of justice.

Chapter 7

Radiation: risks and uses

The issues

Many people have very negative attitudes towards the use of radiation and radioactive substances, particularly in large-scale applications such as nuclear power stations. There is much less opposition to the medical use of radiation (Figure 7.1).

How well founded are people's fears? How can these dangers be balanced against the benefits? How great is the risk to people's health of exposure to radiation? Can the risks be reduced or eliminated altogether? How can non-scientists evaluate the evidence on these issues?

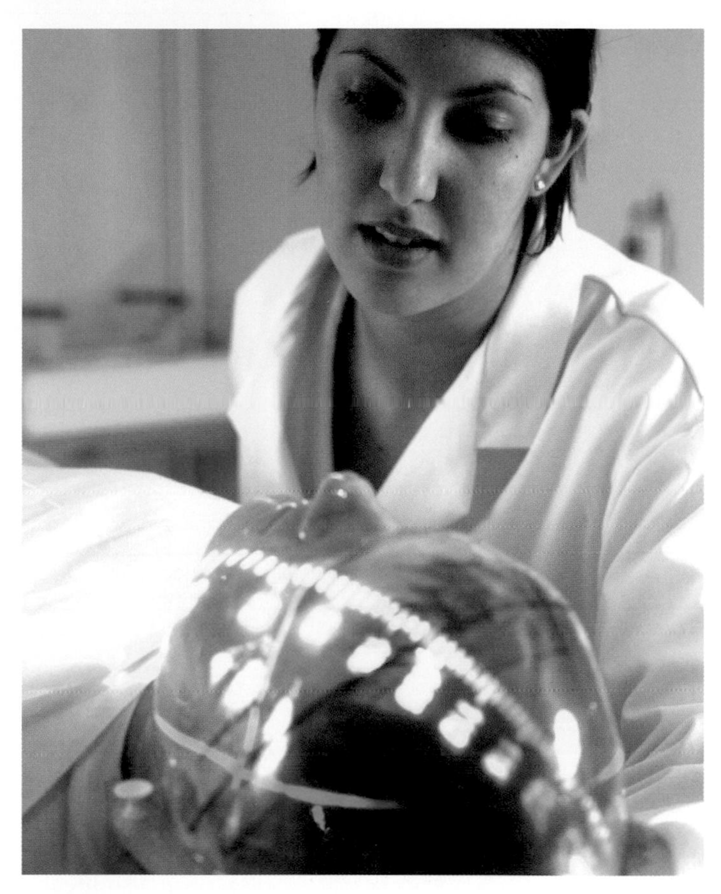

Figure 7.1

A radiographer preparing a cancer patient for radiotherapy using a laser beam to make sure that the invisible ionising radiation will be targeted at the brain tumour. The plastic mask holds the patient's head still and protects healthy cells from the radiation.

The science behind the issues

Radiation is emitted by a source and travels in a straight line until it hits another object. When it is absorbed, it no longer exists as radiation but heats up the absorber and may also cause chemical changes.

Some substances are radioactive. They emit ionising radiation which can cause changes within living cells, changes which can make them grow out of control and cause cancer.

The health risks of ionising radiation increase with the dose a person receives. As well as the risks of irradiation, people and the environment can be contaminated by radioactive material if it spreads from its source.

What this tells us about science and society

Experts and the general public react very differently to estimates of risk. Often the measures used to assess the size of risk have little meaning to individuals. People respond differently to situations where the risk is small but the consequences are very serious, and those where the risk is greater but the consequences are less severe. The way people respond to risks also depends on whether or not the risk is imposed or taken voluntarily.

Electromagnetic radiation and its effects

The light we can see is a very narrow band in the broad spectrum of electromagnetic radiation from the Sun and other sources (Figure 7.2). Our eyes can see because this visible light produces chemical changes in the cells of our retinas. These chemical changes then set off nerve signals to the brain. Other types of electromagnetic radiation with different wavelengths can also bring about physical and chemical changes, but our eyes cannot see them.

Figure 7.2

Types of electromagnetic radiation. The types of radiation which travel through space are forms of electromagnetic radiation. The forms of electromagnetic radiation have different wavelengths.

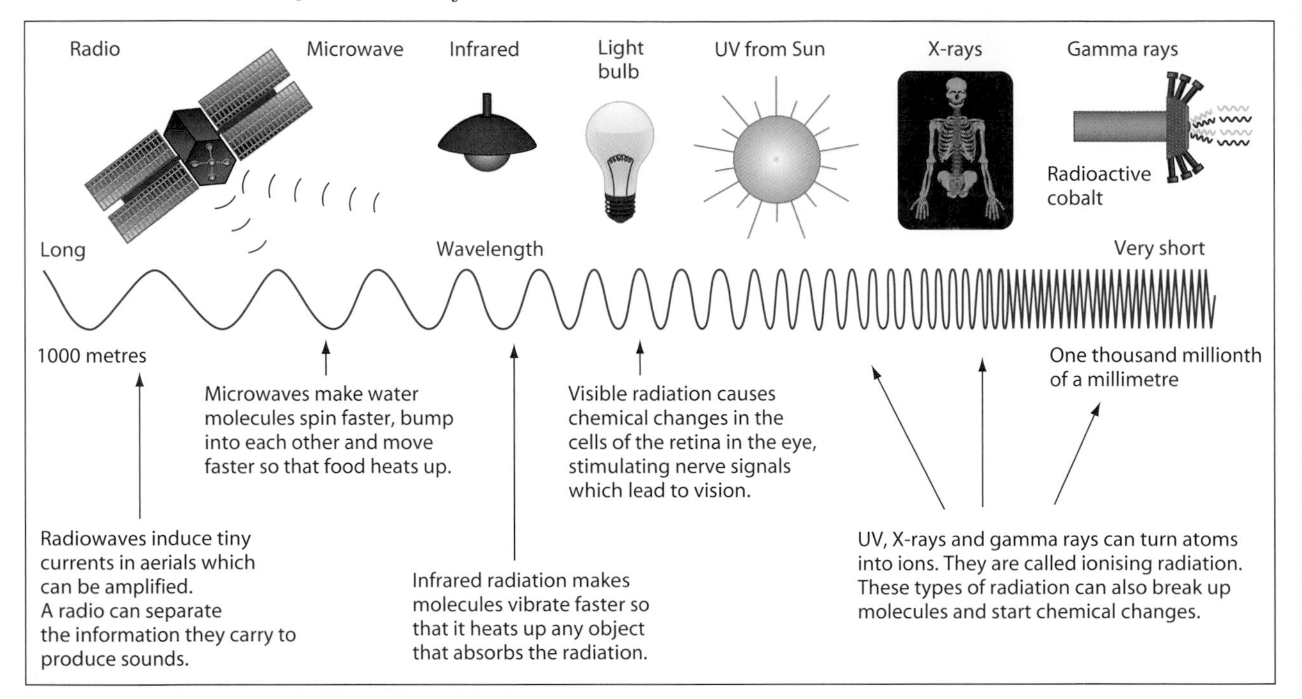

For life on Earth the most important chemical effect of the Sun's visible radiation is photosynthesis, which harnesses the energy from the Sun to make foods that are energy stores. Other effects of radiation are harmful and can kill. When UV, X-rays or gamma radiation is absorbed by cells, the energy carried by the radiation does not just heat up the cells; it can also break up molecules into fragments. Molecules are electrically neutral but some radiations can break them up into fragments which are electrically charged. These charged fragments are ions and the radiation that produces them is called ionising radiation.

High-intensity radiation can damage so many molecules in cells that it destroys them. Cells may be able to repair lesser amounts of damage done by lower levels of radiation. If the DNA in cells is damaged, however (see Figure 7.3), even a small amount of damage can have serious consequences.

If vital genes are damaged, cells can be crippled or killed. If radiation damages the genes that control growth and replication, cells can start to divide uncontrollably to produce tumours of rogue cells. This is cancer (see Figure 6.8, page 85). There is often a time delay of several years between the initial damage to cells and the development of a cancerous tumour.

Scientists are confident in claiming that ionising radiation causes cancer. This is because there is, for all but very low radiation doses, a correlation between radiation dose and the risk of cancer. Also there is a well-established mechanism explaining how it happens.

Key term

Ionising radiation is radiation with enough energy to break up molecules and turn them into electrically charged ions.

Questions

1. Identify three sources of electromagnetic radiation other than the Sun.
2. Give an example of electromagnetic radiation that:
   a) has a heating effect when absorbed
   b) is used to carry information.

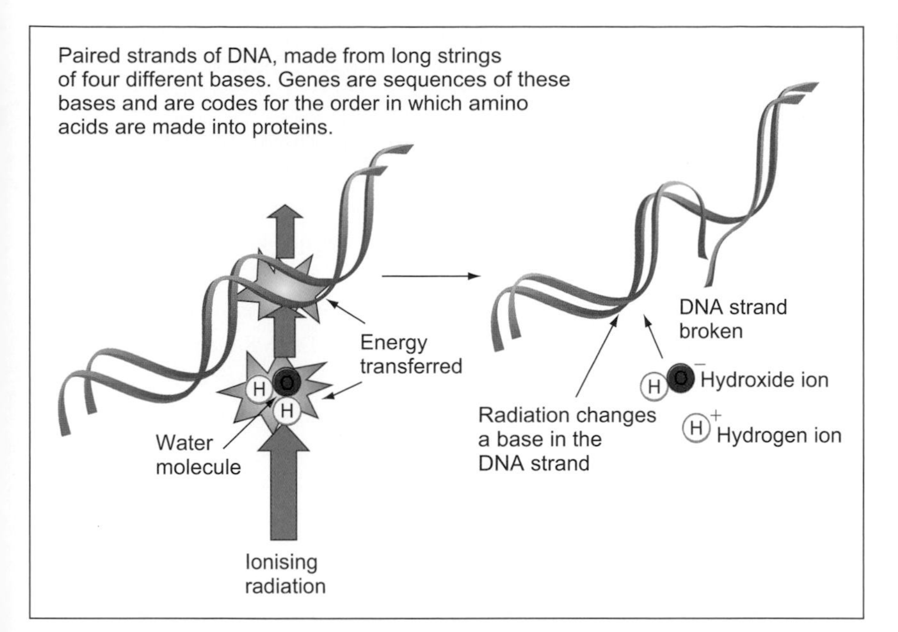

Figure 7.3

Ionising radiation can break up molecules into ions. If the radiation affects DNA molecules this can damage genes.

Are mobile phones a health risk?

There has been a huge increase in the number of mobile phones in use in Britain since they were first introduced in the early 1990s. Because mobile phones are now so common, and so popular, any 'scare' story about their possible dangers is sure to be big news.

Reviewing the evidence

In 1999 the UK government set up an independent expert group of scientists to review the evidence about the health risks of mobile phones. The report, *Mobile Phones and Health* (the *Stewart Report*), was published in 2000. The scientists decided that, on the balance of the evidence, the public was not being harmed by the microwave radiation from mobile phones and base stations (Figure 7.4). However, they accepted that there was evidence that there could be biological effects from the radiation even at exposure levels below those permitted nationally and internationally. The scientists recommended that a precautionary approach should be adopted to the use of mobile phone technologies

Questions

3 a) Suggest reasons why the Stewart Report advised that children might be more vulnerable to any effects arising from the use of mobile phones.
b) Why are research studies to examine any possible effects of mobile phones on children limited for ethical reasons?
c) In the light of the limited evidence, what advice should the Department of Health give to parents about children and the use of mobile phones?

4 What does it mean to 'adopt a precautionary approach' to the use of mobile phones?

Antenna on mast – radiates with a power of 100 watts
Beam of radiation
Mobile phone
50–300 m
The handset transmits the signal to the base station – radiates with a power of ¼ watt

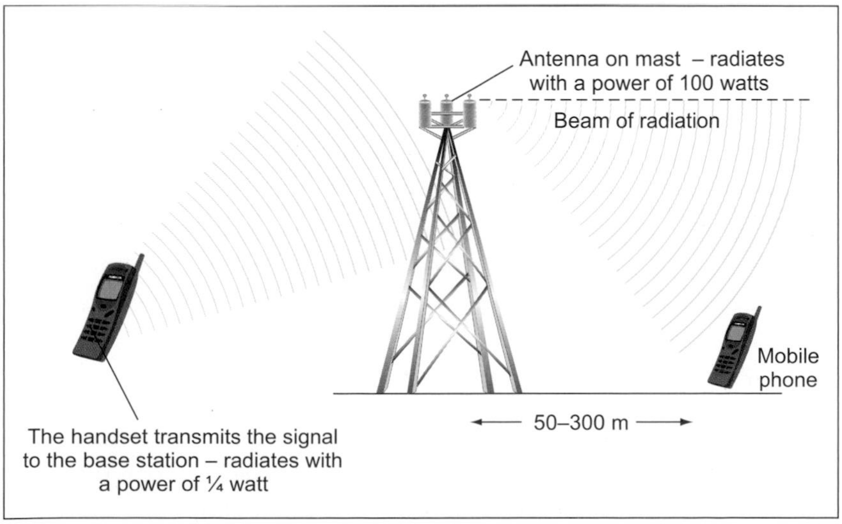

Figure 7.4

Microwave radiation used with mobile phones.

Figure 7.5

Scientists still do not know for sure what the risks are if people use mobile phones for many years.

Questions

5 a) What was the hypothesis being tested by the mobile phone study in England?
b) What type of epidemiological study was used for the research?

6 Write two newspaper headlines based on the results from the mobile phone study in Finland:
a) one headline to reassure readers
b) one headline to scare readers.

(see Chapter 8). At the time there were enough gaps in knowledge for it to be impossible to conclude that mobile phones are free from harmful effects on health.

Since the Stewart Report was published, the growth in the use of mobile phones has continued (Figure 7.5). The public is not behaving as if it considers the technology to be hazardous. In the short term no hard evidence has emerged to suggest that mobile phones are harming people's health.

The National Radiological Protection Board, which is now part of the Health Protection Agency, carried out a review of the *Stewart Report* in 2004, taking into account further evidence that had come to light. This review emphasised that mobile phone technologies are still so new that it has not been possible to get results from long-term epidemiological studies and evaluate the findings.

Newer epidemiological studies

In 2007, teams from three English universities and the charity Cancer Research UK published results of a large four-year study of mobile phones and cancer in the *British Medical Journal.* The researchers collected data from 678 people with acoustic neuroma. In this condition, benign tumours grow in the nerve that connects the ear to the brain. The tumours are not cancers but they cause loss of hearing. They grow slowly and do not spread to other parts of the body.

The data from the group of people with the condition were compared with findings from a control group of 3553 people who did not have acoustic neuroma. The scientists concluded that there was no link between the occurrence of this type of tumour and the number of years of mobile phone usage, the number of calls made, or the total number of hours spent on the phone.

However, a study by researchers from the Radiation and Nuclear Safety Authority in Finland has found evidence connecting brain cancer with the long-term use of mobile phones. This work was also published in 2007. The study compared the mobile phone use of 1521 people with a type of brain cancer with that of 3301 people without the cancer.

Looking at the whole group, the scientists found no link between phone use and brain cancer. However, when they looked only at people who had used a mobile phone for 10 years or more, they found that they were 39% more likely than average to get a cancer on the side of their head where they held their handset.

Possible explanations

Even if a correlation between microwave radiation and brain cancer was established, scientists would still need to provide a plausible story of how microwaves could affect DNA before they could claim that microwaves cause cancer.

One possibility is that microwaves damage cells by making them hotter, just as food becomes hotter by absorbing the radiation inside a microwave oven. This, however, is not very likely. Microwave phones emit less than a thousandth of the amount of energy per second that is used to cook food in an average microwave oven. Even allowing for a possible focusing effect of the brain tissue, creating 'hot spots', the microwave radiation from a mobile phone is unlikely to heat up brain cells as much as they are heated up when someone takes mild exercise.

The radiation from microwave masts is, of course, more powerful than from the phones themselves, but it is many times further from anyone's head and the intensity of the radiation falls off in proportion to the square of the distance (Figure 7.6). In other words, if the distance from the source of radiation is increased by a factor of 10, the intensity of the radiation reduces by a factor of 100.

Scientists investigating the possible link between mobile phones and brain cancer have, therefore, been looking for other evidence of how microwaves might damage cells, usually by using animals or *in vitro* cultures of cells. Various claims have been made, for example that microwaves disrupt the mechanism in cells that repairs damage to DNA. If so, this would make it more likely that they will become cancerous. To date, however, no one has come up with a story that a majority of other scientists find convincing about how low-level microwave radiation could cause brain tumours.

Figure 7.6

Mobile phone masts on top of a block of flats in London. Masts on or near buildings where people live are controversial because of fears that low-level microwave radiation from the masts may harm people's health.

Living close to a high-voltage power line

Pylon lines are a feature of our landscape, both in cities and in the country. They carry electricity from power stations to consumers. Pylons carry safety warnings about the dangers of high-voltage cables, but until around 1980 no one had suggested there was any other danger from power lines. Then a study in the USA suggested a higher incidence of cases of childhood cancer among children living near a power line (Figure 7.7).

Mains electricity in the UK is an alternating 50 hertz (Hz) supply so that the current changes direction and back again 50 times each second. This means that the electrical and magnetic fields around cables

Questions

7 Suggest a reason why cells are more likely to become cancerous if the mechanisms that repair DNA are disrupted by microwaves.

8 a) Do you think that the balance of evidence suggests that mobile phones can lead to brain tumours or do you think that the evidence shows that mobile phones are safe?
b) What additional information would you most like to have to help to reach a decision or to check your current view?
c) Could a study be set up to obtain this information? If so, how could it be done?

Figure 7.7

Power lines over people's homes.

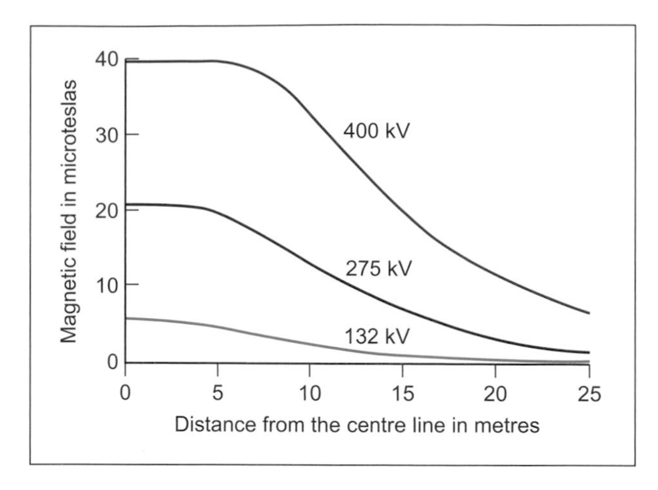

Figure 7.8

Magnetic field strength near a power line.

and power lines are also alternating with the same frequency. As a result the cables emit electromagnetic radiation.

The radiation from wires and cables carrying mains electricity has a very low frequency compared to other types of radiation in the electromagnetic spectrum and is sometimes referred to as extra-low-frequency (ELF) radiation.

ELF radiation, like microwave radiation, is non-ionising. In fact, its only known effect is that it can, like microwaves and radio/TV waves, cause small alternating currents to flow in materials that conduct electricity, such as aerials and, to a lesser extent, living cells. It also sets up small electric and magnetic fields within materials. The size of this magnetic field decreases rapidly with distance from the power line (Figure 7.8).

Ordinary household electrical appliances also emit ELF radiation and generate small magnetic fields around them when they are switched on. As Figure 7.9 shows, these can be larger than the field 25 metres from a power line. On the other hand, many appliances are only switched on for short periods of time, whereas fields due to power lines are always present. As there have never been suggestions that domestic electrical appliances pose a cancer risk, scientists were inclined to dismiss the claims about power lines.

Figure 7.9

Exposure to magnetic fields from household appliances.

Household appliances (30 cm away)	*Magnetic field strength/microteslas*
Vacuum cleaners and drills	2–20
Food mixers	0.6–10
Hair driers	0.01–7
Dishwashers	0.6–3
Washing machines	0.15–3
Fluorescent lamps	0.5–2
Electric ovens	0.15–0.5

Questions

9 How does the magnetic field due to a high-voltage power line 25 metres from a house compare with that due to the appliances inside the house (see Figures 7.8 and 7.9)?

10 On the basis of Figures 7.9 and 7.10, how close to power lines do you think that houses should be built?

11 From Figure 7.10 calculate the ratio of cases to controls for the higher exposure group, and for the lower exposure group. How much bigger is this ratio for the higher exposure group?

12 What features of the Swedish study made other scientists take its findings seriously?

Convincing epidemiological findings

It was a Swedish study, reported in 1993, that caused the scientific community, and the bodies concerned with health and safety regulations, to take the suggestion of a link between power lines and health more seriously. The researchers identified all 127 383 children in Sweden (under the age of 16) who had lived within 300 metres of a power line between 1960 and 1985. In this sample, they found 141 cases of cancer, including 39 of leukaemia. They looked for four controls for each case. These were children who were the same in all respects except that they had not lived near a power line. They took great care over this matching, looking for children from the same area, of the same sex and age. They eventually identified 554. Using records from the electricity companies they then worked out the exposure of each child to ELF radiation and magnetic fields.

In one analysis of leukaemia cases, they compared children exposed to average magnetic fields over 0.2 microtesla with those exposed to average fields below 0.1 microtesla (Figure 7.10). Children exposed to the higher field were 2.7 times more likely to develop leukaemia. This rose to 3.8 times for children exposed to a magnetic field of average strength 0.3 microtesla.

	Cases	Controls
Exposed to larger fields (>0.2 microteslas)	7	46
Exposed to smaller fields (<0.1 microteslas)	27	475

Figure 7.10

Cases of leukaemia among Swedish children under 15 exposed to larger and smaller magnetic fields.

Because of the care with which the samples were chosen and the information collected, and the size of the study, these findings were generally felt to be more reliable than many epidemiological studies. It was unlikely that other environmental factors were responsible for the results. On the other hand, the numbers of people contracting cancers was small. The researchers estimated that power lines account for 3.5 cancer cases per year in the whole of Sweden.

A controversial theory

In 1996, and again in 1999, Denis Henshaw and his colleagues at the University of Bristol published a theory to explain the health effects of power lines. The scientists had studied the effect of ions in the electric field around power lines. The field is strong enough to ionise molecules in the air, producing so-called 'corona ions'.

The idea is that the ions in the air join up with small particles of air pollution so that these particles become electrically charged. The electric charge on these particles means that they are more likely to stick in the lungs. So, in this theory, the rise in cancers is caused by the increased effect of air pollutants. Most scientists do not accept this theory, pointing out that it does not account for the observed increase in leukaemias near power lines but would be more likely to cause a rise in lung cancer, which has not been detected. In 2004, an advisory group of the National Radiological Protection Board decided that this was not a plausible explanation.

Newer results from a study in England

The results of a major case-control study of power lines and childhood cancer were published in 2005 (see page 80). The researchers identified 29 081 children with cancer, including 9700 with leukaemia. The children were aged 0 to 14 years and were born in England and Wales between 1962 and 1995. The researchers also matched a control group of children with respect to sex, approximate date of birth, and the district where their birth was registered. They worked out the distance from the nearest power line of all the children born within 600 metres of the electricity grid network.

This large study shows that children living within 200 metres of power lines had a significantly greater risk of contracting leukaemia. There was a greater risk too for those living between 200 metres and 600 metres away (Figure 7.11). However, for the other childhood diseases studied

Distance	Relative risk	95% confidence interval
0–200 m	1.69	1.13 to 2.53
200–600 m	1.23	1.02 to 1.49

Figure 7.11

The relative risks of childhood leukaemia for two ranges of distances from power lines.

Questions

13 Identify differences in the scientific basis of the investigations by the researchers in Sweden and the researchers in Bristol.

14 Explain why the Bristol group's theory might predict that cases of leukaemia would be higher on the downwind side of power lines.

15 In the context of the case control study of powerlines and childhood cancer, explain these terms in Figure 7.11:
a) relative risk
b) 95% confidence interval.

16 How do the findings of the 2005 English study differ from those of the earlier Swedish study?

17 a) Do you think that the balance of evidence suggests that living near power lines increases the risk of cancer or do you think that the evidence shows that power lines pose no risk?
b) What additional information would you most like to have to help to reach a decision or to check your current view?
c) Could a study be set up to obtain this information? If so, how could it be done?

there was no increased risk. The results could not be explained in terms of the magnetic fields round power lines because they are too low to have an effect beyond 200 metres. According to the research team, the results did not support the 'corona ion' theory of the Bristol group either because there was no evidence that the effect was greater downwind of the power lines.

This study suggests that about 5 of the 400 cases of childhood leukaemia every year could be linked to power lines – which is about 1% of cases. However, the scientists also acknowledge that since there is no known biological mechanism to explain the higher risk. The results, although statistically significant, may be due to chance.

What do we mean by 'risk'?

Science and technology have done much to make our lives less risky. People in the industrialised countries live, on average, longer, healthier lives than in the past. But there are also growing concerns that science and technology may introduce new health risks – from radiation or from the chemicals used in agriculture and food processing or from industrial wastes. This makes it important to be able to assess how big different risks really are. Knowing the risks we can make more rational decisions about whether to accept them or try to reduce or eliminate them.

Life is uncertain. We might catch a cold in the next month, or we might not. The car journey we make this weekend might go uneventfully – or we might have an accident (Figure 7.12). There is a definite risk of these events happening. It is impossible to predict what will happen to an individual but we can calculate the probability of the event occurring by looking at larger groups of people. To measure the size of a risk, scientists look at the probability of the event occurring. So, for example, in a town of 6000 people, the evidence might show that 150 people caught influenza during one winter. Their risk of catching it that winter was therefore 1 in 40. In the same period, it might be that 600 people in the town caught a cold. The risk of this would then be 1 in 10 – rather higher than the risk of catching influenza.

Figure 7.12

Road crashes are the single biggest killer of 15- to 24-year-olds in industrial countries. Death rates among under-25s are double those of older drivers. Young men have three times the death rate of young women. The more young passengers a young driver carries, the greater the risk of an accident.

Risk estimates

Risk estimates like these are never exact. One reason is that someone might be diagnosed as having influenza when they only have a heavy cold. For this reason, the risk of death from different causes is easier to measure – as the outcome can be detected with certainty. Figure 7.13 shows the risk of a person in the UK dying in any one year from a number of different causes.

Cause of death	*Risk of death (in one year)*
Smoking 10 cigarettes a day	1 in 200
Influenza	1 in 5000
Road accident	1 in 16000
Accident at home	1 in 26000
Accident at work	1 in 43000
Radiation (for a worker in the nuclear industry)	1 in 50000
Murder	1 in 100000
Train accident	1 in 500000
Lightning	1 in 1000000
Radiation leak from nuclear plant	1 in 10000000

Figure 7.13

Risk of an individual in the UK dying in any one year from various causes.

All statements about risk are estimates. Often it is difficult to measure the size of a risk. This may be because the outcome is hard to detect with certainty, or because the risk is low and only a few people are affected, even when data is collected from a large population.

The size of a risk may also be very different for different groups within a population. For example, the risk of death from influenza in Figure 7.13 is an average for the whole population. The risk for people over 65 is much higher than 1 in 5000. The same is probably true about the risk of accidents in the home. This may be important when comparing risks.

Question

18 a) Which are the big voluntary risks in your life which you could do a great deal to control if you wanted to do so?
b) What are the small involuntary risks which you worry about but cannot do anything to change?

19 The risk of developing lung cancer in the UK is estimated at 209 per 100000 people per year for smokers and 14 per 100000 people per year for non-smokers. What proportion of people are likely to contract lung cancer over a period of 25 years if they live that long and are:
a) smokers
b) non-smokers?

Different ways of expressing a risk

Sometimes a risk is expressed in terms of the number of people likely to experience the event in question. For example, the population of the UK is 60 million, so the numbers in Figure 7.13 suggest that around 60 people will be killed by lightning over the next 10 years (Figure 7.14). This might seem higher to some people than a risk of 1 in 10 million.

After the accident at the nuclear power station in Chernobyl in 1986, a spokesman for the National Radiological Protection Board said in a radio interview that the accident would probably result in around 30 cancer deaths over the next 50 years. This caused a considerable public reaction – whereas an equivalent statement a few days earlier that the accident would only raise the risk to an individual by less than 1 in 10 million passed without comment.

The size of risk may also depend on exposure to the hazard. For example, the risk of dying in a road accident is greater for an individual who drives 20000 miles a year than for one who drives 2000 miles a year. For this reason the death rate among car drivers is often expressed in relation to distance travelled. In Britain in 1996 it was three deaths per 1000 million kilometres travelled.

Figure 7.14

The risk of being killed by lightning is very low.

How people react to risks

People often do not know the actual risk for many of the activities they choose. They have to use their own perception of risk to decide whether it is safe to do something. However, research shows that people are not very good at judging risk from experiences.

Figure 7.15

Some activities such as scuba diving have a higher risk of death (1 in 200 000 dives) than others such as rock climbing (1 in 320 000 climbs) or canoeing (1 in 750 000 outings).

People tend to overestimate the risk of unfamiliar or rare events, and underestimate the risks of familiar or common ones. If there has recently been an accident or a health-scare story in the news, people are likely to see the risk of this as larger than it really is.

Sometimes people are willing to accept the risk of an activity because of its benefits. So concerns about possible risks from mobile phones have not slowed the rapid growth in their use. And people still engage in 'dangerous' sports, despite the risks – because they enjoy them (Figure 7.15).

People are more willing to accept the risk from something they choose to do voluntarily, as compared to something they have no option about. So a person might willingly ride a powerful motorcycle, but would be horrified about a proposal to build a nuclear power station nearby.

People are more willing to accept a risk that ceases when you stop the activity in question – and more wary about risks that will continue for a long time. So people are concerned about the effects of chemical pollution, or radioactive materials, because the risk these pose may persist for a very long time. Even though hang-gliding might be much more dangerous, it ceases to have any risk once you stop doing it.

The hardest kind of risk to deal with is one which is very small – but with disastrous consequences if it does occur. The risk of a nuclear power station exploding is an example (Figure 7.16). The risk is extremely small. But if it were to happen, the consequences would be enormous. It is much harder to decide whether such risks are acceptable than ones where the risk is higher but the consequences are less severe.

Questions

20 There is some evidence that non-smokers tend to overestimate the risks of passive smoking, while smokers underestimate the risk of their own smoking. Suggest some reasons for this.

21 Some people seem to expect to be assured that an activity is entirely free of risk. Is this a reasonable expectation?

Figure 7.16

Hartlepool nuclear power station in Teesside, North East England

Risk factors

There are often claims that something is believed to pose a risk to health, or to improve our health. For instance, living under high-voltage power lines has been claimed to increase the risk of some types of cancer. On the positive side, it has been claimed that drinking a glass of red wine a day reduces the risk of heart disease. Claims like these, however, are often contested. The findings are questioned. So what is involved in establishing that there is a link between a factor and an outcome – or in showing that there is no connection?

Most things are influenced by several factors, not just one. This makes it difficult to spot a factor that has an important effect. If there is a suspicion that one particular factor might be important, then experts may be able to organise a cohort study to collect evidence by comparing two situations. In one situation the factor is present and in the other it is absent, while all the other factors that might be important are kept the same (see pages 60–61).

Detecting (or eliminating) small effects

If a factor causes a major effect, it is easy to spot – and people quickly recognise that it is important. But often a factor causes, at most, a very small effect; for example, worries about the effect of low levels of ionising radiation or about the use of mobile phones. The number of cases is very small. So if a study compares a group that is exposed to the factor with a control group, the numbers of individuals affected will be small in both groups. And if the numbers are small it is hard to be sure if any difference is real – or just due to random variation.

One way to tackle this is to use large samples. The bigger the samples, the more cases there will be in both groups – and the bigger any difference will be. The larger the difference, the less likely it is that it is just due to chance. But this may not solve all the problems. For instance, imagine that a study collects data about 10 000 people who have not been exposed to a certain factor. The study finds one person who has suffered harmful effects. The study also gathers information from another 10 000 people who have been exposed and finds three cases of people with symptoms. This is three times as many, but it is still just three cases. Maybe the extra two are just chance.

Another option is to use a case-control study (see page 84). This type of study has the advantage that it guarantees that there are enough cases of the ill effects in the data set to reveal a pattern if there really is one. But it is harder to avoid bias. The people making the study are more likely to 'see' what they expect to see – even if it is not really there.

Interpreting information about risk factors

Sometimes claims about risk factors say that the factor increases the risk by so many percent, or by so many times. It is, however, also important to know how many actual cases were observed. Figure 7.17 shows some data on two different risk factor claims.

Questions

22 Suggest reasons why it can be hard to set up a meaningful cohort study to study a risk factor.

23 How would you set up a cohort study to investigate whether taking more than 3 g of vitamin C per day reduces the risk of catching a cold.

24 How would you set up a case-control study to investigate whether having a domestic pet reduces stress?

	Number of cases in samples of 10 000			
	Exposed to factor	*Control*	*Risk increased by (%)*	*Risk increased (number of times)*
Claim 1	3	1	300	3 times
Claim 2	260	200	30	1.3 times

Figure 7.17

Cases found in exposed and control groups of 10 000.

Although the percentage increase is much bigger in the case of claim 1, many people might be more convinced by claim 2, because the number of cases has gone up from 200 to 260. This seems unlikely to be just random variation, whereas an increase from one to three might be.

Question

25 a) Imagine that you have two coins – one is normal and one has a small bias to turn up heads when tossed. What problems would you have in getting good evidence that one was biased?
b) Now imagine that you have two normal coins. How would you collect evidence to show that they are **not** different? Why would this be harder than showing that one coin is biased?

Convincing people there is no effect

Industries that want to persuade people that their activities pose no significant risk to health, or official regulatory bodies who want to establish that something does not pose a risk, have a particularly difficult task. There are always likely to be some differences between any two groups. So it is very hard to persuade people that a factor has no effect. This involves 'proving a negative'.

Media reports – genuine concern or just hype?

Media reports sometimes sensationalise new scientific claims (Figure 7.18). Reports can make them appear to be quite certain even when they are in fact provisional, based on limited evidence, and not yet confirmed by other scientists. Conflicting findings reported by other scientists may be ignored. This can then lead other scientists to enter the discussion, to counter the claims being made. The result is a public disagreement between scientists. If people think of science as 'certain knowledge', this can lead them to conclude that one side or other must be either incompetent, or deliberately trying to mislead – perhaps because of whom they work for or represent. If scientists taking part in such a discussion are shown to have changed their views from the ones they expressed at some time in the past, this can also lead to suspicion.

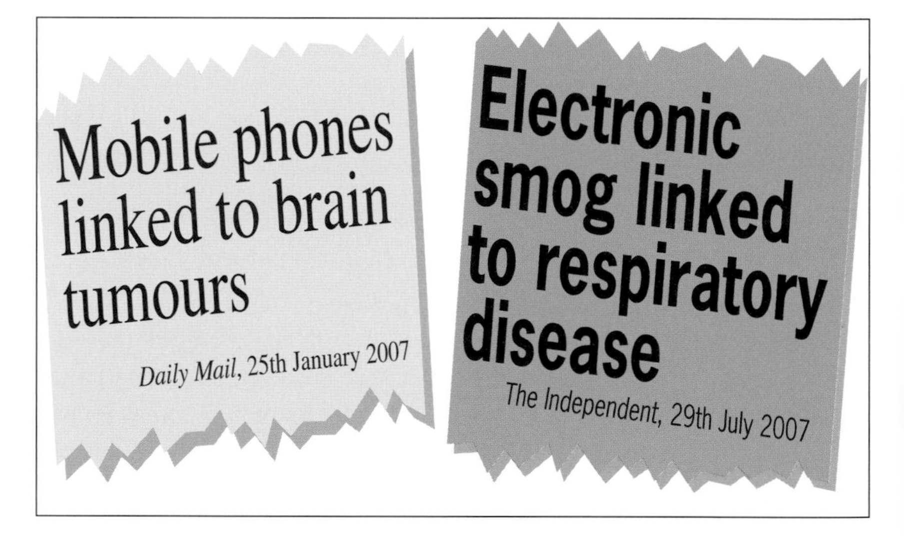

Figure 7.18

Many science stories in the media are about claims that our lifestyles or some feature of our daily environment influences our health.

In fact, much of the uncertainty about such issues stems from the fact that good evidence is hard to obtain. Scientists cannot give a straight 'yes or no' answer. They can state their best judgement at the time, in the light of the (incomplete) evidence available. This may result in considerable resentment, since quite technical matters are often involved and people realise they have no alternative but to accept the view of experts. But this expert view is often more complex and less clear-cut than people would like.

Media reports also have a tendency to give undue weight to case histories of individual people (Figure 7.19). This may be done to add human interest to the story. But it can be very misleading, as it usually implies that there is little or no doubt that the individual's ill health was caused by whatever the 'scare' story is about.

Teachers want an investigation into whether there are any health risks from wireless computer Wi-Fi networks in schools. A teachers' union says: "There's a worry that the potential health risk of this technology hasn't been investigated fully."

The union is "concerned that so many wireless networks are being installed in schools and colleges without any understanding of the possible long-term consequences. It is not saying there is a danger, but that there is enough concern to ask for it to be investigated."

The union highlights the case of a teacher who "had never had any problems before the Wi-Fi. When it was put into his classroom he suffered nausea, blinding headaches and a lack of concentration. When the school removed the Wi-Fi his condition improved."

According to the spokesman the concern is that "if this is having effects on adults, what effect is it going to have on pupils?" The union worries that "Wi-Fi may become like asbestos in the past. Everybody thought it was safe for years and then we suddenly discovered that it wasn't."

Figure 7.19

Reporting of possible health effects of Wi-Fi in April 2007.

Question

26 a) Does the experience of the teacher (Figure 7.19) provide good evidence about the long-term health risks of Wi-Fi?
b) Is the union spokesman making a fair point when comparing the possible health risks of Wi-Fi with the long-term effects of exposure to asbestos?

Scientists have, however, contributed to some of these problems of communication that they now face. Where radiation is concerned, there are very strong historical grounds for the media taking any evidence of the harmful effects of radiation very seriously, as shown by this quotation from Catherine Caufield's book, *Multiple Exposures*: 'Mankind's brief acquaintance with ionising radiation seems almost to have been designed to exacerbate these feelings of ... suspicion. X rays, radium, nuclear fission – each new discovery was greeted with wild enthusiasm, which gave way to alarm when unforeseen side effects appeared. Protection measures were introduced, and always, sooner or later, they had to be strengthened and strengthened again.'

Radiations from radioactive materials

Radioactivity in general, and the nuclear power industry in particular, are often adversely reported in newspapers and in the media. This 'bad press' certainly contributes towards the public's fear and hostility.

Adopting a cautious attitude towards radioactive substances is entirely justified on the basis of the available evidence. There is little doubt that people exposed to high doses of radiation have suffered dreadfully. This has been shown by the effects of the US atomic bombs dropped on Hiroshima and Nagasaki in Japan at the end of the Second World War, and by the consequences of the explosion of the Chernobyl nuclear power station in the Ukraine in 1986. Figure 7.20 shows the immediate effects on people in relation to the amount of radiation they received.

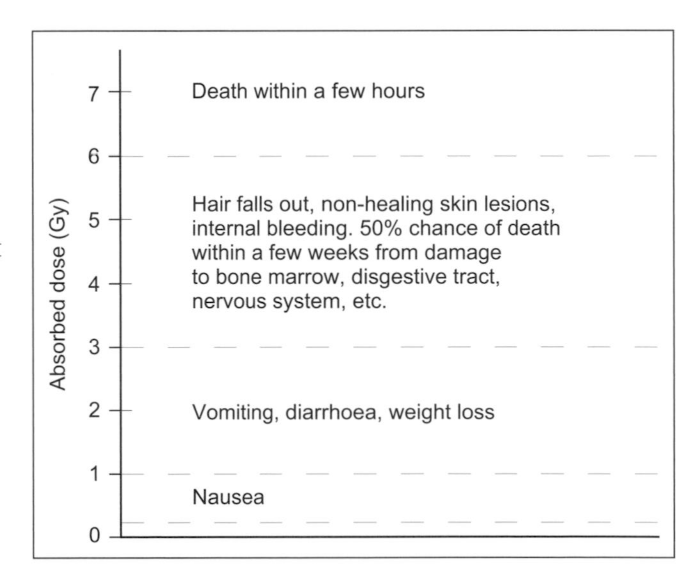

Figure 7.20

Effects of different radiation doses. For doses below 0.25 Gy there is no immediate effect but there is a risk of cancer developing in the long term. This risk is proportional to the radiation dose.

Though people are right to be cautious about anything that might increase their exposure to radiation, not all of the fears concerning radiation that are common among the general public are

Key terms

The **activity** of a radioactive source is measured in becquerels. Radioactivity of 1 becquerel (Bq) is equal to the disintegration of one radioactive atom per second.

The **unit of radiation dose** is the gray (Gy). It is a measure of the amount of energy transferred by radiation to each kilogram of body tissue. ($1\ Gy = 1\ J\,kg^{-1}$)

The **half-life** of a radioactive isotope is the time taken for the rate of emissions from a sample of the isotope to fall by half. So this is also the time for half the unstable atoms in a sample of a radioactive chemical to decay. The half-life is a constant which is characteristic of the particular isotope. The half-lives of radioactive isotopes vary from a fraction of a second to billions of years.

justified. Anxieties about different sources of radiation are often completely out of proportion to the actual risk. For example, many people are far more concerned about the radiation risk of nuclear power stations than they are about the radiation risk in their own homes. In fact, however, for most people the radiation risk from their own homes is far greater. Furthermore, one very commonly held belief about radiation is incorrect – things which are exposed to radiation from radioactive substances do not themselves become radioactive.

Radioactivity

Figure 7.21

The nucleus of a carbon-12 atom. The '12' refers to the total number of protons and neutrons. The nucleus of an atom is tiny – much, much smaller than the atom itself.

Figure 7.22

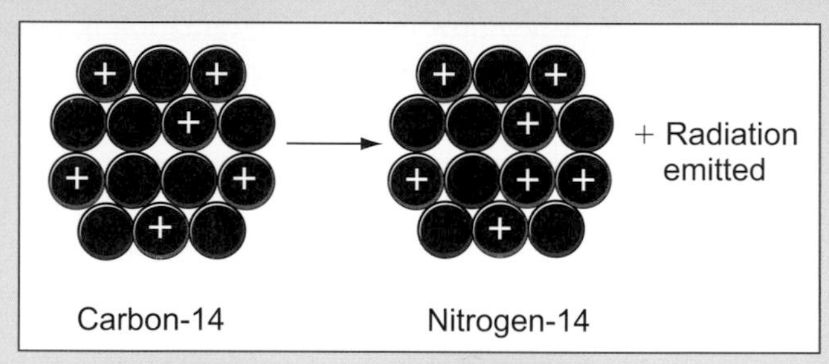

The nucleus of a carbon-14 atom emits radiation when it decays to produce the nucleus of a nitrogen-14 atom.

Everything is made of atoms. There are about 100 different elements, each made from a particular type of atom. Atoms of different elements can, however, join together in different ways to make millions of different chemicals.

Most atoms are stable, but the nuclei of atoms of radioactive substances are unstable. They decay to produce different atoms and emit radiation as they do so. There is nothing that we can do to speed up, slow down or stop this process – it just happens. Even joining radioactive atoms to other atoms to make new chemicals has no effect on the rate at which the unstable atoms decay. Each type of unstable atom has its own particular rate of decay.

All atoms consist of a small central nucleus containing most of the atom's mass together with electrons in the space around the nucleus. Electrons have very little mass but have a negative electrical charge. The nucleus is made up of protons (each of which has a positive electrical charge) and neutrons (which are electrically neutral). Protons and neutrons have equal mass. Atoms have the same number of protons and electrons; this means that they have no overall electrical charge.

Atoms of the same element always have the same number of protons (and hence electrons), but may have different numbers of neutrons. Atoms of the same element with different numbers of neutrons are called isotopes (or nuclides) of that element. Unstable isotopes are called radioisotopes (or radionuclides).

Radioactive decay is a change to the nucleus of an atom; it is a nuclear change. During a nuclear change an atom of one element becomes an atom of a different element.

Questions

27 a) What is the difference between atoms of
(i) carbon-12 and carbon-14
(ii) uranium-235 and uranium-238?
b) Explain why the above pairs of atoms are isotopes of the same element.

28 What happens to the nucleus of an atom of carbon-14 when it decays to produce an atom of nitrogen-14?

Three types of ionising radiation

Figure 7.23

Though all of the radiation emitted from radioactive substances is ionising radiation, it is not all the same type. One of the types of radiation emitted is called gamma (γ) radiation. This is electromagnetic radiation from the short-wavelength (high-frequency, high-energy) end of the electromagnetic spectrum (look back at Figure 7.2, page 94).

Two further types of 'radiation' emitted by radioactive substances are not electromagnetic waves at all but are, in fact, fast-moving particles of matter which have a mass.

- Beta (β) particles are electrons, so they have a negative electrical charge and only a very small mass.
- Alpha (α) particles have a positive electrical charge and are several thousand times more massive than beta particles. (They are, in fact, helium nuclei, that is they consist of two protons and two neutrons.)

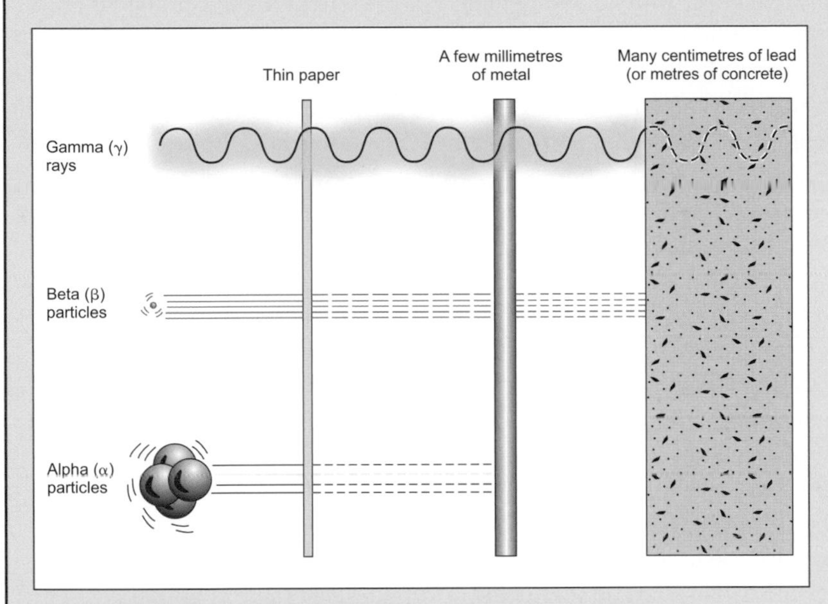

How different types of radiation are absorbed.

Because they have very short wavelengths and no charge or mass, gamma rays can pass through most materials very easily. In other words, gamma radiation is very penetrating and can only be significantly reduced by sheets of dense metal (e.g. lead) many centimetres thick or by concrete several metres thick.

Beta particles pass reasonably well through many materials, being gradually absorbed in proportion to the thickness of the material. They are, however, more or less completely absorbed by a sheet of any metal just a few millimetres thick.

Alpha particles, being larger and more massive than beta particles and carrying a greater charge, are much more easily absorbed. They are more or less completely absorbed by just a few millimetres of air or by a thin sheet of paper.

The ionising effect of radiation is caused by the transfer of energy to molecules of the absorbing material, as the radiation is absorbed. The more readily radiation is absorbed, therefore, the more its ionising effect is concentrated in a short distance.

Questions

29 Surgeons' instruments and some foods are both treated with gamma radiation to kill bacteria. Many people approve of the former but disapprove of the latter. Are they being reasonable? Give reasons for your answer.

30 As it passes through a material, which type of radiation produces:
a) the least ionisation
b) the most ionisation?

31 Radioactive chemicals can be used to monitor the thickness of paper or metal foil as it is being manufactured. Why are beta emitters more suitable for this purpose than alpha emitters or gamma emitters?

How dangerous are different types of radiation?

Since alpha particles are very easily absorbed by cells (see Figure 7.23) and so cause a lot of ionisation in a short distance, it might seem that they are more likely to damage cells than beta particles or gamma rays. This is certainly true if the unstable atoms that are the source of the alpha particles are close up against living cells or even inside them. However, for sources of radiation outside of a person's body, alpha radiation is the least dangerous. This is because alpha radiation from a source more than a few centimetres away from a person's body will not even reach their body. But if the source is inside the body, alpha particles cause a lot of ionisation.

To take account of the greater damage done by any alpha radiation that is absorbed by body cells, the absorbed dose of this type of radiation (in gray, Gy) is scaled up by a factor of 20. This is then added to the absorbed dose of beta and gamma radiation to give the equivalent dose.

In addition, the different types of cells in the different tissues and organs of the body are differently affected by radiation; reproductive cells in the testes and ovaries, for example, are more strongly affected than the cells in the hard outer parts of bones. When this has also been taken into account the resulting figure is then called the effective dose, measured in sievert (Sv). Whenever you see a figure giving 'the dose of radiation' it is the effective dose that is being referred to.

Effective dose is an important quantity because it is related directly to the risk of dying from cancer. This risk can, of course, only be a best estimate in the light of all the statistical knowledge that is available.

This method of calculating risk assumes that there is a proportional relationship between effective dose and cancer risk and that all radiation doses, however small, carry some risk. That is to say that there is no threshold level of dose needed to cause cancer (Figure 7.24). Although the proportional relationship is well established for moderate doses, the absence of a threshold is only assumed and is not accepted by all scientists. Some argue that below a certain level there is no risk.

The average radiation dose for an individual in the UK from all sources is 2.6 mSv. For comparison, a dose of 1 Sv (400 times larger) is high enough to mean that, on average, 5 in 100 people exposed to the dose will die. This is a measure of the annual risk of dying from an annual radiation dose of 1 Sv.

Key term

The **effective dose** of radiation is a measure used to estimate the risk resulting from an exposure of ionising radiation. The units are sieverts (Sv) or millisieverts (mSv).

Questions

32 Express the annual risk of dying from a dose of 1 Sv:
a) as a percentage
b) as a ratio.

33 What is the annual risk of death:
a) for a radiation dose of 1 mSv?
b) for the average UK radiation dose of 2.5 mSv?
c) How does your answer to (b) compare to the risks in Figure 7.13?

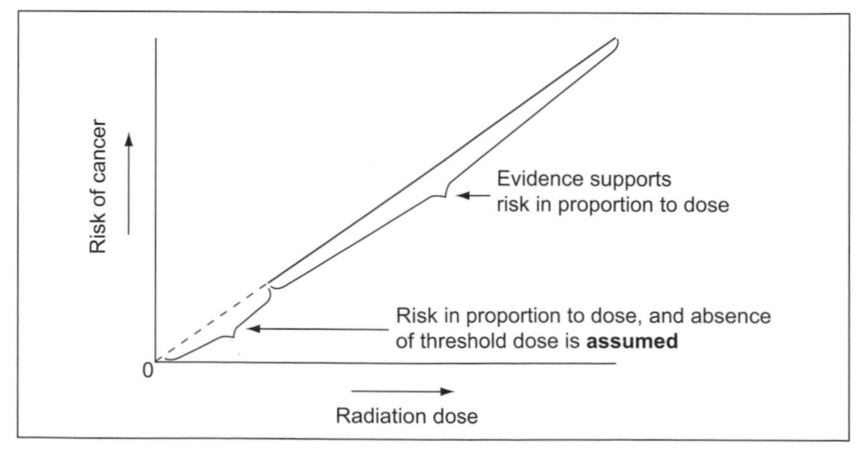

Figure 7.24

The relationship between radiation dose and cancer risk.

Irradiation and contamination

A source of radiation that is outside a person's body can harm that person's body by the radiation it emits: it can irradiate the person's body. This irradiation can be reduced by screening the source or moving the source and the person apart.

The source of radiation may, however, be actually on, or inside, the person's body. We then say that the person is contaminated by the radioactive substance that is the source of the radiation. It may then be difficult to remove the source and so it will continue to add to the person's radiation dose for as long as it continues to emit radiation (Figure 7.25).

Figure 7.25

Irradiation and contamination.

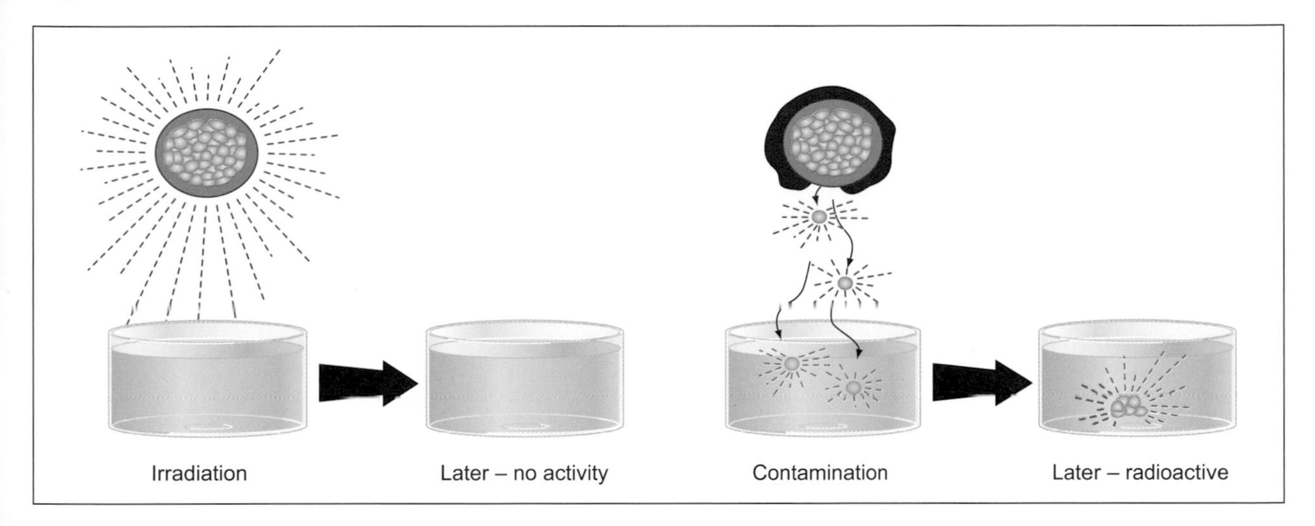

The ideas of irradiation and contamination can also be applied to the environment. If a source of radiation is kept securely contained, it can be used to irradiate things without contaminating them or their environment. There is then less chance of people being accidentally irradiated and no risk at all of their becoming contaminated. If, however, a radioactive substance is not securely contained and is allowed to contaminate its immediate surroundings, it may then spread out over a very wide area.

How do radioactive substances get moved around the environment?

Radioactive materials that are allowed to escape into the immediate environment may then be transported elsewhere by many different processes. The radioactive substances which escaped from the accident at the Chernobyl nuclear power station in 1986, for example, were carried long distances by winds and contaminated much of Europe.

If soluble radioactive substances are dumped on landfill sites they can be carried by rainwater into the soil on agricultural land (Figure 7.26) or into drinking water supplies.

Contamination is especially serious if the radionuclide involved emits radiation at a high rate and/or if it has a long half-life.

Figure 7.26

Flow diagram for the movement of a radioactive element through the environment.

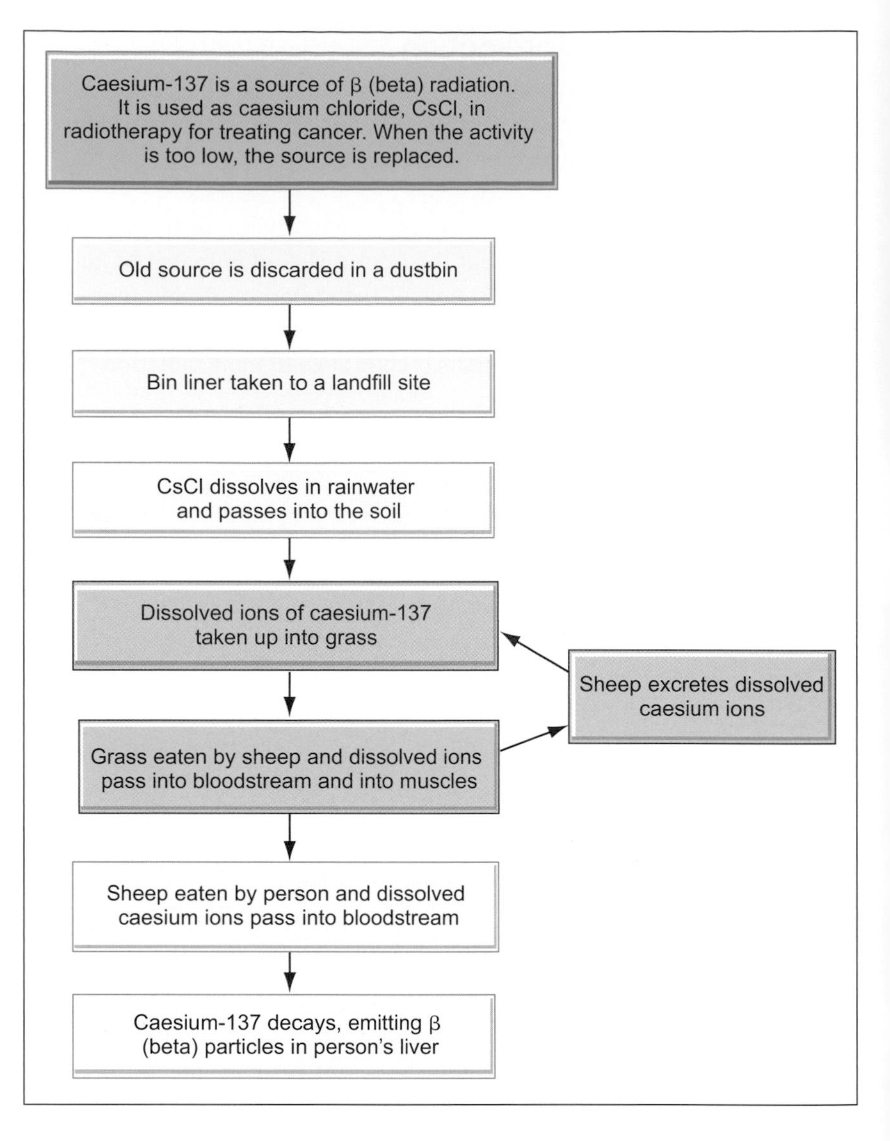

Questions

34 Some smoke detectors contain a very weak source of americium-241, an alpha emitter.
a) Why are smoke detectors unlikely to be a health risk in normal use?
b) Why might they be a health risk if they eventually end up in a landfill site?

35 Identify examples of irradiation and radioactive contamination in Figure 7.25.

Where does our radiation dose mainly come from?

For most people, the largest single source of the radiation that they are exposed to is the air that they breathe in their own homes, due to a gas called radon. The average radiation dose in the UK from radon is estimated as being 1.3 mSv per annum; this is quite close to the estimated world average.

Radon is a radioactive gas that is produced in a series of radioactive decays that begin with unstable uranium and thorium atoms. Since there are small quantities of these elements in almost all types of rock and soil, radon is being produced in the ground almost everywhere. Certain rocks and soils, however, contain far more than average amounts of uranium and/or thorium and so release radon at a rate that is much greater than average (Figure 7.26).

The isotope of radon that results from the decay of uranium-238 atoms is radon-222. This isotope has a half-life of only 3.8 days so that out of doors it quickly disperses and decays and is not a significant health hazard. Indoors, however, radon levels can build up. The highest levels of radon inside buildings are found in areas where the underlying

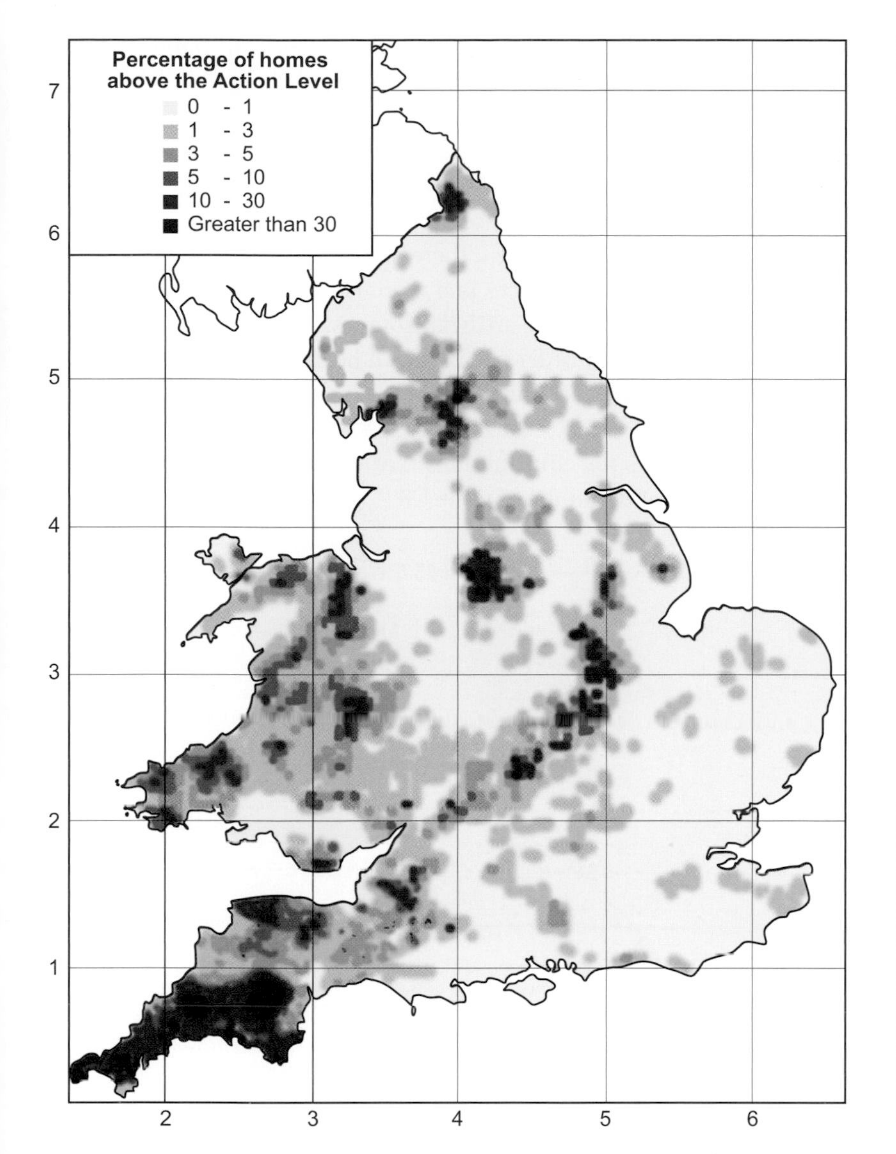

Figure 7.27

Radon doses in England shown by the colours on this map of the country.

rock is limestone and at the edges of masses of granite. The levels of radon and the resulting radiation dose in some buildings can be hundreds of times, and sometimes thousands of times, higher than normal.

Most of the radon inside a building seeps up through the ground immediately below the building. It can be prevented from reaching the rooms inside a building by constantly pumping out the air from the space underneath the ground floor rooms.

The main health hazard from radon is not, in fact, from the decay of the radon itself. Radon is an intermediate stage in a long series of decays (Figure 7.28).

The decay products from radon – the radon 'daughters' – are all solids. When an atom of radon gas decays a speck of a solid substance is produced. Tiny drops of water vapour tend to form around these specks.

Questions

36 The average radiation dose for people in Cornwall is 7.8 mSv per year. Why is this dose higher than average for the UK?

37 Is the decay of radon itself likely to increase a person's radiation dose significantly? Explain your answer.

Figure 7.28

The decay series from uranium-238 to stable lead-210, including radon–222.

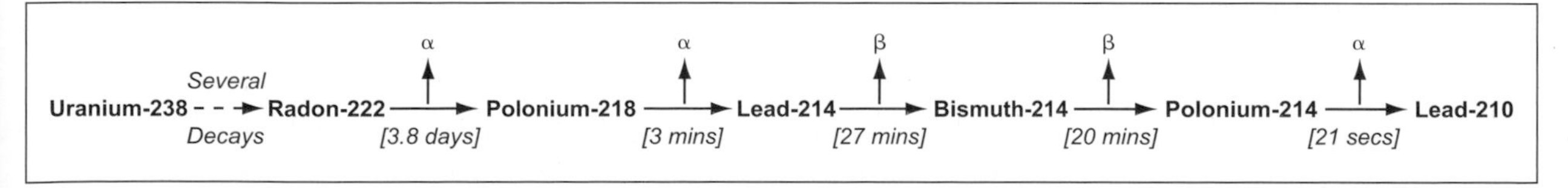

These droplets may then be breathed in and the specks of radioactive material may lodge in, and contaminate, a person's lungs, thereby increasing the risk of lung cancer.

What other sources of radiation are we exposed to?

Though radon and its 'daughters' are the major source of the radiation to which most people are exposed, there are also various other sources. Some of these are natural sources and there is little or nothing that we can do about them. Others are artificial sources. The artificial sources are not, of course, created in order to produce a health risk; the health risk is the price that has to be paid for using the radioactive sources for other, beneficial, purposes.

Figure 7.29 shows the contributions that different sources make to an average person's radiation dose in the UK.

Figure 7.29

Where an average person's radiation dose comes from in the UK.

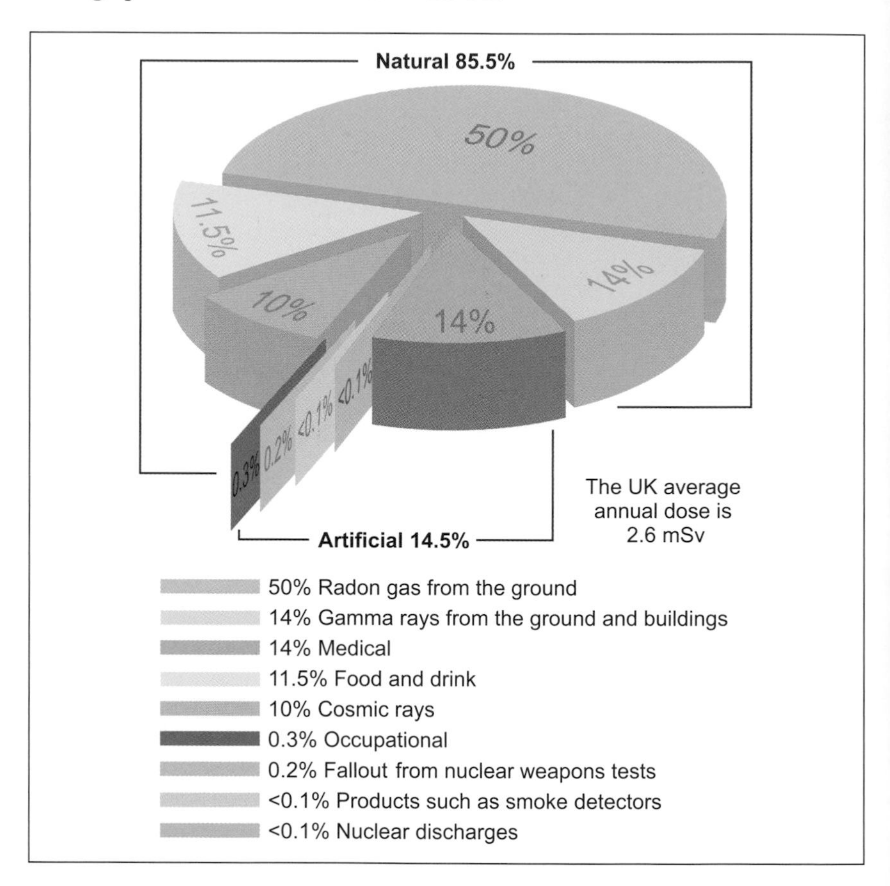

Is low-level ionising radiation a cancer risk?

Most scientists believe that there is a risk of cancer from ionising radiation, however small the radiation dose received. The main evidence is the clear correlation between radiation dose and risk for moderate and large doses. It then seems reasonable to assume that the correlation will hold for smaller doses too. There is also a plausible mechanism: even a single alpha or beta particle, or pulse of gamma radiation, can damage the DNA in a cell in such a way as to make it cancerous.

Other scientists, however, believe that there may well be a threshold level of radiation below which there is little or no risk of cancer

Figure 7.30

Scientists disagree as to whether low radiation doses pose a significant health risk.

(Figure 7.30). Once again, this belief is not just a hunch: it is based on evidence, even though that evidence isn't conclusive. Firstly, scientists agree that cells can, to some extent, repair damage that occurs to their DNA. It might, therefore, require several radiation 'hits' in a short time to cause damage to cells which they are unable to repair. Secondly, there is some evidence that cells can adapt to radiation: exposure to low radiation doses seems to make them less likely to be damaged by later doses.

If scientists who think that there is no risk from radiation below a certain level are right, then many of the estimates made by the International Commission on Radiological Protection (ICRP) and the Health Protection Agency in the UK would be invalid because they are mainly based on low doses.

To try to resolve these differences and reach agreement, scientists conduct two different types of research. Firstly, they carry out laboratory experiments, either *in vitro* (in glass) using cultures of cells or on animals. Secondly, they obtain detailed statistical data from epidemiological studies about, for example, how the incidence of lung cancer is related to the levels of radon that people have been exposed to.

Epidemiological studies of radon and lung cancer

Results from an epidemiological survey published in 1998 suggest that the link between radon and lung cancer is a genuine one. Sarah Darby, at the Imperial Cancer Research Fund's Epidemiology Unit in Oxford, compared the radon concentrations in the homes of 982 lung cancer victims in Devon and Cornwall with those of 3185 controls. After allowing for other variables such as smoking and social class, she found that people exposed to radon with an activity of more than 200 becquerel per cubic metre – the government's safety limit – were 20% more likely to develop lung cancer than people exposed to the average UK level of 20 becquerel per cubic metre.

This finding was confirmed and extended to cover the European Union (EU) by a study funded by Cancer Research UK and the European Commission. The researchers, led by Sarah Darby, combined information

Questions

38 What hypothesis has Sarah Darby been investigating?

39 a) Estimate the lifetime risk of cancer of someone who receives a dose of 10 mSv from radon (the government safety limit for radiation doses from radon). Assume that the risk is proportional to the dose and that the UK average dose of 2.6 mSv gives a lifetime risk of 1 in 300.
b) Compare your answer to the risks from Sarah Darby's research.

from 13 smaller studies across Europe. At the time it was the largest study ever to be conducted into the effects of radon exposure in the home (Figure 7.31). The report, published in 2004, concluded that radon in homes causes about 20 000 lung cancer deaths in the EU each year, of which 1000 are in the UK. This means that radon in ordinary homes is causing about 9% of lung cancer deaths each year in Europe, which is 2% of all cancer deaths. In the UK, where radon levels are lower, radon in homes causes about 1% of all cancer deaths.

Figure 7.31

A monitor for detecting radon levels at home.

The Seascale leukaemia cluster

A hard-hitting TV programme in 1983 drew attention to the unusually high level of childhood leukaemia in the West Cumbrian village of Seascale (Figure 7.32). The village is near to the Sellafield nuclear plant and it was natural to assume that radiation leaks were responsible for the cluster of cases.

In 1990, Martin Gardner, an epidemiologist from the University of Southampton, claimed that he had evidence indicating that children of workers at Sellafield were between six and eight times as likely to develop leukaemia if their fathers had received cumulative radiation

Figure 7.32

The village of Seascale, near to the Sellafield nuclear plant.

doses exceeding 100 mSv. A possible explanation for this was radiation damage to the fathers' sperm, since it is known that reproductive cells are particularly prone to being damaged by radiation. Male reproductive cells in particular are likely to receive higher radiation doses than organs deeper inside the body. The government's Health and Safety Executive, however, pointed out the puzzling fact that Gardner's findings seemed only to apply to fathers living on Seascale, not to fathers who worked at the Sellafield nuclear plant but lived elsewhere in West Cumbria.

A large-scale epidemiological study was also undertaken during the late 1990s in order to test the Gardner theory. This involved 40 000 children born to 18 000 workers in the nuclear industry at several sites in the UK. The report, published in 1999, concluded that the children of fathers who had been exposed to radiation before their conception were more likely to have leukaemia than the children of fathers who had not been exposed to radiation. The risk was greater for the children of fathers whose radiation dose was more than 100 mSv. The study also showed, however, that the overall rate of all types of cancer among the children was not significantly greater than in the UK population as a whole. The study failed, therefore, to settle the issue, concluding only that Gardner's finding was 'not disproved'.

But then, later in 1999, there was a new twist to the story. Evidence was found supporting the quite different 'virus' theory, which had originally been suggested 10 years earlier. This theory suggests that leukaemia clusters are triggered by infections that occur in populations where many of the people are new to the area. This was the case in the village of Seascale, where thousands of new workers had come to the area over 50 years to build and operate the nuclear plants.

The virus theory was first put forward, in 1988, by Leo Kinlen of the University of Oxford. This was a 'population mixing' theory based on the idea that when large numbers of workers move into an isolated area, the immunity passed down by migrant mothers may not prepare their children to fight off the particular strains of viruses and bacteria found in the area.

Evidence to support the theory came from the work of Heather Dickinson and Louise Parker of the University of Newcastle. They scrutinised the health records of 120 000 children born in Cumbria between 1969 and 1989. They found that if both parents came from outside Cumbria, the chance that their child would develop leukaemia during the first six years of their life was 2.5 times higher than when one or both parents came from within Cumbria (Figure 7.33). In areas where the population influx was heaviest, and up to 80% of the parents were migrants, the risk was up to 11 times as high as in the areas with the fewest migrants.

Questions

40 Why did many people start by making the assumption that the Seascale leukaemia cluster was caused by radiation or radioactive contamination?

41 Identify findings from the epidemiological studies at Seascale that cast doubt on the theory that radiation was responsible for the higher levels of childhood leukaemia.

42 Suggest further research questions that arise from Professor Kinlen's explanation of leukaemia clusters.

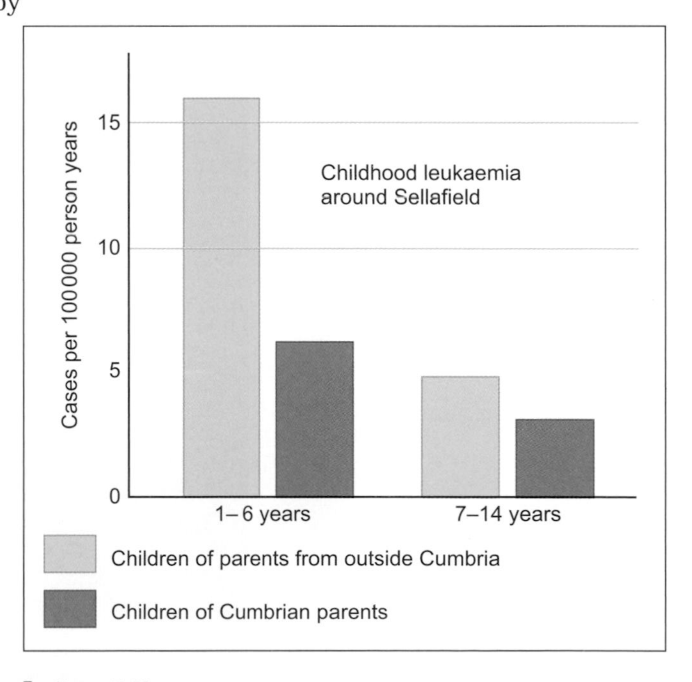

Figure 7.33

The incidence of childhood leukaemia in Seascale.

Professor Kinlen has found more support for his theory, which he published in 2006. The report was based on a research study of a leukaemia cluster in the desert town of Fallon, Nevada, USA. Around 100 000 recruits passed through a military base in the town in 1999 and 2000. The next year, doctors recorded a big increase in the incidence of childhood leukaemia. The disease was diagnosed at more than 30 times the expected rate.

Radiation and health

Ionisation radiations can cause cancer. They can also be used to diagnose and cure many health problems. Doctors and patients have to balance the benefits of using radiation against the possible risks to health.

Diagnostic radiology

Radiologists use X-ray machines to obtain images of bones and other parts of the body. For many years, taking an X-ray was the only way to see inside patients. The X-rays used to examine bones, teeth and the lungs now use very low doses of radiation. A chest X-ray involves a radiation dose of 0.02 mSv.

A CT scan is a more sophisticated way to use X-rays. The patient lies on a narrow table and the X-ray source and detectors rotate around a slice of the body. The signals from the detectors are fed to a computer which builds up a series of images from the data (Figure 7.34). A CT scan, however, involves a much higher radiation dose than a conventional X-ray. A CT scan of the pelvis gives a person a total radiation dose of about 10 mSv.

Research funded by Cancer Research UK suggests that about 700 cases of cancer a year arise from the diagnostic use of X-rays. However, there is no reliable evidence to prove conclusively that the low doses of radiation used for diagnosis do cause cancer.

Questions

43 The lifetime risk of cancer is assumed to increase by 1 in 20 000 for every 1 mSv of radiation dose. What is the increase in the lifetime risk of cancer due to a CT scan?

44 Explain why it is difficult to prove whether or not low doses of X-rays cause cancer.

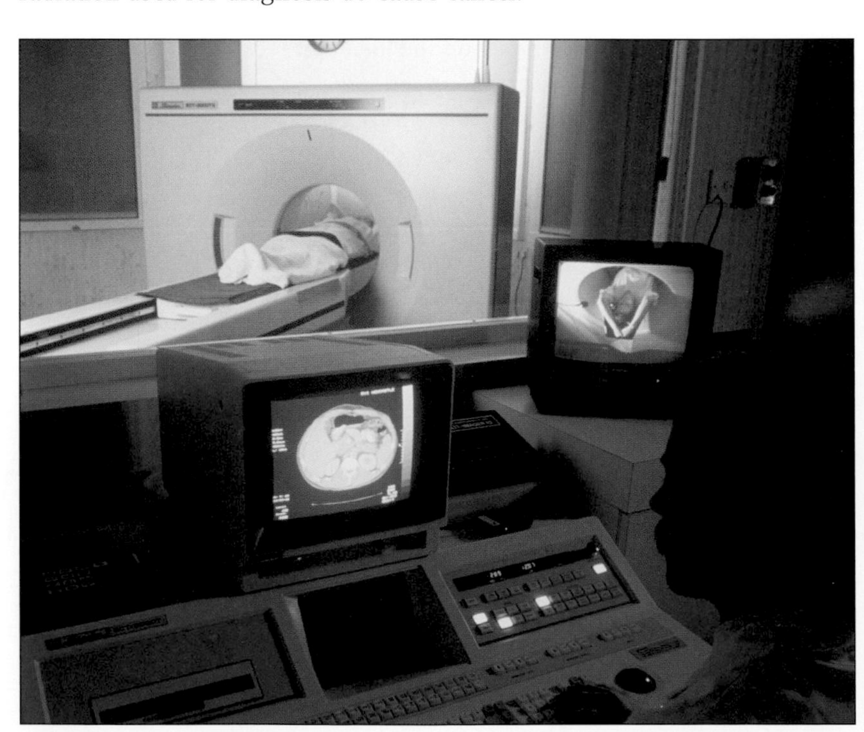

Figure 7.34

CT scan of a patient's abdomen being recorded by a radiographer.

Nuclear medicine

Nuclear medicine uses radioactive substances introduced into the patient for diagnosis or treatment. For diagnosis a small quantity of a radioactive element such as technetium-99 is injected into a vein. The radioactive atoms are part of a chemical known to be taken up by the tissues or organ to be examined.

Technetium-99 gives out gamma radiation. The radiation escapes from the body and can be detected with a special gamma camera linked to a computer. The computer then builds up an image of what is happening inside the body (Figure 7.35).

The radioactivity in the patient's body falls to insignificant levels in a few days because the half-life of technetium-99 is only six hours. In some cases the chemical with the radioactive isotope is excreted in urine, in which case it disappears from the body faster.

An example of radiotherapy with radioactive isotopes is the use of iodine-131 to kill cancer cells in the thyroid gland. The thyroid is a gland in the neck and it is the only part of the body to concentrate iodine which it uses to make a hormone. The patient swallows a capsule containing the radioactive isotope. He or she then cannot eat anything for several hours while the thyroid absorbs the iodine-131.

The iodine-131 gives out beta particles which have enough energy to destroy cancer cells. It has a half-life of eight days.

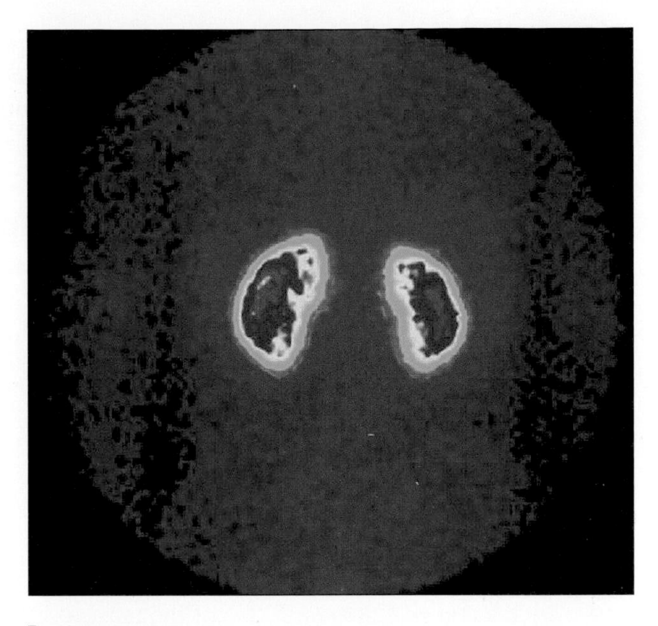

Figure 7.35

False-colour image of a pair of healthy kidneys seen from behind with a gamma camera.

Questions

45 Why is the isotope chosen for medical imaging:
a) a gamma emitter?
b) an isotope with a short half-life?

46 a) Why are visitors only allowed to stay for a short time with patients being treated for cancer with iodine-131?
b) Why is the contact with visitors only restricted for about two weeks?

Radiotherapy

Radiotherapy aims to treat or cure cancers with a powerful beam of radiation from X-ray machines or radioactive sources. The doses of radiation are many thousands of times higher than those used for diagnosis. The aim is to cure the cancer or at least to alleviate the most distressing symptoms. For a cure, the doses have to be high enough too kill all the living cells within the cancer tumour (Figure 7.36).

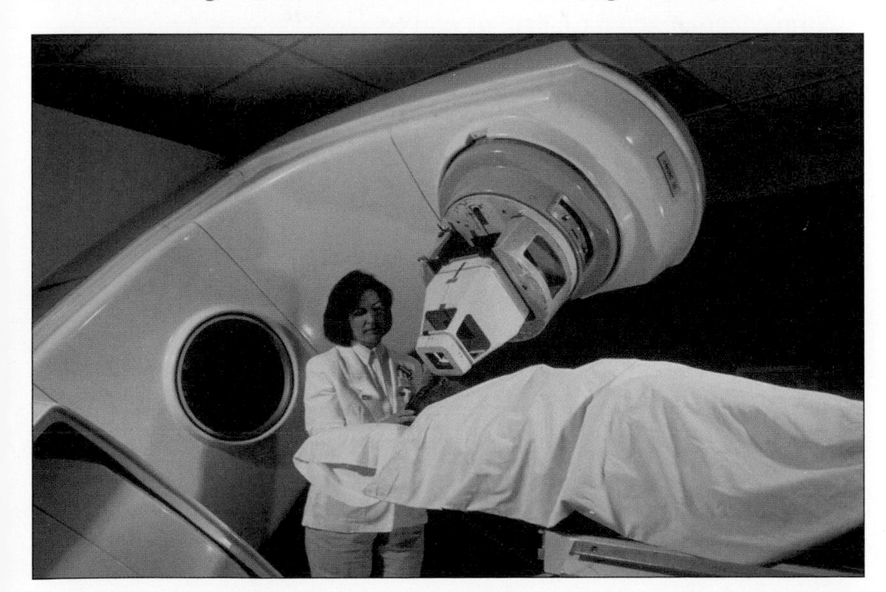

Figure 7.36

Radiotherapy for a brain tumour with a powerful beam of X-rays from an instrument that rotates around the patient's head.

This radiation also causes damage to surrounding healthy tissues. The damage is limited by rotating the beam around the patient so that it irradiates the tumour from many different directions. This means that the radiation does not pass continuously through the same healthy tissues.

Questions

47 Use examples from the medical uses of radiation to distinguish between irradiation and contamination by radioactive chemicals.

48 The Health Protection Agency supports research into health risks from diagnostic radiology and nuclear medicine but not from radiotherapy. Suggest a reason for the difference.

49 What does Figure 7.38 show about the relationship between radiation dose and the cost of reducing the dose?

50 Suggest ways in which the ALARA principle might be applied to decide the acceptable upper limit for the radiation dose from radon in homes.

51 Suggest how the ALARA principle might apply to medical staff who work with radiations and radioactive materials.

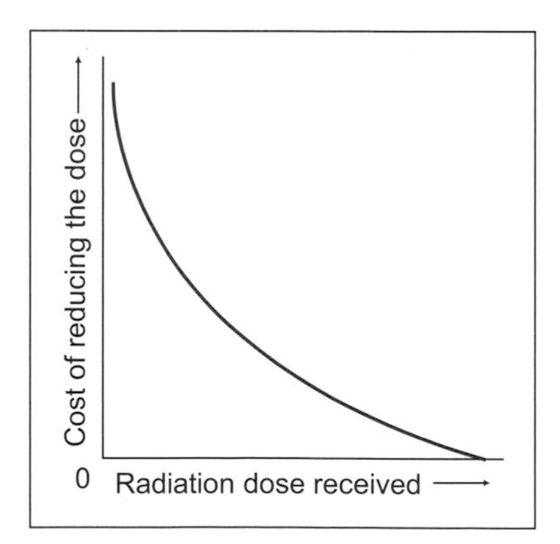

Figure 7.37

The reason for the ALARA principle as opposed to setting a target of zero dose.

How can people be protected from radiation risks?

Since there are natural sources of radiation all around us, and even inside us, we obviously cannot be completely protected against radiation. We can, however, be protected from being unduly or unnecessarily exposed to radiation from artificial sources. Indeed, various international and national bodies have been set up to ensure that people are all adequately protected. This includes people who use sources of radiation in their work, the people on whom these sources may be used and members of the general public.

The International Commission on Radiation Protection (ICRP) publishes recommendations, which are regularly updated in the light of the most recent evidence. National governments then implement these recommendations, modified where necessary to match their particular circumstances. In the UK, the Health Protection Agency advises the government about how best to implement the current ICRP recommendations.

ICRP policy is based on three fundamental principles:

- Justification
  No practice involving radiation shall be adopted unless its introduction produces a positive net benefit.
- ALARA
  All radiation exposures shall be kept As Low As Reasonably Achievable, economic and social factors being taken into account (Figure 7.37).
- Dose limits
  The radiation dose to individuals shall not exceed the limits recommended for the appropriate circumstances by the Commission.
  The UK dose limits for exposure to artificial sources of radiation are:
- 20 mSv per annum for radiation workers
- 1 mSv per annum for members of the general public.

These doses are in addition to the average annual radiation dose, mainly from natural sources, of 2.6 mSv per annum. The limit for the general public is intended to be a five-year average, to allow for any dosage involved in medical treatment which may occur over a short period of time. Furthermore, the recommended dose limits may be exceeded if, on balance, it is in the best interests of patients to do so, for example in the treatment of cancer.

In practice, the average additional dose for people whose work exposes them to a particular radiation risk from artificial sources is about 2 mSv, giving an overall exposure of 4.6 mSv. A small minority of workers, however, are exposed to an annual radiation dose of up to 15 mSv.

Review Questions

52 Use ideas and examples from this chapter to explain:

a) why some radioactive materials are much more hazardous than others

b) the difference between ionising and non-ionising radiation

c) the difference between radioactivity and radiation dose

d) the difference between irradiation and contamination by a radioactive material

e) why ionising radiation can cause cancer.

53 Give at least one example from this chapter to illustrate each of these ideas about how science works:

a) In some situations a causal factor can increase the probability of an outcome but not invariably cause it.

b) Correlation between a factor and an outcome can be investigated using a case-control study.

c) It is very difficult in complex situations to obtain convincing evidence that a factor does not cause an effect.

54 Give at least one example from this chapter to illustrate each of these ideas about risk:

a) New technologies based on science and technology often introduce new risks.

b) The perceived risk of an activity is often quite different from the actual measured risk.

c) People are more willing to accept a specific risk if they can see some benefit from the activity and can choose it voluntarily.

Chapter 8

Lifestyle and health

Figure 8.1

A 2002 study reported that grilling and frying some foods at high temperatures create end-products which are associated with an increased risk of cancer, coronary heart disease and diabetes. However, there are also other studies that suggest there is little or no association. A balanced and informed consideration of risk factors relies on good study design, quality reporting and interpretation of studies.

The issues

Science and technology have done much to make our lives less risky. People in industrialised countries live, on average, longer and healthier lives than in the past. It is now known how to prevent or cure many of the infectious diseases which used to kill large numbers of people. As a result, the main threats to health today in countries like the UK are chronic diseases such as cancer, cardiovascular diseases and type 2 diabetes. Many factors can contribute to the development of these diseases, including the genetic make-up of individuals, their diet, environmental pollutants they encounter, the work they do and the choices they make in how they live.

There is plenty of advice from many sources telling people how to live their lives in ways which will help them to reduce the risks to their health. Often the link between cause and effect is uncertain or remote. Sometimes the advice is contradictory.

The science behind the issues

Cancer is a multifactorial disease. Scientists have identified genes which give people a predisposition to certain cancers, but the health risk can also be increased or decreased by exposure to certain environmental factors, including lifestyle. Large-scale studies help link genes with diseases, by identifying clustering of diseases within families, geographic areas or particular occupations.

What this tells us about science and society

Many illnesses are influenced by several factors and it is sometimes difficult to show convincingly that a particular factor has an effect. This is especially so when the influence of a factor is small or the disease is rare. For example, many of the associations between diet and disease are small, which means that determining whether the effect is real or not is often very difficult. These weak associations may reflect a real effect or may reflect some kind of subtle bias or measurement error that the researchers were unable to detect and eliminate.

Scientists look for correlations, or relationships, between exposure to specific risk factors and disease occurrence. Any uncertainties may be interpreted by scientists, the media and other stakeholders in quite different ways. Their backgrounds, personalities and interests can influence their interpretations. We need to take this into account when evaluating what they say.

Studying causes of disease

The work of John Snow (Chapter 1) set the scene for the birth of epidemiology – the study of how often diseases occur in different groups of people and why. Many diseases, including cancers and heart disease, do not have simple or single causes. Genetics and environmental factors such as diet, lifestyle, exposure to ionising radiation, cigarette smoke and other toxic chemicals can all contribute to the risk of disease. Epidemiologists gather data from large-scale studies to test hypotheses suggesting connections between the incidence of disease and potential risk factors for the disease (Figure 8.2).

Figure 8.2

The effects of smoking on health are well known, but many young people still choose to smoke.

Both morbidity and mortality are studied in epidemiology. The morbidity of a disease is the proportion of the population that becomes ill. Morbidity is easy to record when a disease is simple to diagnose and causes clear-cut symptoms. Accurate diagnosis is more difficult when symptoms are not so clear-cut. An inaccurate diagnosis can lead to errors of classification of the disease and contribute to misleading data about diseases.

Mortality measures the number of people who die of a particular disease. While this is easier to measure than morbidity, it is only useful in diseases serious enough to cause death. Mortality is also subject to misclassification.

Studies of morbidity and mortality data are used in association with social data to discover risk factors for diseases. Data about people in society include income, housing conditions and occupation, extent of smoking and exposure to pollutants. Also significant are biological data such as genetics and recorded observations of changes in cells and tissues.

Scientists are careful to distinguish between 'risk factors' and the 'causes' of a disease. For example, being male cannot be said to cause heart disease, but men have a higher risk of heart disease compared with women. So being male is one of the risk factors for the disease. Epidemiological studies often provide the first clue in the hunt for the causes of a disease.

Questions

1. What was John Snow's hypothesis? What factors did he investigate? What data did he collect to test his hypothesis? (Look back at Chapter 1.)
2. Why can it be useful to doctors and the public to identify risk factors for a disease even if it then takes a long time before the causes of the disease are understood?

Health risks from smoking

In the first half of the twentieth century smoking was freely promoted and often glamorised (Figure 8.3).

Studies in the 1950s by epidemiologists such as Richard Doll changed attitudes. In 1971, for example, The Royal College of Physicians issued a statement:

> Premature death and disabling illnesses caused by cigarette smoking have now reached epidemic proportions and present the most challenging of all opportunities for preventive medicine...

Figure 8.3

A cigarette advertisement in 1939.

The evidence

Research into the links between smoking and disease was prompted by data of the kind shown in Figure 8.4.

Figure 8.4

Deaths from lung disease in England and Wales, 1920–1960.

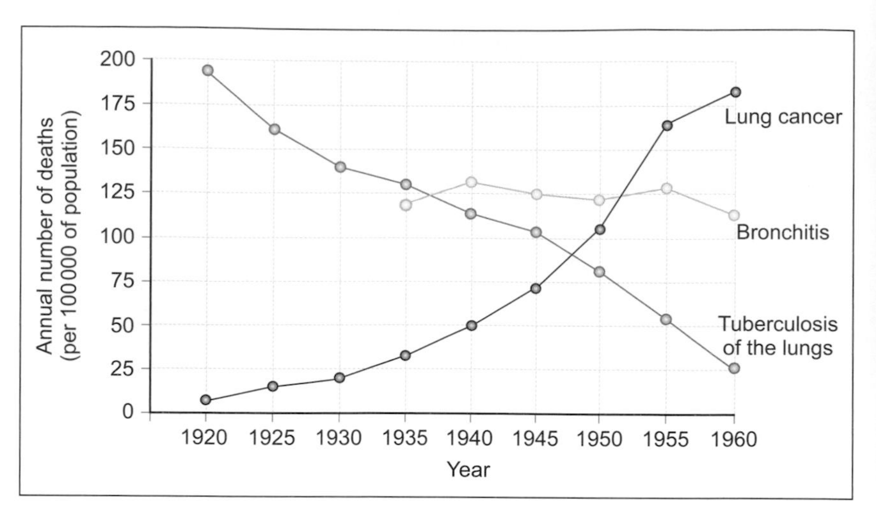

Key term

A **correlation** describes the tendency for two measures or variables, such as the extent of smoking and the number of cases of lung cancer, to vary together.

Notice that deaths from lung cancer started to increase from the early 1900s, increasing more sharply over time. In the same period, deaths from other forms of lung disease such as tuberculosis fell steadily. At first, scientists thought there might be a correlation between the increase in deaths from lung cancer and increasing air pollution because of the growing number of motor vehicles on the road. Cars were invented at the end of the nineteenth century, so making this link seemed reasonable.

The habit of smoking cigarettes became widespread around the same time as the increase in cars on the roads. Attention turned from vehicle exhaust fumes to the possible correlation between lung cancer and smoking. Studies in the 1950s showed that over 90% of patients with lung cancer were tobacco smokers.

Although it was difficult to prove the cause of lung cancer, when doctors saw the early evidence many of them gave up smoking cigarettes. Their swing away from smoking provided further valuable evidence for the dangers of smoking cigarettes (Figure 8.5).

Figure 8.5

Death rates from lung cancer in male doctors and men in England and Wales. Doctors understood the significance of the early research into smoking-related diseases more quickly than other people.

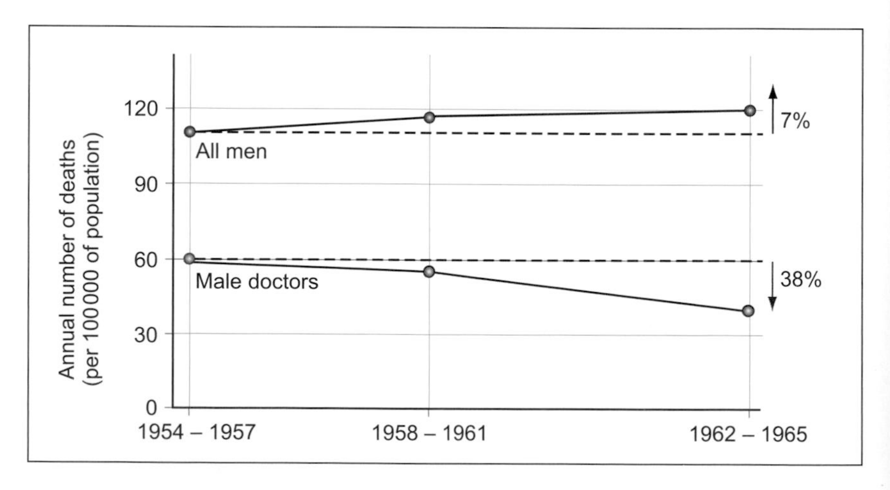

Other studies helped confirm the relationship between the risk of dying from lung cancer and the number of cigarettes smoked – the more cigarettes smoked, the greater the risk (Figure 8.6).

Since 1971, cigarette smoking among men and women in the UK has dropped (Figure 8.7), but it is currently increasing among children and, in

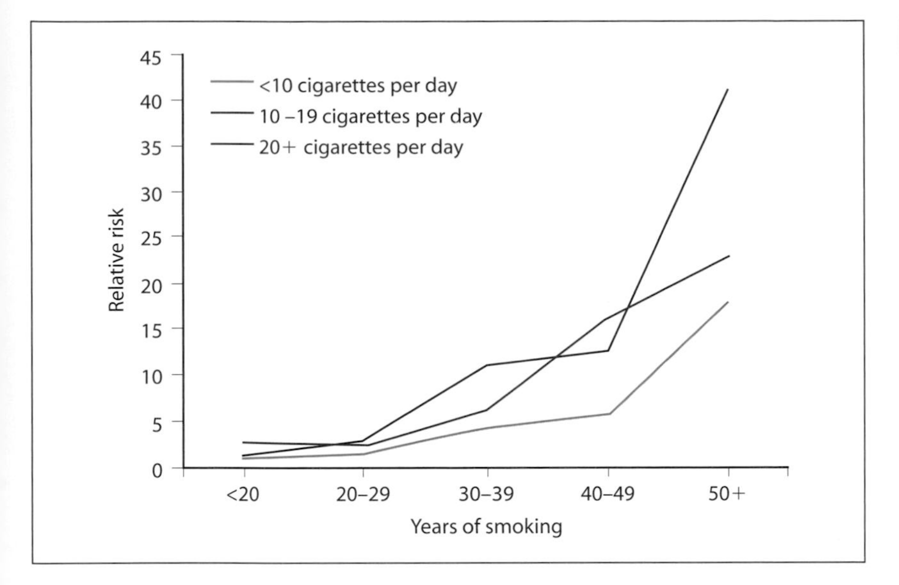

Figure 8.6

The relationship between the risk of men dying from lung cancer and the number of cigarettes smoked daily.

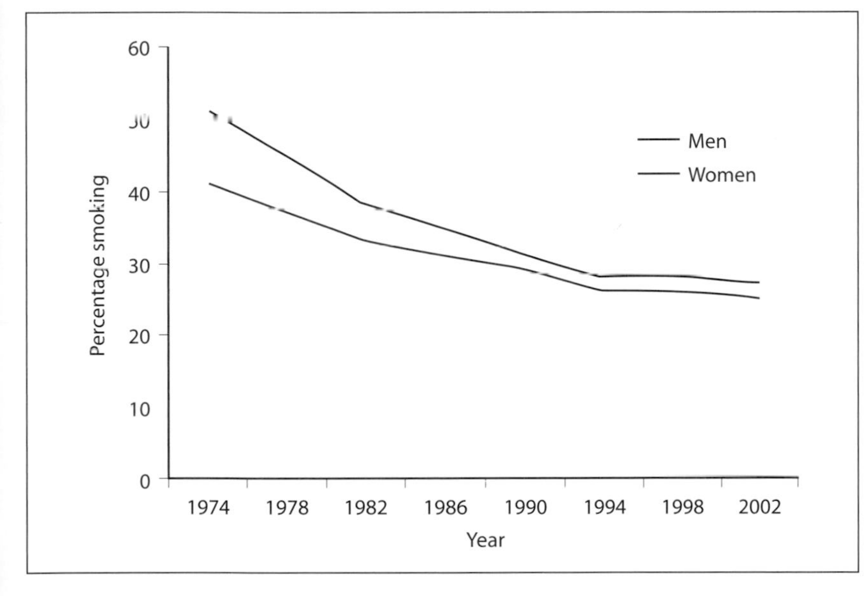

Figure 8.7

Prevalence of cigarette smoking in people aged 16 or over, UK, 1974–2002.

particular, teenage girls. In 2007, 25% of 15-year-old girls were regular smokers, compared with 16% of boys.

In 2007 there were around 10 million adult cigarette smokers in Great Britain – this represents about a quarter of the adult population. Compare this with 1974, when about half the adult population were smokers. The highest rates of smoking are in the 20–24 age group, with 34% of men and 30% of women in this age group recorded as smokers. Smoking declines with increasing age, with only 14% of people over 60 smoking.

The evidence for health risks

Epidemiological studies look for correlations between the risk factors and disease. Finding a correlation, such as the increase in lung cancer as smoking increases, can suggest but not prove that smoking causes cancer.

Scientists can use toxicology studies to test the hypotheses that particular chemicals or mixtures of chemicals actually cause a disease. For example, toxicology studies to measure the effects of toxins on living cells in a laboratory culture contribute to the calculation of risk factors

Questions

3 Explain why, in the early twentieth century, the growth in motor traffic was a plausible explanation for the rise in lung cancer.

4 Why did it take a long time before people began to recognise the risks of smoking?

5 Suggest reasons why doctors were among the first to respond to the evidence linking smoking to an increased risk of lung cancer.

6 What generalisations about risk can you make based on the information in Figure 8.6?

7 How do you account for the fact that people aged 24–34 are most likely to ignore warnings about the risks of smoking?

from environmental pollutants, including tobacco smoke (see Chapter 4). Data from such studies may provide an explanation for the correlations observed in epidemiological studies.

As tobacco contains such a complex mixture of chemicals, the mechanism by which it causes diseases is not clear. Tobacco manufacturers assess the toxicity of their products with the help of *in vitro* and animal studies. They also carry out studies with volunteer smokers to examine effects of smoking on health.

It is now clear that there is a strong association between lung cancer and smoking. The evidence shows that giving up smoking reduces the chance of developing the disease (Figure 8.8) but it is also the case that some people who smoke heavily never develop lung cancer. So this is not a straightforward example of cause and effect. Even so, scientists have their theories of the way in which chemicals in tobacco smoke can cause the disease.

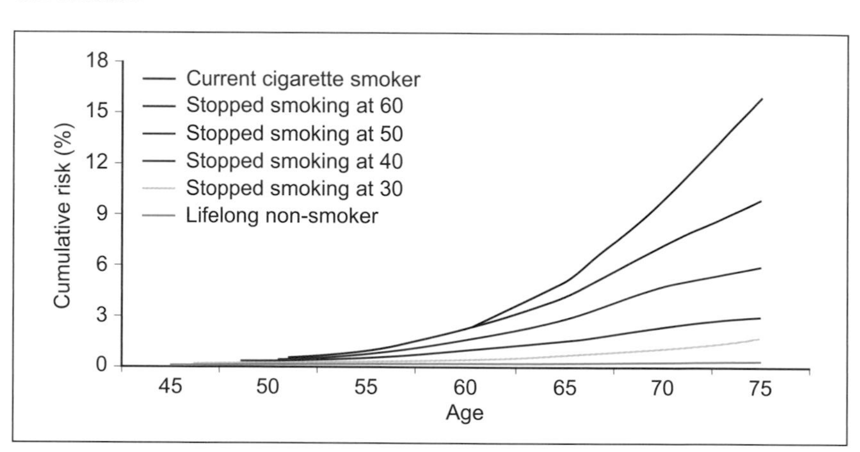

Figure 8.8

Effects of stopping smoking at various ages on the cumulative risk (%) of death from lung cancer by age 75. Risk depends much more on how long people smoke than the amount they smoke. Smoking one packet of cigarettes a day for 40 years is much more hazardous than smoking two packets a day for 20 years.

The chemicals in cigarette smoke

Cigarette smoke contains a complex mixture of chemicals.

- Nicotine is a powerful and addictive drug. It acts as a stimulant to the nervous system, increasing heart rate and raising blood pressure.
- Carbon monoxide is a toxic gas which reduces the amount of oxygen the blood can carry.
- Sticky brown tar is a mixture of around 4000 different chemicals that accumulates in the lungs; about 60 of the chemicals are suspected or known carcinogens.

Carcinogens in cigarette smoke have been shown to cause mutations in the genes which control cell division, resulting in formation of a cancerous tumour. This mechanism probably accounts for 25% of cases of lung cancer.

Tobacco carcinogens may also cause mutations in other genes which help to stop tumours forming. Inactivating these genes can also lead to loss of control over cell division and subsequent development of cancers – especially lung cancer.

Explaining the onset of lung cancer is, however, difficult because there is usually a delay, which may be as long as 10 to 20 years or more, between exposure to the harmful agent and the onset of cancer (Figure 8.9). The sharp rise in lung cancer among men, for example, came 20 years after a similar rise in cigarette smoking. For women the same pattern was repeated a generation later. This delay made it more difficult to establish a link between smoking and the disease.

Question

8 There is a range of tobacco products available – cigarettes with differing tar levels, pipe tobacco, chewing tobacco. How could researchers take advantage of the existence of such products to investigate some of the risks of tobacco use in a systematic way?

9 Why is it difficult to establish whether or not a particular chemical or mixture of chemicals causes cancer (is carcinogenic)?

10 Why are people more likely to be persuaded that smoking causes lung cancer if scientists can come up with an explanation of how it happens?

Mutations and cancer

Figure 8.9

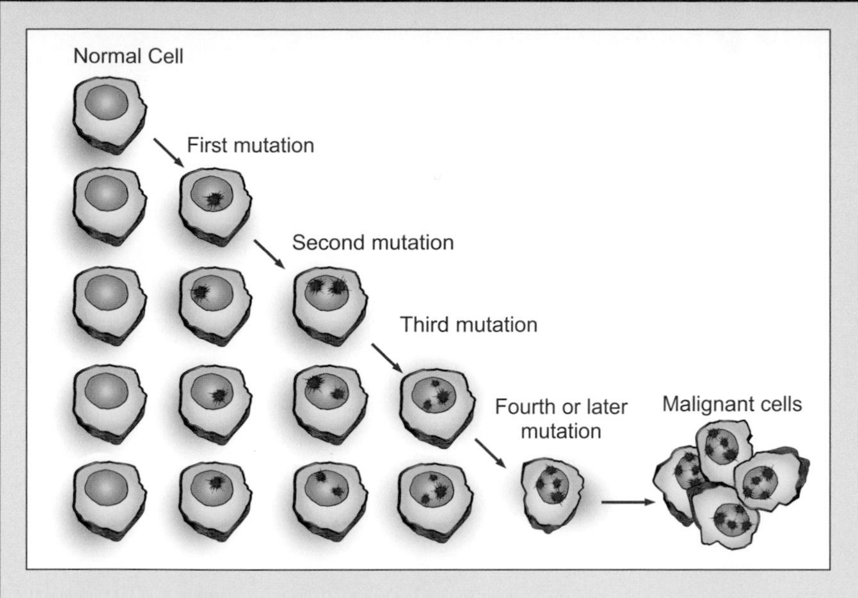

Cancer usually arises in a single cell. This model for cancer risk suggests that the cell's progress from normal to malignant and then to the stage when cancer cells break away and move to other parts of the body occurs in stages (see page 85 in Chapter 6). These stages involve a series of irreversible genetic changes (mutations). Exposure to a harmful agent may trigger just one or all of the stages. Cancer morbidity data is consistent with the theory that only two critical mutations are required, but three or more smaller changes may have the same effect. Anything that increases the rate of cell division also increases cancer risk, by reducing the time available for repair of DNA changes before cell division.

Studying health risks

Case-control studies

Scientists study health risks with the help of large-scale studies, including cohort studies (Chapter 4), randomised controlled trials (Chapter 4) and case-control studies (Chapter 6).

Case-control studies attempt to 'prove' a link between exposure to known risk factors and occurrence of a disease as compared with a control group. The scientists select people to take part in a case-control study on the basis of the participant's health status. This means that there is potential for bias. Any method where participants are not selected randomly introduces the possibility of favouring certain participants for inclusion over others. This can invalidate attempts to infer causal factors.

Genetic epidemiology

Until quite recently scientists had to draw inferences about the genetic basis of disease from patterns of inheritance in families and across groups of families (see Chapter 6). Now they have access to information about DNA and this has opened up new lines of inquiry. Large-scale studies looking for correlations between complex diseases and variations in DNA sequences in large populations have helped to identify genes linked to particular diseases.

If clustering of a disease within families suggests a genetic factor, epidemiological studies may lead to further research to discover the precise location of the gene in the DNA of chromosomes (Figure 8.10). Scientists also try to explain in detail how the gene causes the disease.

Genetic risk factors are calculated by comparing the incidence of the disease among close relatives of affected individuals with the incidence

Questions

11 With the help of page 84 (Chapter 6), draw a diagram in the style of Figure 4.18 on page 62 to summarise the procedure for a case-control study to study risk factors for a particular cancer.

12 A cohort study of risk factors associated with cancer of the colon was based on tests on volunteers attending a Parent's Association lunch in a primary school. Explain how this approach to sampling might introduce bias into the data.

13 A case-control study of the links between diet and cancer selected participants to make up the control group from people waiting for appointments in a hospital outpatient's department. Explain how this might bias the results. How might the bias affect the outcomes of the study?

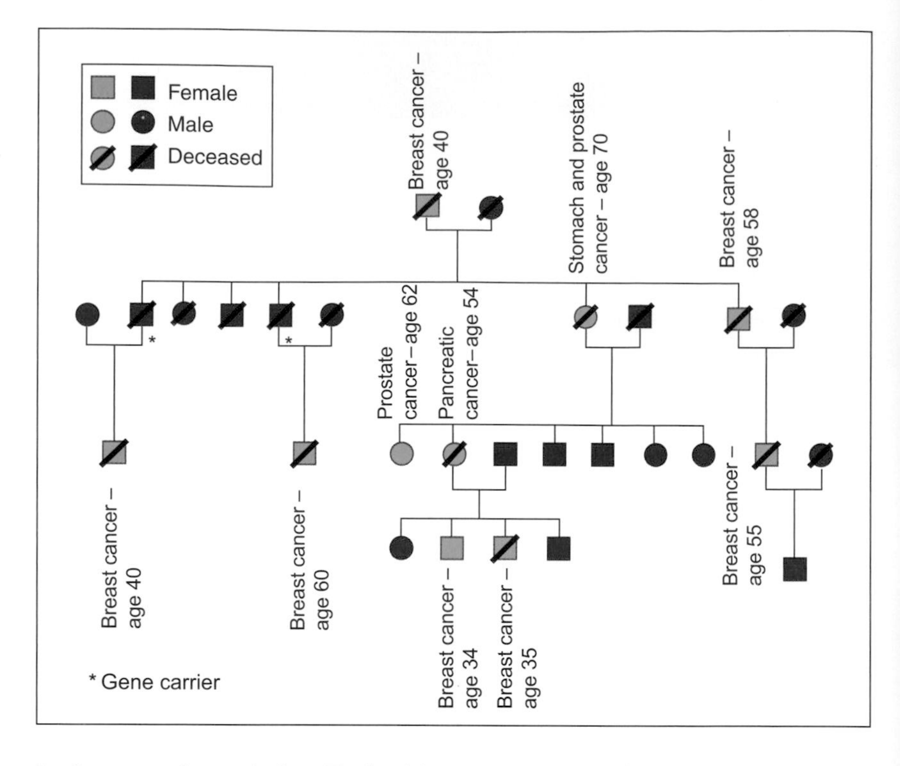

Figure 8.10

A higher than expected incidence of a disease such as cancer in a family may indicate the presence of a genetic factor. This family tree shows inheritance of an increased risk for breast cancer. Those shown in blue have developed cancer.

in the general population. To do this, scientists must first agree how to diagnose the disease, then they can estimate how many people suffer from the disease in a representative sample of the population. With rare genetic diseases, it is difficult to find a large enough random sample to allow valid inferences to take place. This means that people are often recruited to a study precisely because they are affected. This non-random sampling is one of the most common reasons that researchers may think they have discovered an association or correlation which in reality does not exist.

Correlation between a disease and family members may not have a genetic cause. Lifestyle factors are shared by families, so can contribute to any observed correlation. However, for many complex diseases, lifestyle risk factors are only weakly correlated in relatives. For example, known risk factors for breast cancer such as body mass index (BMI) explain less than 5% of the increased risk of breast cancer in close relatives of affected people.

Lifestyle factors are typically measured by interview or questionnaire. This can lead to substantial errors and bias. People are not necessarily good at recalling, or admitting to, risky lifestyle factors. This means that non-genetic contributions to disease can be greatly underestimated.

Questions

14 Explain why clustering of a disease within a family is not necessarily the result of genetic factors.

15 Tracking down the genes involved in disease can take advantage of isolated populations that have lived for generations out of contact with the rest of the world. Studies of this kind have helped to identify genes linked to a number of disorders, including multiple sclerosis, schizophrenia and autism. Suggest why it helps to be able to study isolated populations.

16 Explain why identical twins have the same genes.

17 Why is it helpful to study both identical twins and non-identical twins when exploring genetic risk factors for disease?

Genes and environmental factors

Studies involving twins help separate the contributions of environmental and genetic risk factors. Identical twins have the same genes, so any differences between them are likely to be due to the environment.

One team of scientists studied twins to investigate risk factors for melanoma, which is the most serious form of skin cancer. They counted freckles, moles and other pigmented areas on the skin of adult twins

(Figure 8.11). They collected data from 127 pairs of identical twins and 323 pairs of non-identical twins.

From their data, the scientists concluded that for twins less than 45 years old, it was exposure to sunlight rather than genes that did more to determine the total count of pigmented areas. However, the relative importance of genetic effects increased as the twins got older. It rose from 36% for twins under 45 and to 84% after this age. This percentage describes the proportion of the observed pigmented areas that are due to the expression of genes rather than the environment (in this case, exposure to the Sun).

The results of this study show that genetic and environmental factors interact. People with a genetic predisposition to form freckles or other pigmented skin areas only do so when exposed to sunlight. When considering genetic risks it is important to take into account any environmental factors which may affect the way that the genes are expressed. It is also important to appreciate that some environmental risk factors will become relatively more important as people get older, as they have been exposed for longer.

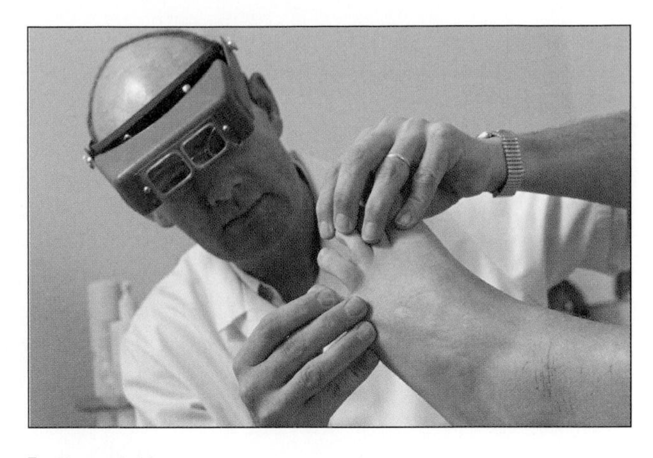

Figure 8.11

Checking a patient's skin for moles and other pigmented areas. The doctor is wearing magnifying lenses to help him see small moles. Pigmented areas of the skin are risk factors for a type of skin cancer called melanoma. The genes for pigmented areas are affected by exposure to the sun, so the total risk is a combination of genetic and environmental factors.

Diet and cancer risks

The experts that advise Cancer Research UK suggest that about a quarter of all cancer deaths are caused by unhealthy diets and obesity. What we eat may affect our risk of many cancers, including cancers of the bowel, stomach and mouth. Typically doctors recommend people to cut their cancer risk by eating a healthy, balanced diet that is high in fibre, fruit and vegetables, and low in red and processed meat and saturated fat (Figure 8.12). However, the links between health and diet are very complex and epidemiological studies have to involve very large numbers of people to provide reliable results.

Questions

18 Suggests reasons why the incidence of melanoma has shot up in the UK recently.

19 Suggest an explanation for the growing importance with age of the genes that determine the total count of pigmented areas on skin relative to the effect of exposure to sunlight.

Figure 8.12

In 2001 the Department of Health initiated its '5-a-day' programme to encourage people to eat more fruit and vegetables. Many food producers have used this idea in promoting fruit and vegetable products.

Question

20 Suggest reasons why studies of diet and health that are based on one population are likely to miss:
a) important risk factors for disease
b) dietary factors that can cut the risk of disease.

There have been many studies exploring the connections between diet and cancer. Many of them have been flawed for a variety of reasons, including:

- they were too small
- they focused on one population with limited variation in diet
- they did not measure dietary intakes accurately, often because people were asked to recall what they ate months or years in the past.

Scientists in Europe are now taking part in a long-term study called the European Prospective Investigation of Cancer (EPIC). The researchers have designed the cohort study to be as reliable as possible. Data is being collected from over 500 000 people in 10 European countries. This includes around 90 000 British men and women, including about 30 000 vegetarians.

Key term

A **biobank** is a database of genetic, clinical, environmental and lifestyle information on individuals, together with corresponding clinical specimens.

UK Biobank

Another long-term 'prospective' project is the UK Biobank (Figure 8.13). This £61 million project was given the full go-ahead in August 2006. The aim is to provide the richest source of health-related data and samples for researchers from around the world. By measuring many different exposures (not just genes) in large numbers of people, this study will assess the impact of a wide range of factors, alone or in combination, on many different health conditions.

Figure 8.13

The logo of the UK Biobank.

In its first few years, the UK Biobank team will recruit 500 000 adults aged 40 to 69. That is nearly 1% of the UK population. Volunteers attend an assessment centre where they complete a detailed lifestyle questionnaire and are interviewed about their medical history. Nurses take measurements of a range of health indicators such as blood pressure, body size and lung function. They also donate blood and urine samples. About 15 million samples will be stored for decades at ultra-low temperatures. Work on this scale requires a robotic system for handling the samples from up to 1000 participants every day.

In the years following the initial assessment, information about participants' health is being collected, with permission, from medical and other health-related records. As follow-up continues, researchers will compare the lifestyle, genes and other factors among participants who develop a particular disease with participants who remain healthy. For common conditions, such as heart disease and diabetes, this will be possible within 5 to 10 years of starting the project; for less common diseases it is likely to take much longer before there are sufficient cases for reliable analysis.

Questions

21 The EPIC and UK Biobank studies have started with healthy people. Their lifestyle and health are being followed for many years. Suggest reasons why this type of study is more accurate than one that asks patients who already have cancer, or other diseases, to recall their lifestyles before the onset of the disease.

22 Why is it important that an independent Ethics and Governance Council has been set up to monitor the UK Biobank?

Promoting healthier lifestyles

Claims about health risks and benefits are often based on complex studies with many variables and interacting factors. Research on human lifestyles is never simple; there is no such thing as an 'average' person, and there are many variables that cannot be controlled.

Even when the evidence of the risks is very strong, as it is with smoking, people may not change their behaviour. Despite sustained education programmes and anti-smoking campaigns (Figure 8.14), large numbers of people still smoke.

Changing human behaviour is never easy. One campaign that has perhaps proved more successful than many is the 'Slip-Slop-Slap' campaign in Australia. Faced with a frightening level of cases of skin cancer, the authorities encourage people to slip on a shirt, slop on sunscreen and slap on a hat. The campaign started in 1981, with an advertising jingle and seagull mascot called Sid.

Despite the campaign, new cases of skin cancer still outnumber all other forms of cancer in Australia by more than three to one. The country continues to suffer from the highest incidence of skin cancer in the world but there are some recent trends which are encouraging. The rate of diagnosis of skin cancer in younger people, especially among women, has begun to level off and, in some age groups, decline. This is the first time such a trend has been noted anywhere in the world. It has taken 20 years of campaigning to bring about the changes.

This campaign has been copied in other countries. Cancer Research UK has been running a SunSmart campaign since 2004 aimed not just at holidaymakers but also at people who work out of doors (Figure 8.15).

Figure 8.14

Health warning on a cigarette packet in 2007.

Human behaviour and health risks

Theories from behavioural science suggest that human behaviour is often irrational when judged by logic or scientific fact. If behaviour was governed by scientific evidence, no-one would smoke and everyone would eat sensibly and exercise enough to maintain a healthy weight; this would mean that current trends in obesity would not be an issue.

Valuing the outcome of a behaviour (Figure 8.16) depends on the person's point of view, their individual beliefs and their previous experience. The final value attributed to the behaviour depends on all the positive and negative values of each possible outcome (Figure 8.17).

Generally, tangible and immediate outcomes are more likely to have a large influence on behaviour than more theoretical or long-term outcomes, such as skin cancer.

Figure 8.15

If you are SunSmart you:

- ***S**pend time in the shade between 11 and 3*
- ***M**ake sure you never burn*
- ***A**im to cover up with a T-shirt, hat and sunglasses*
- ***R**emember to take extra care with children*
- ***T**hen use factor 15+ sunscreen*

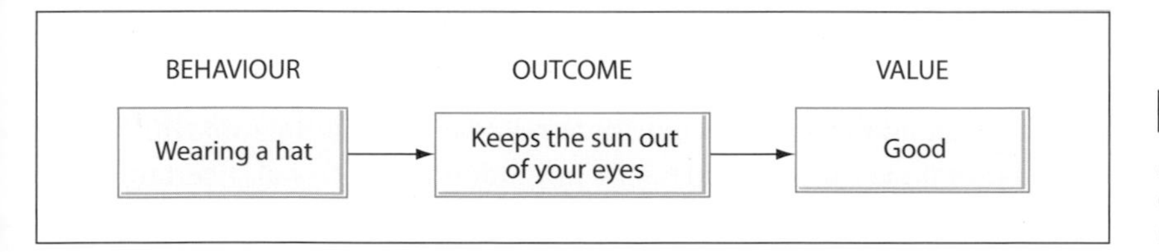

Figure 8.16

Behaviour is influenced by the likely outcomes of that behaviour combined with the perceived value of the outcome.

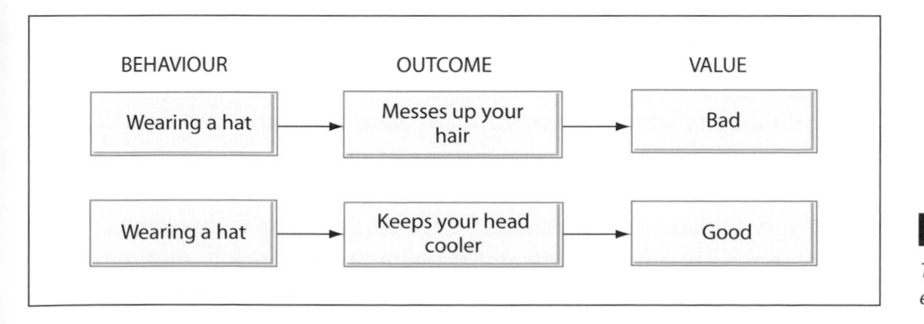

Figure 8.17

There may be many possible outcomes for each behaviour.

Figure 8.18

These tourists can control the amount of protection they use, but may be less able to control behaviour such as using shade.

Questions

23 Why might simply recommending people to use sunscreens when outside do more harm than good?

Sun protection behaviour is complex; it involves many different actions, for example avoiding exposure by seeking shade, wearing sunscreen and wearing clothes which cover the skin. There is a time component to the behaviour – when you apply the sunscreen, when you stay indoors or go outside, and how long you wear your hat for. The context also affects the behaviour; an uncomfortably hot day may prompt some of the actions which avoid sun exposure, or being on holiday might make a difference (Figure 8.18). Behavioural sciences show that the more complex behaviour is, the more difficult it is to change.

Sometimes, behaviour depends on the degree of control a person has over their actions. An adult's control over children's sun protection decreases as children get older. Avoiding exposure or applying sunscreen are under an adult's control, but there may be external factors which are not. Children may have to take part in outdoor activities at a certain time of day, or there may be no shade available. Social norms also have an influence, in this case tans being fashionable or not fashionable.

Key term

The **precautionary principle** states that when an activity raises threats of harm to human health or the environment, precautionary measures should be taken even if there is scientific uncertainty about the cause and effect relationships.

Public policy and the precautionary principle

Judging what is an 'acceptable' level of risk for society is a political responsibility, not simply a scientific decision. When risks to human health become known, it is often the role of government to determine the appropriate action or inaction. Governments set up regulatory or advisory bodies to help make the difficult decisions in situations where the scientific evidence may be incomplete or open to conflicting interpretations.

When there is a new issue, scientists may disagree about the meaning of the data available. In these circumstances many would argue that the lack of scientific certainty should not be used as an excuse to delay action to deal with the possible risk. In other words, regulators should adopt a precautionary approach and give priority to protecting public safety. Often the public supports the notion that new technologies should not go ahead unchecked, requiring evidence that the benefits outweigh the risks. In other words, the policy should be 'better safe than sorry' (Figure 8.19).

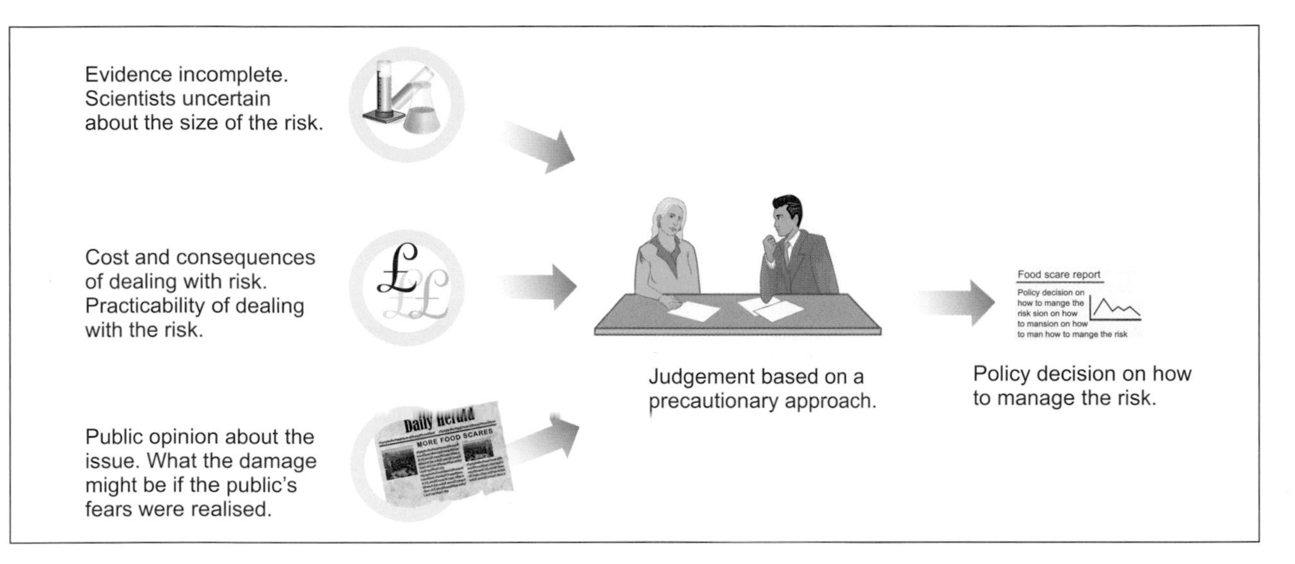

Figure 8.19

Taking a precautionary approach to making decisions about food safety. Regulators do not make judgements on their own. They have to consult both experts and the public. Decisions do not only involve scientific evidence but also have to take into account the costs and consequences of dealing with the risk and public opinion.

At first the precautionary principle was put forward as the basis for policies to prevent damage to the environment. Now it is also applied much more widely to cover a broad range of contexts where there is threat to human, animal or plant health.

Applying the precautionary principle is often controversial. Some people welcome regulations aimed at avoiding risk and removing hazards from our lives. Others take the view that we can learn to manage risks and that hazardous activities should only be banned as a last resort.

Campaigning pressure groups and other opponents of new technologies can take the precautionary principle to extremes when they attempt to stop new developments. In their opposition to technological change, they can use the 'better safe than sorry' principle to argue that just because there is no evidence of harm, that does not mean that something is not harmful and so it should not be allowed. However, when the risks are small and harmful effects are rare it can be very hard to assess the risks (see Chapter 7).

Questions

24 Critics of the English smoking ban in public places, which became law in 2007, pointed out the lack of evidence to justify the ban, and the lack of regard for balancing the legislation with civil liberties. The Parliamentary Under Secretary for Public Health at the time commented 'It is clearly the case that, in relation to deaths from smoking and second-hand smoke, the most serious aspect of that is smoking in the home. 95% of deaths are related to smoking in the home.' Comment on whether the ban on smoking in public places showed a justifiable application of the precautionary principle.

Review Questions

25 a) Explain how mutations in genes sometimes lead to cancer.
b) Give examples of risk factors for cancer which probably have their effect by causing mutations.

26 Give at least one example to illustrate each of these ideas about how science works:
a) Identifying a correlation between a factor and an outcome may *suggest* that the factor is the cause of the outcome, but does not *prove* a causal link.
b) Scientists are much more willing to accept that a risk factor causes a disease if there is a convincing theory to explain how the factor has its effect.
c) New technologies and processes based on science often introduce new risks.
d) Estimates of risks are more accurate if the sample size is large.
e) If the probability of harm from a risk factor is low it can be very hard to make an accurate assessment of the risk.
f) Regulators sometimes take a precautionary approach to a new product or process.

27

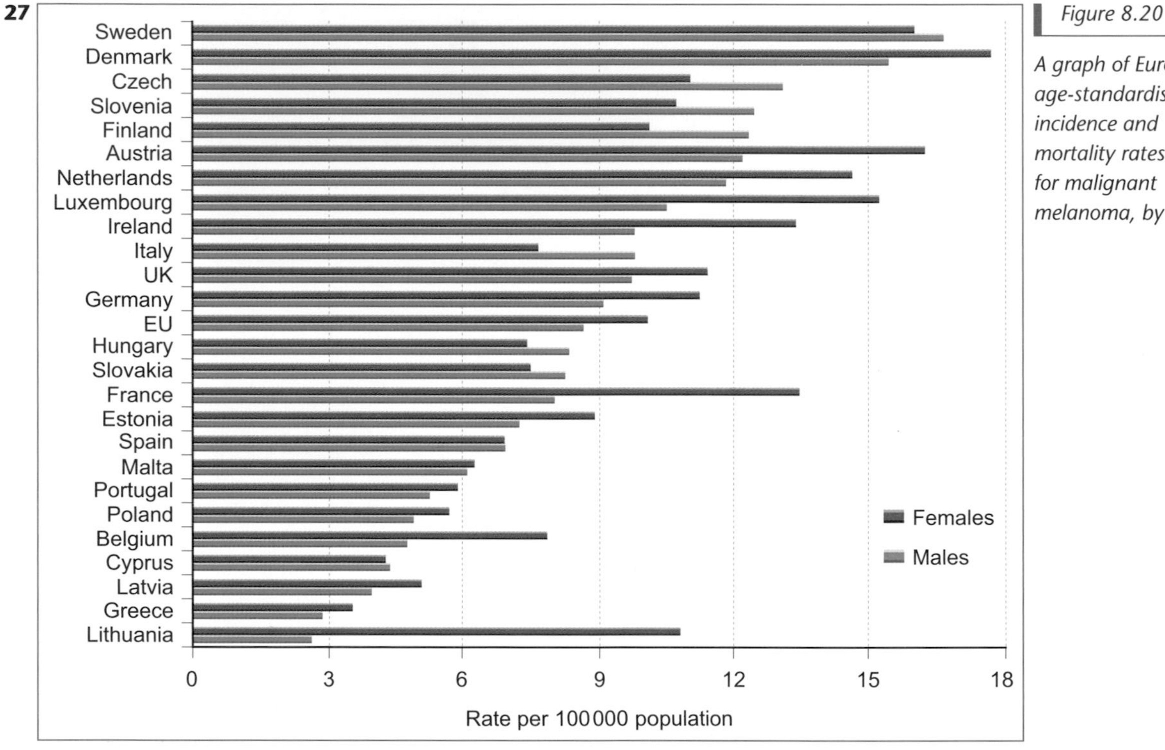

Figure 8.20

A graph of European age-standardised incidence and mortality rates for malignant melanoma, by sex.

a) Describe the trends shown by the data in Figure 8.20.
b) Suggest explanations for these trends.
c) What measures could a primary school take to reduce the risks of melanoma for the pupils?
d) Which do you think will have more effect on reducing the incidence of skin cancer: an education programme in schools or a public promotional campaign with advertising?

28 Out of 13 published epidemiological studies of the link between the use of sunscreens and melanoma risk, three showed a decreased risk of melanoma, four showed an increased risk and six were inconclusive. What might account for these inconsistencies?

29 a) Suggest two areas of behaviour linked to health risks (other than sun exposure) which are complex, possibly involving frequent or regular actions.
b) Suggest two others which involve single or infrequent events, explaining why these are more likely to become part of people's routine behaviour than the more complex events described in (a).

30 One supermarket chain has voluntarily banned over 20 synthetic food colours in response to consumer concerns even though the colours are permitted for use in foods by regulators in the EU. Instead they are using colours from natural sources. Discuss whether you think the supermarket was right to make the changes.

Chapter 9

Evolution

The issues

The theory of evolution has major implications for how we see ourselves. If biologists are correct, humans and all other forms of life on Earth shared a single common ancestor some 3500 million years ago.

Yet theories about evolution have always been controversial and, to some people, still are. The general public in some countries is deeply divided over whether evolution really has taken place. Many people doubt that it has and some religious believers maintain that the world is only 10 000 or so years old or even younger. Many people see no overlap between religion and science – for them science and religion deal with different questions. Others happily combine their religious faith with an acceptance of the theory of evolution.

The science behind the issues

Scientists now know a great deal about how individuals pass on genetic information to their offspring, and about how changes take place in the genetic make-up of species. Changes begin with mutations – permanent alterations to an individual's genes which occur constantly and more or less at random. Some mutations are harmful, and individuals with these are less likely to survive and reproduce, while other mutations appear to have no effect. Occasionally, though, mutations prove beneficial. The process by which the frequencies of genes change over time as a result of the accumulation of mutations is known as natural selection.

Natural selection is widely accepted as a major element of the process of evolution, although explanations of the mechanisms of evolution have been extended as more evidence emerges from molecular biology.

What this tells us about science and society

Evolution is a scientific explanation for the huge diversity of life on Earth. Evolution is, in part, a historical science, although experiments on living organisms play an important part. Scientists cannot observe the lives of the dinosaurs and other extinct organisms directly, but fossils give strong hints of how they lived, their structure and habitat. The processes of fossilisation and natural selection are open to experiment, but any evidence from existing fossils or contemporary experiments is open to interpretation, as with any scientific data (Figure 9.1).

Figure 9.1

Rodolfo Coria dusting the teeth on the fossil jawbone of the largest carnivorous dinosaur discovered so far. He unearthed the 110-million-year-old bones of this dinosaur in 1993 at El Chocon, Argentina.

Figure 9.2

A portrait of Charles Darwin in his early 20s.

Charles Darwin's life and work

One of the most famous biologists who ever lived is Charles Darwin (Figure 9.2). Darwin is famous for two reasons. Firstly, he produced a mass of evidence that succeeded in convincing scientists and much of the public that evolution had occurred. Secondly, he came up with the idea of one major agent of evolutionary change – natural selection.

Darwin's life

Charles Darwin was born in 1809 while England was at war with France. His father had high hopes that Charles would be a doctor like his own father, but as a medical student Charles realised that he couldn't face operations and the suffering of patients. He moved from Edinburgh to Cambridge, where he began to prepare for being ordained as a minister in the Church of England.

As a child, Charles collected shells, postal franks, birds' eggs and minerals. As a student, his collecting instincts focused on insects. He became passionate about beetles and found nothing more thrilling than finding a new specimen. However, the question as to his future career remained. Remarkably, at the age of 22, he was offered the chance to spend two years on a small ship, the HMS *Beagle*, that was to sail round the world making maps (Figure 9.3).

Charles was employed both as the ship's naturalist and as a suitable dining companion for the ship's captain, Robert Fitzroy. Fitzroy was an eminent geographer, and made an important contribution to the science of meteorology. Later in life, after marrying a pious lady, he held a firm belief in the literal truth of the bible and the impossibility of evolution. It is interesting to note that on the voyage Fitzroy expressed doubts about the biblical flood at a time when Darwin still had an entirely orthodox trust in the book of Genesis (the story of Adam and Eve). Captain Fitzroy was a man of great temper, and he and Darwin must have made a remarkable pair. Fortunately, Darwin was very good at getting on with people and he managed to maintain a fairly calm relationship with Fitzroy.

Figure 9.3

Map showing the route taken by the HMS Beagle *on its voyage.*

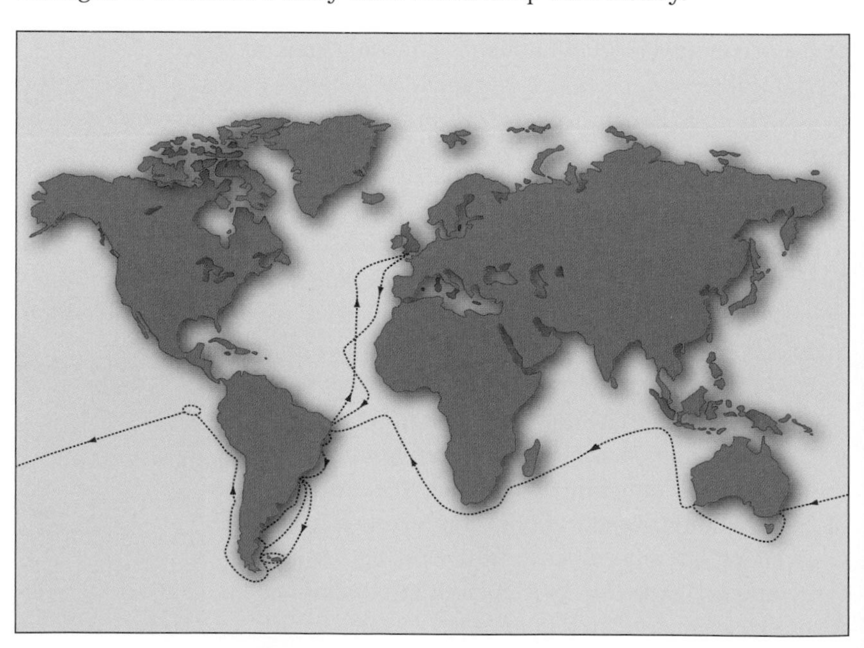

During the five years of the voyage, Darwin was struck with some of the similarities and differences between animals and plants living in different parts of the world. On his return, he slowly became convinced that evolution explains the huge diversity of living species on Earth.

Key terms

A **species** is best defined as a group of organisms that are able to breed with one another to produce fertile offspring. So, for example, orang-utans and gorillas belong to different species; they can't interbreed.

A **population** is a group of organisms belonging to the same species.

What was Darwin's evidence for evolution?

During many years of study, Darwin became convinced that species were not fixed, but that they had evolved from their ancestors over the course of extremely long periods of time. For example, fossils often resembled species currently alive in the same place, but were clearly different from them. Darwin saw that the easiest explanation of these similarities and differences between the present and past species in a geographical region was that the present inhabitants had evolved from the past ones.

Another reason for Darwin's change of mind came from the study of living species. Darwin visited the Galapagos Islands (Figure 9.4). These are tropical islands about 900 km west of Ecuador, off the coast of South America. The *Beagle* spent a month in the Galapagos, and the islands provided Darwin with evidence that isolated groups of organisms can change.

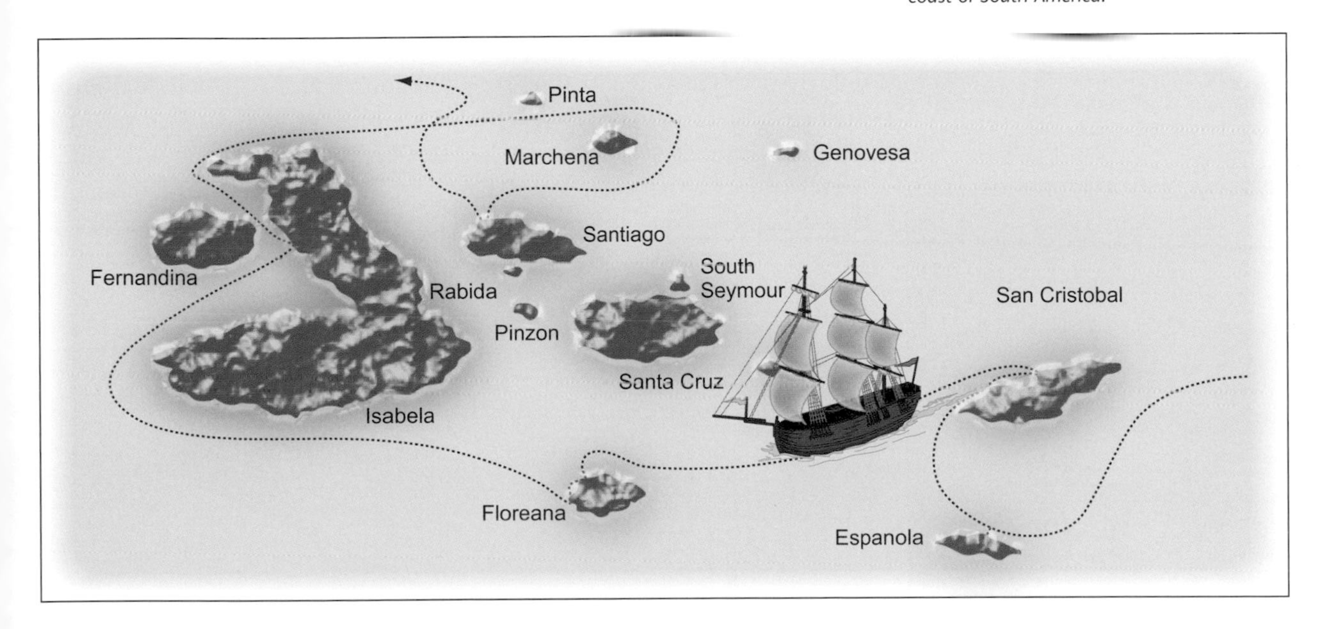

Figure 9.4

Map showing the voyage of the Beagle *through the Galapagos Islands off the west coast of South America.*

Darwin noted that one of the most striking things about the organisms on the Galapagos Islands is that few are found on the mainland of South America yet all resemble organisms found on the mainland. For example, we now know that 13 species of finch are unique to the islands but resemble finches found on the mainland.

Darwin was intrigued by the giant tortoises on the islands (Figure 9.5). A chance remark was made to him at dinner by a certain Mr Lawson, vice-governor of the Islands. He happened to say to Darwin that he could tell by looking at a tortoise which island it had come from. This set Darwin thinking. How could this be? Was each sort of tortoise created independently on each island? Or had each type of tortoise evolved independently from common ancestors separated on the different islands by deep water, so that over time they became recognisably different?

Figure 9.5

An 1886 illustration of the giant tortoises from the Galapagos Islands. These tortoises give their name to the islands – the Spanish for tortoises is galápagos. They can reach 1 m in height and weigh up to 225 kg.

Figure 9.6

A Galapagos mocking bird similar to the ones Darwin recorded on his voyage.

On the first island, Darwin found a mocking bird like the South American ones he knew but clearly different (Figure 9.6). On the second island, he found another kind of mocking bird. He was intrigued by the difference and labelled the two specimens carefully. On island three, he found yet another and labelled that. As the *Beagle* sailed away from the Galapagos, he sorted out his specimens and added those of all his friends who had been collecting with him. He then remembered Mr Lawson's remark about the tortoises and asked himself 'Is this the same pattern – different islands, different species?'

Natural selection

Once home, Darwin tried to come up with a mechanism that could account for evolution. For some scientists, there is a definite moment when a new theory takes shape in their mind. For others, the process takes much longer. Darwin's theory of natural selection appears to have taken a long time coming. But, all the evidence suggests that the key idea came in a flash on 28th September 1838, when he read an essay written in 1798 by the English clergyman, Thomas Malthus. Darwin read a passage about human populations, saw how it might apply equally to animal populations, and there was the explanation for evolution.

Malthus argued that human populations always increase more quickly than the available food supply (Figure 9.7). As a result, Malthus predicted, there will always be a global shortage of food. Darwin saw that Malthus's argument could be applied to all species. The key idea he got from Malthus is that there is a struggle for existence. Malthus thought of this struggle as competition for food. Darwin realised that it could take various forms.

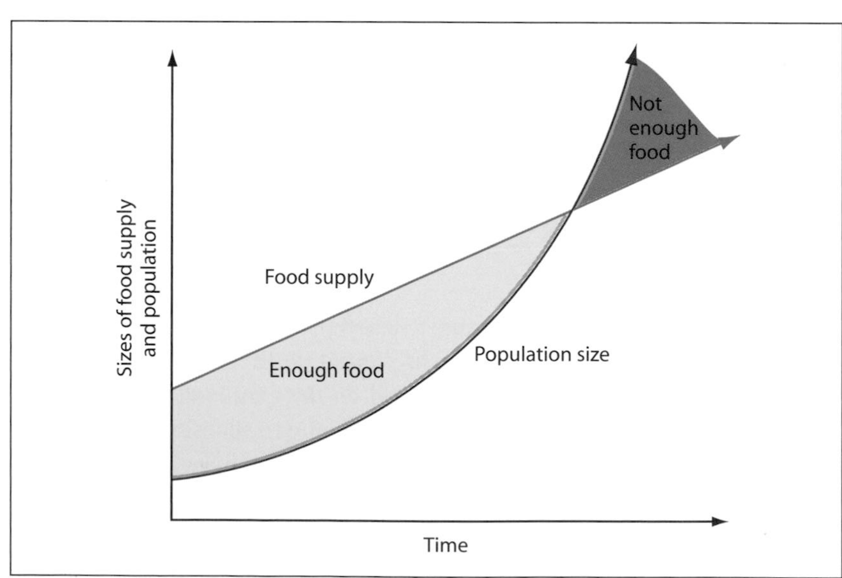

Figure 9.7

A plot of food supply and population size showing the growth of food supply and population. Malthus maintained that, in the long run, population growth would always outrun food supply. Malthus's reasoning helped both Darwin and Wallace (see page 143) to come up with the theory of natural selection.

Eventually Darwin came up with four conclusions:

- Individuals within a species differ from one another.
- Offspring generally resemble their parents.
- In all species more offspring are born than can survive to adulthood and reproduce.
- Some individuals are better suited to their environment than others, and they survive and reproduce more effectively than others.

Darwin suggested the following statements, which together make up the theory of natural selection:

- Individuals that survive to reproduce pass onto their offspring the characteristics that enabled them, the parents, to survive.
- Over time, a group of individuals belonging to a single species may give rise to two different groups that are sufficiently distinct to belong to two separate species.

Question

1 a) Describe two of Darwin's observations that led him to question creation as an explanation for the diversity of life.
b) How did he think these observations could be explained by evolution?

2 Explain why natural selection would not work if:
a) individuals within a species were all the same
b) offspring did not resemble their parents
c) all offspring survived to reproduce
d) all individuals in species had the same number of surviving offspring.

Artificial selection as evidence for natural selection

The theory of natural selection was Darwin's attempt to explain the observations he made. Darwin spent years thinking about this theory and obtaining evidence for it. Scientific theories do not just emerge from data and observations, but involve creative imagination and conjecture on the part of the scientist.

Some of the best evidence for natural selection came from what is known as 'artificial selection'. Artificial selection is intentional breeding for particular characteristics. Darwin spent many hours talking to people who bred farm animals and varieties of fancy pigeons. He was particularly struck by one pigeon breeder's claim that he could produce a new variety within only a few years (Figure 9.8). Darwin realised that if artificial selection, at the whim of the pigeon breeder, could produce a distinct new variety within a few years, perhaps natural forces could produce entire new species given enough time.

Darwin became so interested that he built a pigeon house in his garden, bought several varieties of fancy pigeon and set to work breeding them. To his delight, Darwin found surprisingly large differences between the many fancy pigeon varieties he bred, observed and measured.

Figure 9.8

Breeds of pigeons exhibited in 1864 by one of the London pigeon clubs to which Darwin belonged. All pigeons are descended from the rock dove, including the ordinary pigeon found in towns. The great diversity of pigeon breeds is due to artificial selection. This engraving appeared in The Illustrated London News *in January 1864.*

The differences were not just external; even their red blood cells were differently shaped. To Darwin's fascination he found that pigeon fanciers had produced so much variety between their breeds that if specimens had been shown to ornithologists (bird experts) as wild birds, the ornithologists might have reckoned them to be around 20 separate species.

Evolution before Darwin

Darwin was not the first person to suggest that species had evolved over time. Indeed, the idea had been suggested by the Ancient Greeks more than 2000 years previously. Nor was Darwin the first person to come up with a theory for evolution.

Darwin published his book *The Origin of Species by Means of Natural Selection or The Preservation of Favoured Races in the Struggle for Life* in 1859 (the Victorians liked long titles for their books). Exactly 50 years earlier, a Frenchman called Jean-Baptiste de Lamarck had put forward his own, very different theory of evolution.

Lamarck's theory proposed that organisms responded to their environment by changing the way that they used a particular structure or organ, and that in response to this change in use they physically adapted to become better suited to their environment. He also proposed that these changes could be passed on to the next generation. Lamarck backed up his theory with evidence, such as his observation that the sons of blacksmiths tended to have larger and stronger arms than the sons of weavers. According to Lamarck's theory, blacksmiths exercised their arms as they worked and therefore developed big muscles, a trait that was then passed on to their sons.

Both Lamarck and Darwin would presume that today's long-necked giraffes have evolved from ancestors with shorter necks (Figure 9.9). However, they would explain the presence of long necks in today's giraffes very differently.

Figure 9.9

Lamarck and Darwin came up with alternative theories to explain why giraffes have long necks.

Lamarck's theory sounds a bit like one of Rudyard Kipling's *Just So* stories, for example the one in which the elephant gets its trunk by having its nose pulled by a crocodile (Figure 9.10). Lamark's theory suggested that giraffes reached up to eat leaves growing higher on trees. Their necks stretched as a result. In other words, characteristics which were 'needed' were acquired during an animal's lifetime so the animals progressed towards an ideal form. An acquired characteristic – such as long necks – is then passed on to any baby giraffes.

A Darwinian explanation would be rather different. It would rely on the idea that there are always more giraffes than the food supply can support. As a result, those giraffes that just happen to have taller than average necks would be most likely to survive and produce baby giraffes. The characteristic of having long necks would then be passed on to the next generation, and so the average length of giraffes' necks would increase over time.

Darwin did not agree with Lamarck's explanation of the mechanism for evolution – that animals strive towards some perfect form – and no internal planning mechanism is involved in Darwin's explanation. There is no reason to suppose, though, that had Lamarck been alive when

Figure 9.10

This is the Elephant's Child having his nose pulled by the Crocodile in Rudyard Kipling's Just So *story. He started off with a short nose, but got tricked into going close to the river, and the crocodile pulled and pulled at his nose until it grew longer. It never returned to its normal length. Ever since then, elephants have had long trunks.*

Darwin proposed his theory of natural selection, Lamarck would not have been interested and even accepted the idea. Lamarck is credited with attempting to explain the evolution of higher species from simpler ones before Darwin, even though his explanation did not match with the evidence from simple observations of inheritance.

During the twentieth century, philosophers of science realised that there are no simple rules that make it possible to proceed from observations and data to general theories; there is always the possibility that new observations will contradict an established view. For example, observing a series of white swans might lead to the general theory 'all swans are white'. A single observation of a black swan will then falsify this theory.

This is one of the important ideas in the way that scientific theories develop. According to this view, no amount of observation can *prove* a theory to be correct. On the other hand, observations can prove a theory to be incorrect. Science is, then, based on a method of bold conjectures followed by vigorous attempts to falsify them. The important thing about the scientific method is not how the conjectures and theories arise, but how we test them.

Questions

3 Explain how a) Lamarck and b) Darwin might have explained the ability of birds of prey, such as peregrine falcons, to fly very fast.

4 Describe one observation that could have been made by Lamarck or Darwin that would have shown that Lamarck's theory of evolution is not correct.

Wallace's theory of natural selection

On 18th June 1858 Darwin was stunned by a 12-page letter arriving out of the blue from a relatively unknown naturalist called Alfred Wallace. The letter more or less exactly described the key process in Darwin's own theory of natural selection. Looking back, we can see that this isn't the amazing coincidence it might seem to be. After all both Darwin and Wallace had read and thought about Malthus's arguments about there never being enough food to keep up with growth in a population.

Wallace came from a much less privileged background than Darwin. He had left school at the age of 14 because of his family's financial difficulties and became a land surveyor. A stroke of good fortune led

Questions

5 'Human beings have evolved on planets other than Earth' is not falsifiable, because it is impossible to make an observation that proves humans have not evolved on other planets. The statement 'humans have only evolved on Earth' is falsifiable; a single observation of a human on Mars, for example, would disprove it.

a) Which of the following are falsifiable statements relating to the evolution of different types of tortoises on Galapagos?

(i) Tortoises were sorted into different types and placed on the different islands by pre-historic human travellers.

(ii) Groups of tortoises were isolated from each other on the different islands because they are unable to swim between the islands.

(iii) If you moved a group of tortoises to a different island, they would not survive as well as the tortoises which were already there.

(iv) The different types of tortoises on the different islands were created at the same time as each other and as all the other living things on the islands.

b) Explain exactly what observations would falsify these statements.

6 List the similarities between Darwin and Wallace's life experiences that may have led to them independently thinking of the theory of evolution by natural selection.

to him spending four years on an expedition to the Amazon. Despite appalling conditions, he collected nearly 15 000 animal species, some 8000 of which were new to science.

Then he had a terrible misfortune. On the way home the ship caught fire and all 15 000 species were lost, together with most of Wallace's notes. Despite this, Wallace set off in 1854 for an eight-year expedition to Asia. It was there, as he lay ill with fever in 1858, that he thought about Malthus on human populations and wondered about animals: 'Why do some die and some live? ... from the effects of disease, the most healthy escape; from enemies, the strongest, the swiftest, or the most cunning; from famine, the best hunters or those with the best digestion.'

Later, Wallace wrote up his theory and sent it to Darwin. Although Darwin realised it was the same theory he had thought of 20 years earlier, he and Wallace were happy to credit each other for the idea. A joint paper was presented to scientists at the Linnean Society in 1858, so it really should be known as the Darwin–Wallace theory of natural selection. This collaboration was mutually beneficial to the two scientists. Wallace lacked Darwin's supportive network of influential friends in the scientific community, and Darwin benefited from Wallace's ability to write clearly and logically.

How long does evolution take?

Breeds of domestic species, such as dogs, sheep and domestic pigeons, differ greatly from the breeds of just a century ago. However, prehistoric paintings of animals up to 30 000 years old, and drawings, paintings and carvings of animals from 4000-year-old Egyptian tombs show no recognisable differences from today's wild species. It seems that, in the wild, most species stay remarkably constant (Figure 9.11).

So just how long does evolution take? Evolutionary biologists think that many species take periods of millions of years to evolve into new species. Darwin didn't know how long it took species to evolve, but he appreciated that his theory only made sense if the world was very old. Fortunately for Darwin, a growing number of geologists also believed that the world was very old, possibly even thousands of millions of years old.

Figure 9.11

A prehistoric painting of big-horn sheep carved into the rock by Native Americans around 12 000 years ago. The rock face is now part of the Capital Reef National Park in Utah, USA. Such paintings suggest that evolution of this species, if it has occurred, has taken place only very slowly.

Nowadays, radioactive dating of rocks suggests that the Earth itself is about 4.5 billion years old (i.e. 4500 million years old). The first fossils have been dated at 3.5 billion years old. Not surprisingly, such fossils are of very simple, single-celled organisms. Multicellular organisms don't appear until about 700 millions years ago (Figure 9.12).

Key term

Fossils are the remains of organisms that have died, and have left an imprint in rock or have become replaced by minerals. They only form in very unusual circumstances. Many fossils faithfully preserve details of the structures of organisms that lived long ago.

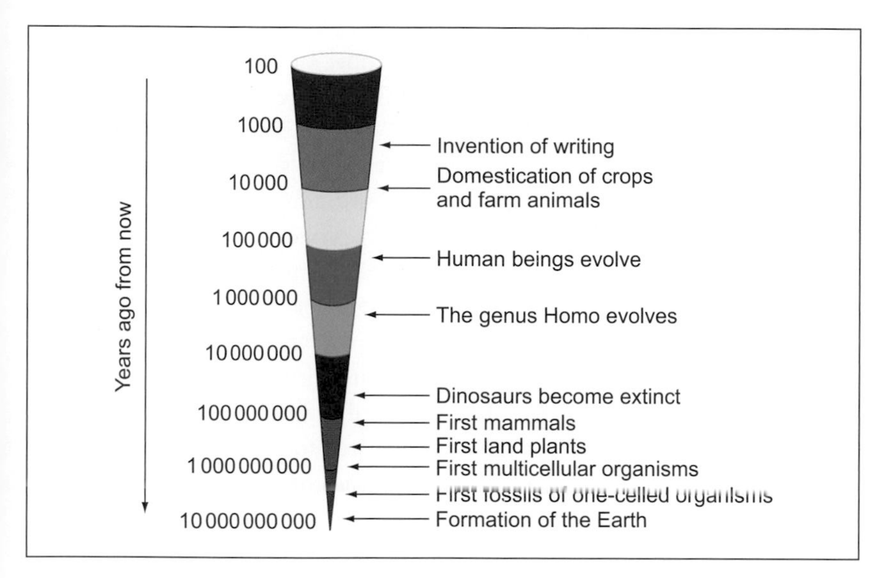

Figure 9.12

A time chart showing the main events in the history of life on Earth. Note that each unit on the scale is 10 times larger than the one before. This is a logarithmic scale, often used when the range of the scale is very large.

Within 10 years of the publication of the first edition of *The Origin of Species*, Darwin was shaken by the pronouncements of the physicist William Thomson (later Lord Kelvin). Thomson calculated that the Earth was only 100 million years old. Darwin was aware that 100 million years was too short either for evolution to occur or for all the thousands of feet of sediments visible in the fossil record to have been laid down. Nevertheless, Darwin spent the winter of 1868–69 making lots of alterations for the next edition of *The Origin of Species*, to fit with Thomson's calculation.

Decades later, long after Darwin had died, physicists realised that while Thomson's mathematics was correct, his model based on cooling of the Earth was faulty, making his calculations incorrect. The actual age of the Earth – 4.5 billion years – gave plenty of time for evolution to have taken place. Darwin needn't have worried about Thompson's estimate of the age of the Earth.

Reactions to the theory of evolution

In 1838 Darwin worked out that his theory of natural selection could explain evolution. Yet it was 20 years later that he published his ideas, after receiving the fateful letter from Wallace. Why did Darwin delay so long? Perhaps he was simply being cautious, preferring to gather enough evidence to support his theory. Certainly, the sheer size of his 500-page book *The Origin of Species*, which he described as a 'sketch' of his theory, along with the weight of evidence it revealed, helped his work to gain acceptance. By 1859 he was a well-known scientist with an excellent scientific reputation. Many of his published ideas had already withstood critical scrutiny.

Questions

7 Why was Darwin worried by the idea that the Earth might be much younger than he had thought?

8 a) Despite the problems posed by Thomson's estimate of the age of the Earth, Darwin did not abandon natural selection as an explanation for evolution. Suggest some reasons why scientists may not abandon a theory because of such a problem.
b) Darwin did, however, modify his explanation to include some elements of Lamarck's theory. Suggest why he did this.

Figure 9.13

Once Darwin's theory of evolution by natural selection was published, it was widely read and excited great controversy. This cartoon, published in the London Sketch Book *in 1874, shows an ape-like Darwin holding a mirror up to an ape to show how alike they are.*

Possibly Darwin's reticence was due to fear of public reaction (Figure 9.13). It is difficult for us, nowadays, to imagine how revolutionary his ideas sounded to Victorian times. At that time most people in England accepted the literal truth of the Christian and Jewish scriptures. These imply that the Earth could not be more than about 6000 years old. Yet Darwin's theory needed the Earth to be many hundreds of millions of years old.

Perhaps more importantly, the Bible tells how all humans are descended from Adam and Eve, who were created by God in the image of God out of the dust of the Earth. This conflicts with the Darwinian vision, which has all of us descended back through primitive mammals, through invertebrates looking rather like present-day earthworms, through single-celled creatures barely visible without a microscope, back to a primordial soup.

Interestingly, some of the theologians of Darwin's day were quicker to accept his views than were many scientists. Nevertheless, some of the clergy were vehemently opposed to Darwin's views. They felt that the theory of evolution was questioning their authority and treading on their area of expertise or, if evolution had occurred, that there could be no valid base to morality and people would be free to behave 'like animals'.

To this day, the theory of evolution remains controversial in some places (Figure 9.14). In 1999 the Oklahoma State Textbook Committee in the USA produced a statement which defined evolution as 'unproven' because 'no one was present when life first appeared on Earth'. The statement went on to say that 'evolution should be considered a theory, not fact'. In the same year the Kansas State Board of Education voted to remove evolution from its science curriculum.

In 2007 the Kansas State Board of Education finally deleted a phrase from their teaching guidelines that challenged the validity of evolutionary theory, after international ridicule and pressure from moderate voters. The redrafted guidelines no longer suggest that it is controversial to the scientific community to suggest that all life on Earth comes from a common ancestor, or that species are able to change into new species over time. Interestingly, the board also redefined science as 'the search for rational explanations of what occurs in the universe'.

A human need to make sense of the organisation of living organisms on Earth is seen right back in ancient writings. The Ancient Greek philosopher, Aristotle, described a form of 'intelligent design', suggesting all species could be placed in a hierarchical chain, from worms on the bottom to humans at the top (followed by angels). In Aristotle's view, the Great Chain of Being was perfect; there were no empty links in the chain, and no link was represented by more than one species.

Figure 9.14

Young Earth Creationists believe that the Earth is at most 10 000 years old and that God created all species in much the form that they exist today.

Later, the philosopher William Paley described an 'intelligent designing mind' in his book *Natural Theology*, published in 1802. Paley's most famous analogy compares a watch with works of nature. Paley argues that if he kicked a stone on the heath and wondered where it had come from, he might answer that it had been there for ever. Suppose he had found a *watch* on the ground, and wondered how the watch had got there. The answer would not be the same as for the stone – that it had been there for ever. He would be able to tell that the watch was made from many parts, designed and constructed for a purpose.

Paley used the watch analogy to reason that the existence of order in design implies that there must be a designer. It then follows, according to Paley, that the universe must have a designer. A few years later, Darwin's theory of natural selection challenged these ideas; natural selection explains evolution with no need for a designer.

There is still a cultural argument about the origins of life on Earth. On the one hand there are people who believe in 'creation'. They believe that an 'intelligent designer' (God) has had a role in determining the course of evolution. On the other hand there are those who see scientific theories and other interpretations of evidence as distinct from religious belief. Religious ideas about how living organisms originated may be 'based on evidence', as they are alternative interpretations of the same sort of observations that Darwin made. However, they are not science, because they can't lead to predictions that could be falsified through testing. Religious interpretations are usually based on an authority such as religious texts, and are not susceptible to change in the face of new evidence.

Questions

9 a) Explain the difference between a religious belief and a scientific theory.
b) Use your answer to explain why evolution is a scientific theory and creationism and intelligent design are not.

10 Do you feel that more scientific evidence for evolution would help creationists accept that it had occurred?

11 Discuss whether creationist views can be considered 'rational'.

The genetic basis for evolution

Gregor Mendel was born in 1822 (Figure 9.15). He entered a Moravian monastery in what is now the Czech Republic. The monks were encouraged to study – Mendel went to Vienna University, where he became interested in plant breeding. Over the course of a decade, from 1856 to 1866, Mendel bred, examined, described and counted over 28 000 garden peas. He discovered that the various characteristics of the peas – such as their height and the colour and shape of their seeds – were determined by 'factors' or particles that passed unchanged down the generations and did not blend. He also appreciated that each adult pea plant has each factor in pairs whereas the pollen and egg cells have only one copy of each factor. We now know Mendel's factors as genes. Mendel's explanation of inheritance was right; in most species there is just a single copy of each gene in the sex cells (sperm, pollen and eggs) whereas all other cells have two copies of each gene (see Chapter 6).

Figure 9.15

Gregor Mendel, the founder of modern genetics.

Unfortunately Darwin, along with many other scientists at the time, was probably unaware of Mendel's work, and certainly failed to appreciate its significance.

Mendel's work was rediscovered, and its significance realised, in 1899, 33 years after he had first published it. Soon the hunt was on to understand more about genes. Much of the work in the early decades of the twentieth century was done on organisms that breed much more rapidly than we do – such as maize, mice and tiny fruit flies. Scientists soon appreciated that Darwin had been right when he saw variation between individuals as lying at the heart of evolution. We now know that mutations are the source of this variation.

Mutations are changes in the structure of an organism's genetic material. They can be caused by spontaneous errors when DNA is copied during cell division, by certain chemicals, including some found in cigarette smoke, and by ionising radiation (X-rays and gamma rays; see Chapter 7). If a mutated gene happens to be in a sex cell, copies of it can be handed on to offspring. It is only these mutations that occur in the sex cells that have a role in natural selection. Most mutations are probably

The development of ideas about inheritance

Figure 9.16

Until the nineteenth century, people were unaware of the detail of how babies were conceived and how heredity worked. The Ancient Greek philosophers speculated that small portions of each body part were transmitted into sperm. During the 1600s, the Dutch mathematician Nicolas Hartsoeker thought he had discovered miniature 'animalcules' in the sperm of humans and other animals.

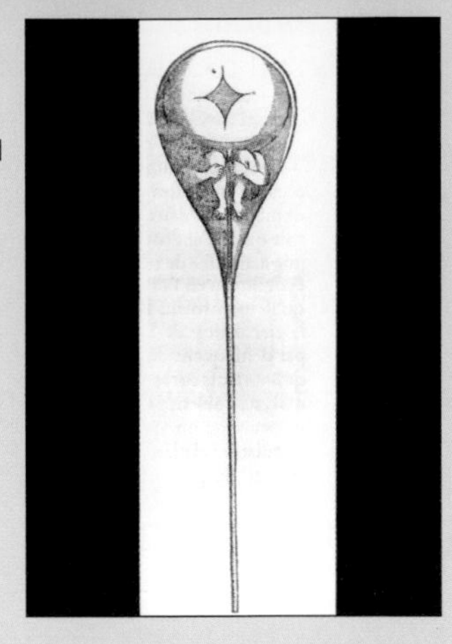

Hartsoeker's Homunculus.

Scientists in the nineteenth century realised, from evidence from plant breeding, that both parents contribute to the characteristics of offspring. They thought hereditary materials from male and female parents blend to form the offspring, and were then inseparable. But this theory did not fit with the observation that some characteristics are absent in one generation, then present in the next.

Modern genetics began in the 1860s, when Gregor Mendel discovered the fundamental principles of inheritance. He realised that the 'factors' of inheritance do not blend together, but are inherited independently.

Nicolas Hartsoeker, along with other scientists of his time, 'saw' little humans inside each sperm. This led Hartsoeker to formulate the 'spermist' theory of conception. The only contribution of the female in this process was to provide a womb in which the homunculus grew.

neutral, having no effect at all on the characteristics of an organism. Sometimes, though, they are harmful. Very occasionally they may be beneficial, but whether a mutation is beneficial or not depends on the environment.

Examples we have of beneficial mutations include human skin colour (Figure 9.17), and, more recently, mutations that allow organisms to survive better as a result of human changes to the environment. For example, mutations have enabled many disease-causing bacteria to become resistant to various antibiotics. People who don't accept the theory of evolution conclude that there is no evidence that mutations ever help organisms in the natural environment. Examples such as heavy metal tolerance in plants demonstrate that organisms do evolve to be able to tolerate conditions that their ancestors could not survive (Figure 9.18). Evolutionary biologists often use examples of evolution due to human activities, but this is because most of the recent changes to the environment have resulted from human actions.

The theory of evolution by natural selection

The fundamental principle behind the theory of evolution is that all the Earth's present-day life forms have evolved from common ancestors, and ultimately from self-replicating molecules which happened to develop under the conditions then prevailing on the Earth.

Questions

12 Mendel observed and counted the characteristics of peas. He then proposed a model of inheritance to explain his findings. Describe the model proposed and explain the differences between data, which must be observed and reported, and a model, which may involve imagining things that cannot be observed directly.

13 Why do scientists use fast-breeding organisms to study evolution?

14 Why does the general public nowadays accept the gene theory of inheritance when the leading scientists of Mendel's time failed to appreciate the significance of what he had found out?

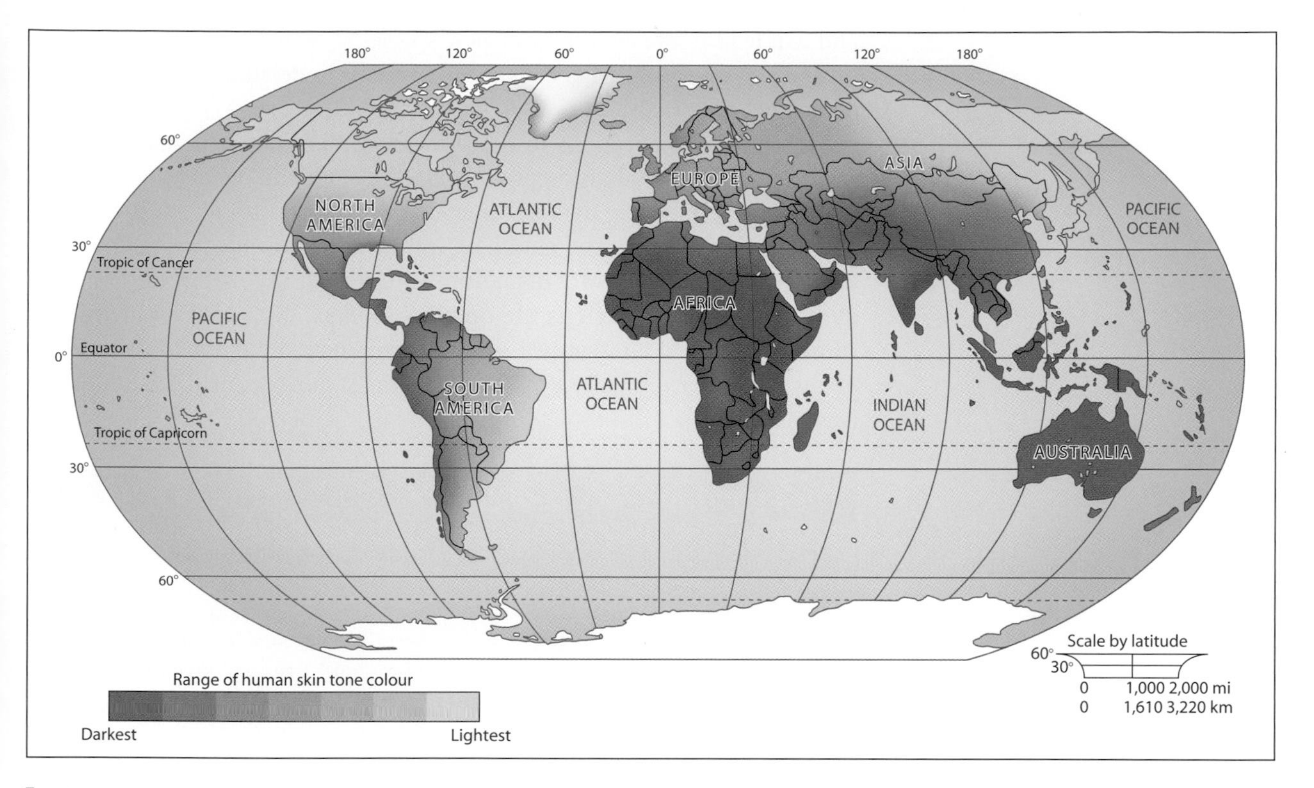

Figure 9.17

This map shows the range of human skin colours as they are thought to have evolved by natural selection, led by climate. The map is an interpretation of information from various sources, including drawings. Regions where there is a lot of variation in skin colour are difficult to represent on a map like this.

Figure 9.18

Studies of the formation of new species by natural selection include case studies of closely related species living in different environmental conditions. It is known that some plant species have evolved tolerance to heavy metals over short periods of time. Metal contamination can be due to natural causes or to human activity. Under conditions of low metal contamination, grasses such as Agrotis tenuis *(colonial bentgrass) show a low frequency of heavy metal tolerant individuals. In contaminated areas, such as around mines, the frequency of metal tolerant individuals is high. The frequency of genes in a population which shows this type of variation can change by natural selection over just one generation. The change is led by environmental pressure.*

The mechanism which best explains evolution is natural selection. Chance mutations are the source of changes in individuals. Within a species, some individuals are better able to survive and reproduce in the environment in which they find themselves than are others. Over time, changes to the genetic make-up of individuals in a species accumulate, and can lead to the evolution of new species. The enormous length of time required for this is consistent with our understanding of the geological time scale.

Among the implications of the Darwin–Wallace theory of natural selection is the idea that the process is led by random events (mutations), and does not have any overall direction or aim. As a result, the path which evolution has taken could, with minor changes in circumstance, have been quite different. This point is interesting in relation to the question of whether there is life on other planets; even with an identical starting point and an identical set of environmental factors, life would almost certainly not have evolved along the same lines as it has on Earth (see Chapter 11).

The importance of classification

Being able to identify and classify species accurately is essential for the study of the relationships between species. Biology in the eighteenth and nineteenth centuries consisted largely of collecting, describing and classifying living organisms, mainly on the basis of their external appearance and structure. Naturalists did not look for explanations for the patterns of similarities between organisms because they believed that all organisms had been created in their present form. Current classification systems in contrast are based on theories about evolutionary relationships between organisms. New discoveries can cause classifications to change.

In addition to advancing our understanding of evolution, precise systems for classification allow scientists to recognise new species, to map biodiversity and track extinctions over time. Understanding the relationship between species also helps humans to exploit organisms; a useful chemical found in small quantities in a rare plant may be found in more useful concentrations in a closely related species.

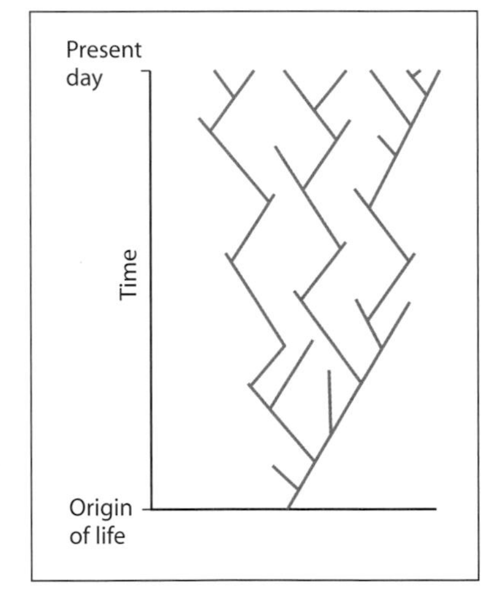

Figure 9.19

Evolutionary biologists think that all of today's life shared a single, common ancestor. The vertical axis shows time, with most recent time at the top. The horizontal axis has no scale but is an indication of variation between species. The more lines there are on the diagram, the more species there were at the time. A point where two lines diverge from a single point indicates a time when one species split into two. A line that suddenly stops before today indicates that the species became extinct.

Evolutionary biologists believe that all the species which now exist, and all those which have existed in the past, can be linked in a single 'branching tree' structure (Figure 9.19). Many of the branches of this tree are 'dead ends' because most of the species which existed at various times in the past have become extinct.

The popular view of life as progress, with humans at the top of the evolutionary tree, is misleading, and certainly Darwin did not believe in this ranking of species. We may be more intelligent than other species but we lack the vision of an eagle, the hearing of a dog, the speed of a swift and the hardiness of many insects. In evolution, there is no upward inevitable ascent towards perfection. Instead, species merely respond to the environmental pressures they face, for example changes in climate or food supply. Humans have lost most of their ability to smell or to climb trees but they have gained the ability to imagine what life is like for other creatures, and they have a far richer system of communication than any other species.

A person's ideas about animal rights and our responsibility to other species may reflect their views on the relationship between species. A view that sees humans at the top of an evolutionary tree can still acknowledge duties towards 'lesser' species, but as 'managers' of their protection and conservation. Other people see the role of non-human species as serving human needs (e.g., for food, natural chemicals, transport and leisure).

This human-centred view of the world contrasts with a more holistic approach which sees humans as one part of a whole, interdependent system of species interacting with their physical habitats.

Environmental pressure

The theory of natural selection relies on the existence of variation among individuals in a population, but also on environmental pressure. If organisms evolve to be perfectly adapted to a particular environment, any new mutation is unlikely to give them a survival advantage.

Competition for resources is just one type of environmental pressure. Competition may be between members of the same species or between members of different species which need the same resource. The end result of competition and natural selection is often extinction, as new species evolve and eliminate their relatives. Ecosystems are often able to withstand minor disruptions that are due to extinctions and environmental changes, but the knock-on effect of a single species becoming extinct is difficult to predict due to the complex interdependence of species.

Does natural selection explain all of evolution?

Generally, scientists are more confident about theories which provide a good explanation for what is being observed. It is also important that new theories are consistent with existing well-accepted theories in related areas of science, and are supported by evidence. Darwin produced a new way of explaining the diversity of living organisms on Earth, and scientists recognised it as a plausible mechanism for evolution. The theory was also consistent with new ideas around at the time about geology, populations and competition for resources.

Most biologists today accept that natural selection is the prime mover behind evolution. However, as more evidence from molecular biology becomes available and our understanding of genetics improves, it is possible to describe the mechanisms of evolution in more detail and to describe other mechanisms important in evolution, building on Darwin's theory of natural selection (Figures 9.19 and 9.20).

Modern evidence for evolution uses changes in DNA and proteins in cells as evidence for natural selection, and this can be demonstrated in much shorter time periods than fossil evidence takes to accumulate. Chapter 2 described how antibiotics apply environmental pressure for bacteria to evolve by natural selection into new, antibiotic-resistant strains over a period of days.

Much of the modern debate about the mechanisms of evolution is around the unit of selection – whether this is genes, cells, individuals or whole species. Richard Dawkins has championed the idea of the selfish gene, arguing that the gene is the only unit

Key term

Interdependence of species describes how all living organisms in an ecosystem rely on each other; for example for food, shelter, pollination or distribution of seeds.

An **ecosystem** is an interacting unit consisting of all the living organisms in a particular environment together with the non-living components such as water, soil and oxygen.

Questions

15 Explain why climate change might result in more rapid evolution of species in some regions.

16 For what resources, apart from food, are organisms in an ecosystem likely to be competing?

17 Why might the extinction of a single species have little effect in some ecosystems, but result in the complete destruction of an ecosystem in others?

Figure 9.20

A chimpanzee drinking from a handful of leaves used as a sponge. Chimpanzees are probably our closest evolutionary relatives. We share about 99% of our evolutionary history and about 98.5% of our genes with them.

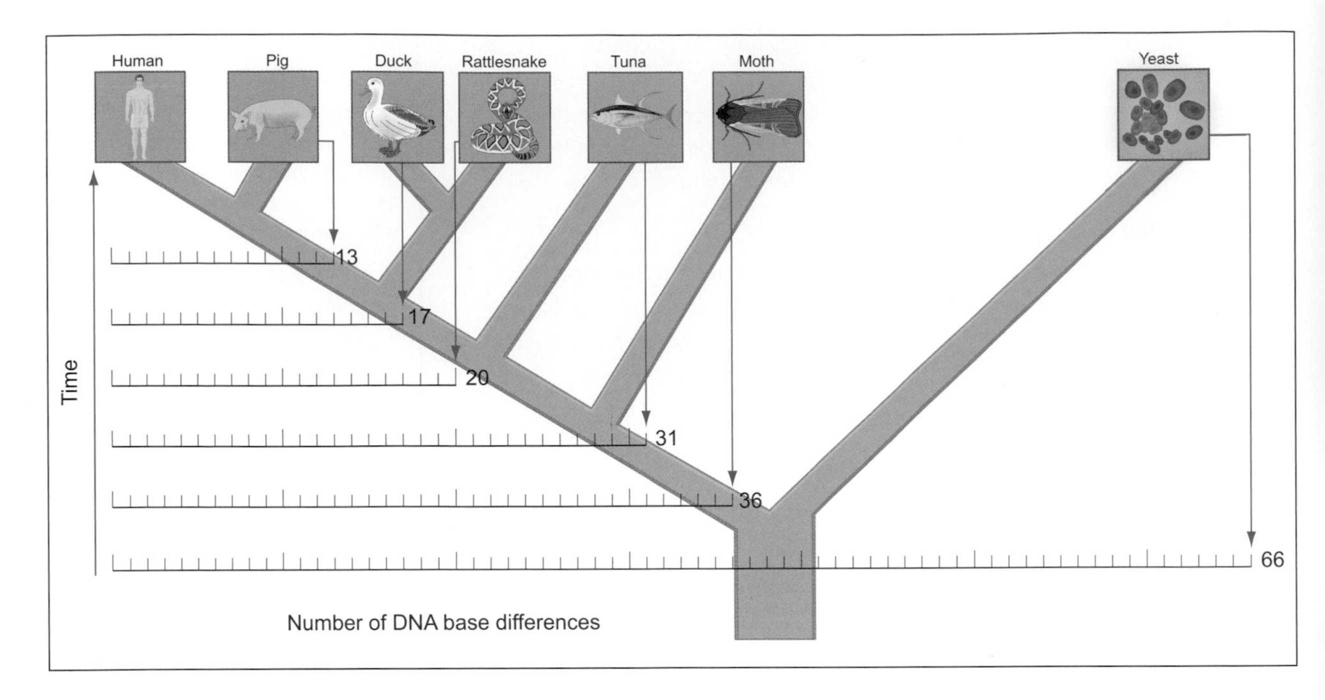

Figure 9.21

Biologists can construct evolutionary trees based on the similarities between organisms. This evolutionary tree is based on molecular biology; it shows the number of base differences in the DNA structure of one molecule – the enzyme cytochrome C – which is found in almost all living organisms. The fossil record supports the idea that the greater the number of differences between two species, the longer ago they separated from one another. Evolutionary trees are increasingly being constructed on the basis of similarities and differences between the DNA of organisms.

that natural selection works on. Consider, for example, the way in which a worker honeybee never breeds and may even lose her life defending her hive. Clearly she herself benefits neither from her sterility nor from her helping tendency to protect the hive, but her genes live on via the other members of the hive.

Origins of life

There are various theories about how life originated on Earth, but most involve the idea of molecules evolving into self-replicating lifeforms. Natural selection could explain the origin of life if it is accepted that molecules can be the units of selection. 'Survival of the fittest' in this case may have involved the ability to make cross-links with other molecules, the ability to self-catalyse and the ability to replicate.

Review Questions

18 Sharks have evolved very little over millions of years, whereas disease-causing bacteria are capable of producing new strains within days if treated with antibiotics. Use the theory of natural selection to suggest reasons for the difference.

19 Explain the importance of the deep water between the Galapagos Islands in allowing tortoise populations to evolve into new species.

20 Identify features of Darwin's theory of evolution by natural selection which required conjecture and creative imagination.

21 Organisms in an ecosystem can be shown to be interdependent through food webs and other interactions. They also need to show some degree of specialisation, to avoid competing for exactly the same resources as other organisms living in the same area.

Use the ideas of competition, avoiding competition and natural selection to explain the diversity of life on Earth using any one ecosystem as an example.

22 Is it possible to believe in evolution but not believe in natural selection? Explain the difference between the two.

23 The Tulsa Interfaith Alliance called a press conference to announce its opposition to a disclaimer that was required to appear in science textbooks. The disclaimer challenged the theory of evolution and asserted the creationist point of view. A heated debate occurred during the news conference between supporters and opponents of the disclaimer. One conservative political and Christian activist stated, 'All scientists do not agree on evolution. This disclaimer is responsive to what the majority of Oklahomans believe.' To this a member of the Alliance responded 'Science isn't done by vote.'

a) Comment on the statement 'Science isn't done by vote', and explain how science *is* done.

b) Explain how the evidence for evolution is different from evidence available to support most other scientific theories.

24 Read the descriptions below, and decide whether the groups described belong to the same species.

a) Group A has 98.3% of its DNA in common with group B.

b) Mules (group C) are the offspring of male donkeys (group D) and female horses (group E). They are normally sterile.

c) Group F is darker in colour than group G. The genetic fingerprints of groups F and G are identical.

25 The eighteenth century botanist Linnaeus carried out the description of living organisms to demonstrate the unchanging order of biblical creation. During his life, Linnaeus realised, through breeding experiments, that new types of plants and animals arose.

a) Suggest why Linnaeus stopped short of accepting that one species could evolve into another.

b) Explain how current ideas about classification have a less static view than that of Linnaeus.

c) Explain why it is useful for biologists to be able to identify and classify living organisms.

Chapter 10

The universe

Figure 10.1

Stonehenge has an astronomical significance, perhaps like an ancient observatory. There are alignments of some stones with the rising Sun, while other stones may be aligned with certain positions of the Moon.

The issues

From earliest times, people have asked questions about the universe. The Sun, Moon and stars are a source of wonder as well as curiosity – they have inspired stories, poetry and music. Our ideas about the universe are important because they profoundly affect the way we see ourselves (Figure 10.1). As our understanding of the universe has grown, the Earth has been steadily pushed away from the centre of things. We have also had to come to terms with the evidence that the universe is vast, with almost unimaginable distances between the stars and the galaxies, and that it is an expanding universe.

The science behind the issues

Common sense tells us that we are living on a large and fairly flat surface, which is fixed and stationary. Science tells a different story: we live on the surface of a sphere which orbits our Sun while also spinning on its axis. The Sun is a star, and the Earth is one of the planets of the solar system. The nearest star to our Sun is over 4 light years away – about 1000 times further from the Sun than Neptune, the outermost planet. Most stars are much further away still. The Sun is towards the outer edge of a spiral galaxy, the Milky Way. This, in turn, is one of millions of galaxies which make up the universe. The distant galaxies are moving away from us at high speeds. Scientists think that the universe began in a 'big bang' around 15 billion years ago.

Figure 10.2

The domes of the Keck observatories which house optical telescopes high up on Mauna Kea in Hawaii. Each telescope has a large mirror with a diameter of 10 metres, made up of 36 smaller hexagonal mirrors.

What this tells us about science and society

The growth of our understanding of the structure and size of the universe depends on accurate measurements and observations – and imagination in devising possible explanations for what we see (Figure 10.2). Patterns and explanations do not simply 'emerge' from the data. A close match between what our explanations predict and what we actually observe is important in identifying fruitful theories. However, astronomers are often working at the limits of their instruments, so many measurements are estimates which may have substantial errors. Some advances were only possible when better telescopes and new techniques like photography and spectroscopy became available. As a result, there have been debates about the data, and how to interpret them, and estimates of distance have often had to be revised.

Ancient astronomy

The idea that the Earth is flat has never been an accepted scientific theory. The Ancient Greek astronomers knew that the Earth was a sphere (Figure 10.3). They saw that the Earth's shadow, during an eclipse of the Moon, was always circular. They also noticed that the north pole star was higher in the sky the further north you went. And they observed that, as ships sail out to sea, the hull disappears below the horizon first, and the masts and sails later.

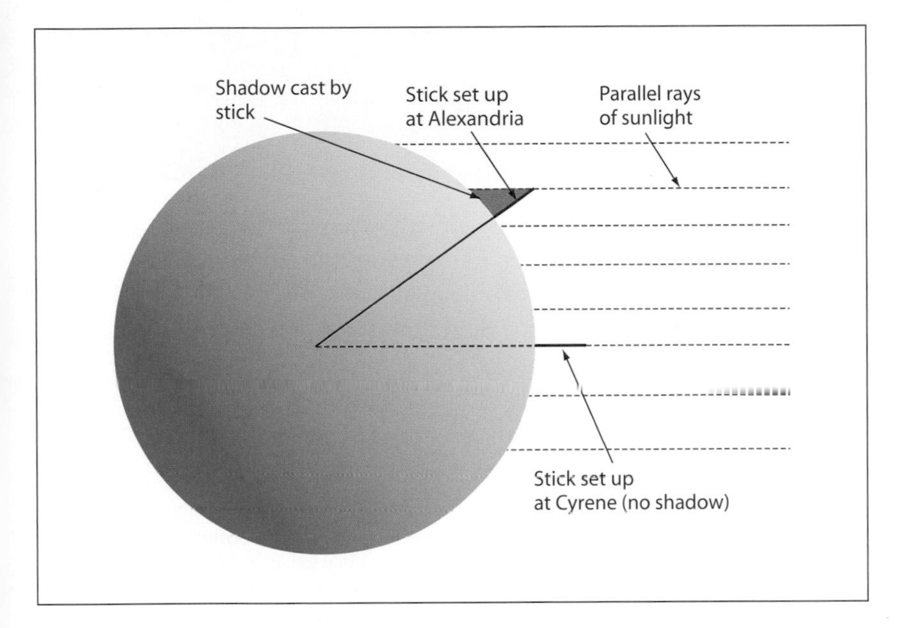

Figure 10.3

Around 200 BCE, Eratosthenes estimated the diameter of the Earth, by measuring the length of the shadow of a vertical post at noon on midsummer day at his home in Alexandria. He knew that a similar post in the town of Cyrene, 800 km away to the south, cast no shadow at that moment, because the sun was directly overhead. The difference is due to the curvature of the Earth – and from his measurement he calculated a value for the Earth's diameter which is remarkably close to the value of 12 756 km that we accept today.

Everything we see in the sky seems to move round us daily. The Sun, the Moon and the stars all follow circular paths across the sky (Figure 10.4). The stars move together, so their patterns (the constellations) remain the same. A few objects in the night sky, however, appear to move among the stars – not fast enough to see them move, but quite clear if you look at them over a series of nights. The Ancient

Figure 10.4

A long-exposure photo of the night sky. The stars appear to move in circular paths, around the north pole star.

Greek astronomers called these objects 'wanderers'. We call them planets. We can see five with the naked eye: Mercury, Venus, Mars, Jupiter and Saturn.

Most of the time, a planet moves steadily along a fairly straight line, relative to the background of stars. But, from time to time, it appears to move backwards, and retrace its steps (retrograde motion) (Figure 10.5).

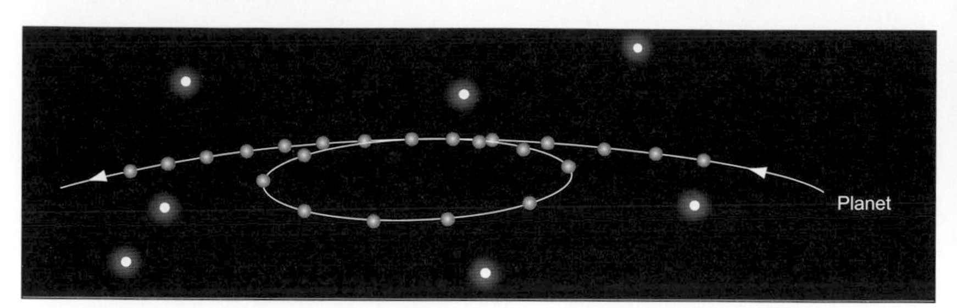

Figure 10.5

Planets move slowly against the background of stars. Occasionally a planet's path changes direction and for a time it moves backwards relative to the stars.

All of these observations were known to the Ancient Greek astronomers. Around 500 BCE, the followers of Pythagoras suggested that the Earth was at the centre of a set of crystalline spheres which moved round, carrying the stars, Sun, Moon and planets with them. This explains the basic motions we observe, but doesn't account for the details, especially the retrograde motion of the planets. Around 370 BCE, however, Eudoxus came up with a way of explaining this (Figure 10.6).

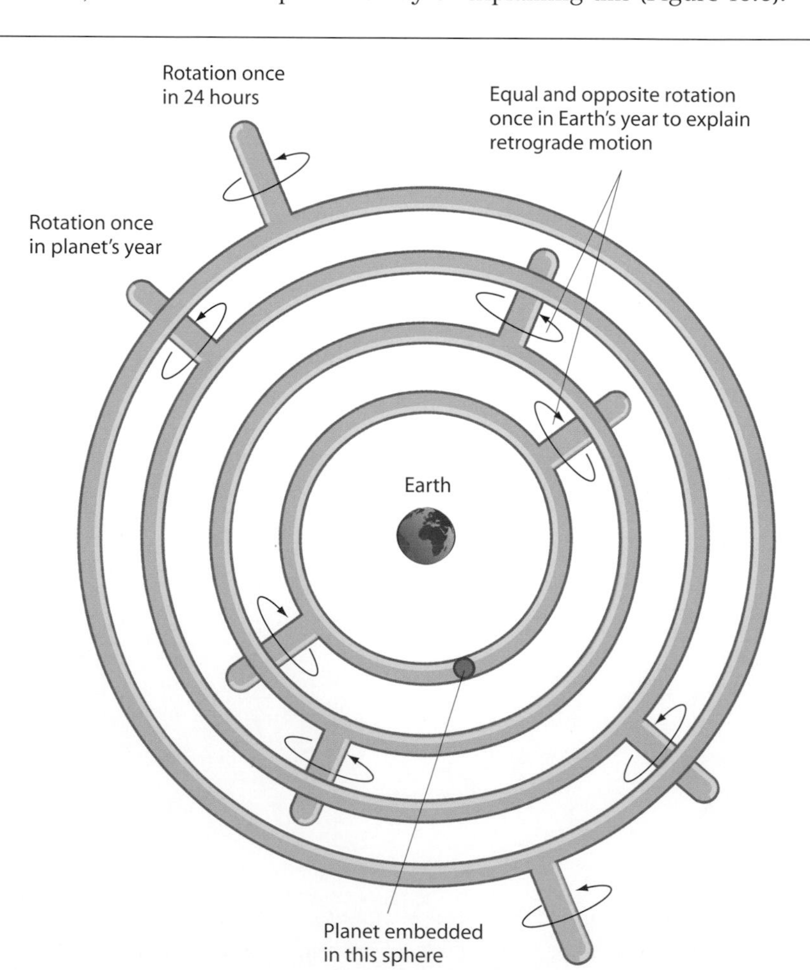

Figure 10.6

Eudoxus's model has sets of invisible spheres, mounted one inside the other. Each sphere rotates independently, and together this produces the motions we observe.

Questions

1 What evidence would you use to convince someone:
a) that the Earth is a sphere?
b) that the Earth goes round the Sun (and not the Sun round the Earth)?

2 What observation(s) suggest to us that the planets are not stars?

3 Aristotle thought that the objects we see in the sky are made of a different kind of matter from that on Earth. What might he have seen as evidence of this?

Aristotle, who lived from 384 to 322 BCE, was the most famous philosopher of his day – and one of the most influential ever. Starting from the ideas of Eudoxus, he developed a model with 55 spheres. In his model, the Earth was fixed and at the centre. All the heavenly bodies were carried round on invisible crystalline spheres, as the sphere was regarded as the 'perfect' shape. The heavens, according to Aristotle, are made of a different kind of matter from that on Earth. They are perfect and unchanging whereas, on Earth, there is imperfection and change.

Even with 55 spheres, this model still could not predict the exact positions of the planets. Some astronomers felt there must be a simpler explanation. Around 230 BCE, Apollonius of Perga suggested that each planet moves in a small circle, while the centre of this moves round a larger circle centred on the Earth (Figure 10.7). The motion of the planet would then be like that of a person on a fairground 'waltzer' – in a car that is spinning around its own centre point, whilst being carried round on a larger rotating platform.

Key terms

Aristotle's model is a **geocentric** one – the Earth is at the centre.

A **heliocentric** model is one which puts the Sun at the centre.

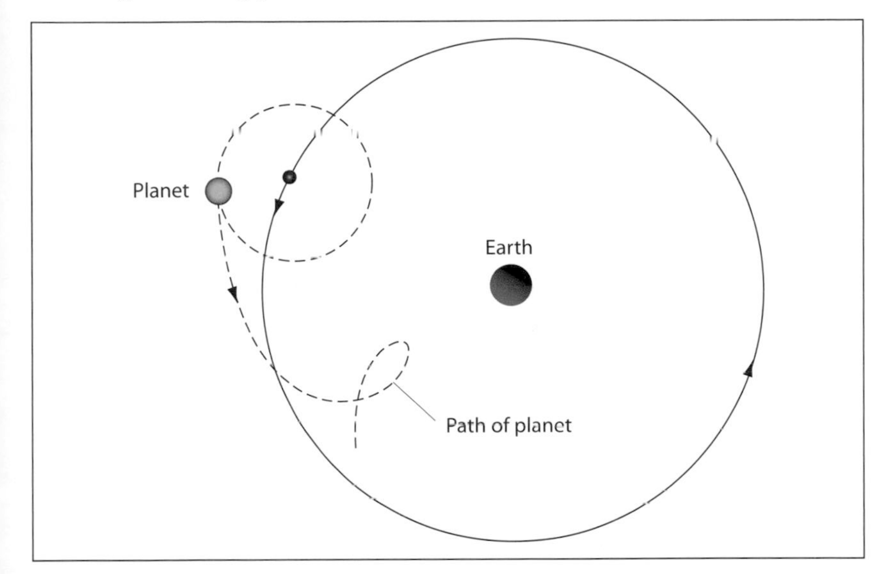

Figure 10.7

In Apollonius's model, which was later developed by Ptolemy, each planet moves round a small circle, whose centre circles the Earth. The combination of these two motions produces the effects we observe – including retrograde motion. (Note that in this diagram the small circle is drawn much larger than it should be, relative to the size of the larger circle, in order to show the model more clearly.)

Apollonius's idea was picked up and developed by Claudius Ptolemy, who lived from 100 to 170 AD. He made some modifications (like having some circles a little off-centre and allowing the planet's speed to change as it moved round its small circle) and added some extra circles. In a famous book, the *Almagest*, he described this very complex model. It had become so complex that most people no longer believed that these spheres and circles really existed – they were just a way of calculating the positions of the planets. It was said to 'save the appearances'. But it worked – and remarkably well. For over 1000 years, it was the most accurate way of predicting the positions of the planets . Although we now have a better way to look at these motions, Ptolemy's model was a remarkable achievement.

Question

4 Explain how the model shown in Figure 10.7 accounts for the fact that a planet usually moves forwards against the background of stars, but sometimes for short periods appears to move backwards.

The Copernican revolution

A new model

The ideas of Aristotle, and Ptolemy's model of the universe, were widely accepted and became very influential. Around 1260 AD, Thomas Aquinas, a Christian monk, developed a philosophy which combined these ideas with Christian teaching. This became the accepted viewpoint of the Church. He argued that the heavens were the realm of God, and hence perfect. It was God who kept the spheres in Ptolemy's model moving.

However, astronomers were becoming increasingly aware that Ptolemy's model did not lead to completely accurate predictions. Also it seemed impossible to explain such complex motions: what could possibly make things move in this way? Early in the sixteenth century, Nicolaus Copernicus (a Polish mathematician and Catholic priest) wrote a book which started a revolution in thinking about the universe. He suggested that the motions of the stars and planets would look much simpler if we imagined the Sun at the centre, with the Earth and the other planets circling around it (a heliocentric model) (Figure 10.8). In this model, the Earth is a planet. It spins on its axis, which causes day and night and explains why we see all the celestial bodies moving across the sky (Figure 10.4).

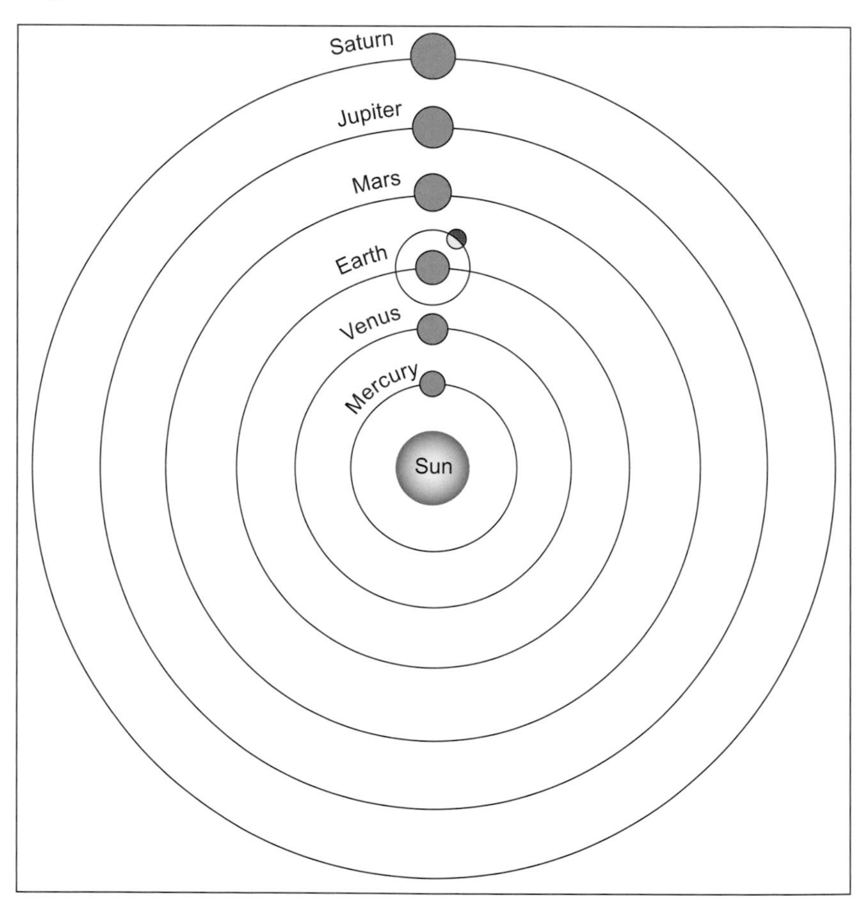

Figure 10.8

Copernicus's model of the solar system. This is a simplified version of the diagram in his book De Revolutionibus.

In Copernicus's model, the retrograde motion of the planets (Figure 10.9) is due to the fact that the planets closer to the Sun are moving faster than those further away. As the Earth overtakes Mars, for example, Mars will appear to move backwards against the background of the more distant stars.

Publication

Copernicus thought long and hard before publishing his ideas. He started to write his book (*De Revolutionibus Orbium Coelestium*, or *On the Revolutions of the Heavenly Spheres*) in 1509, but it was only published shortly after his death in 1543. He worried that his ideas would be ridiculed by other astronomers. So he spent many years checking his model against observation, and improving it. In the process he made it much more complicated than Figure 10.8. In fact he added so many extra circles and other modifications that it ended up almost as complicated as Ptolemy's.

Copernicus may also have been concerned about opposition from the Church. He was living in troubled times. It was shortly after the Protestant Reformation, initiated by Martin Luther, and the Catholic Church was actively trying to suppress ideas that might challenge its authority. When Copernicus's book was eventually published after his death, it contained a preface written by a colleague arguing that the model was just a simpler way of calculating the planets' positions and it was not a description of how things really are. Copernicus, however, seems to have believed that the Earth and the other planets really do move round the Sun.

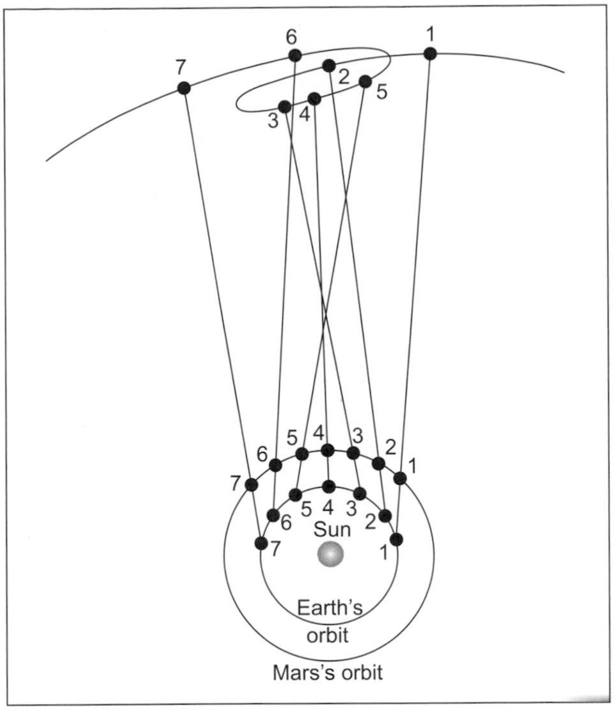

Figure 10.9

The retrograde motion of planets is due to the fact that they orbit at different speeds. So as the Earth overtakes another planet, the position of the other planet against the background of stars changes and it appears to loop backwards.

The response to Copernicus

From our perspective today, it is easy to think that Copernicus's ideas were 'obviously' correct. At the time, however, it was far from obvious. There were some persuasive arguments in favour of a geocentric model, including one (the final one in Figure 10.10) to which the Copernicans did not have a very convincing response.

Figure 10.10

Arguments for and against a geocentric model

Arguments in favour of the geocentric model...	*How a Copernican might respond...*
The Earth feels as though it is fixed, not flying through space and rotating on its axis.	We are moving along with it. If you are in a moving vehicle, you often cannot tell that you are moving.
If the Earth rotates once per day, this means that the whole surface is moving at a high speed. So why do we not feel a strong wind, as the Earth's surface rushes along?	The Earth's atmosphere moves along with it – so we do not experience a wind.
If we throw something straight up, it comes back to our hand. Surely if we were rotating at high speed, it would land a long way away, because we would have moved while it was in the air.	This is based on a common misconception. If you throw a ball upwards while you are running along, the ball does come back to your hand (try it and see...). A projectile released from a moving object has the same forward motion as that object had at the moment of release.
If we observe a star now, and then again in six months' time (when the Earth is at the other side of its orbit), it should be in a different direction (this is called parallax). But we do not observe this.	We don't observe parallax because the stars are very, very far away. In fact, there is a change of direction, but it is very, very small. (It was first observed in 1838.)

Questions

5 What were Copernicus's two main worries about the reception of his ideas?

6 Suggest some reasons why some people in the Church might have opposed a heliocentric model of the solar system.

7 In Copernicus's day, what was, in your view, the strongest argument for the heliocentric model? What was the strongest argument against it? If you had been alive in 1543, do you think you would have accepted the heliocentric model or the geocentric one? Give your reasons.

8 What two observations did Tycho make that challenged the accepted theories of the time? Why were they a challenge?

9 Tycho's evidence did not immediately lead all astronomers to reject Ptolemy's model. Suggest some reasons why.

Copernicus's book did not cause an immediate stir. In fact, few people appeared to notice it, perhaps because it was complicated and technical, and the new ideas it contained were not immediately obvious. It was not immediately banned by the Church. But in the years that followed, social and political events led to a hardening of attitudes, and a greater fear of unorthodox ideas. One major factor was the Counter Reformation – the response of the Roman Catholic Church to the Protestant Reformation. Between 1545 and 1563, a great Council met at Trent in northern Italy. It reaffirmed traditional doctrine, founded the Jesuit order to oppose the reformers, made a list (or Index) of forbidden books, and set up the Inquisition to suppress unorthodox views. This was to have an important influence on the events which followed.

Convincing others

Three people played key roles in getting the heliocentric model generally accepted: Tycho Brahe, Galileo Galilei and Johannes Kepler.

Tycho Brahe: the great observer

In 1559 the 13-year-old Tycho Brahe decided to become an astronomer when he observed a partial solar eclipse. It was not the eclipse that impressed him but the fact that astronomers had been able to predict it. The son of a wealthy Danish family, Tycho was able to pursue his interest in astronomy. On the night of 24th August 1563, something happened that was to shape the rest of his career. There was a conjunction of Jupiter and Saturn – that is, the two planets appeared so close together in the sky that they seemed to be a single bright object. Like the eclipse, this had been predicted – but the prediction was several days out.

This was not good enough for Tycho. It convinced him that Ptolemy's model had to be improved. And for that, he would need more accurate data. By the age of 26 he had set up the greatest observatory in Europe, with instruments provided by the Danish king. Chief among these was a huge quadrant (Figure 10.10), with which he could measure the positions of stars and planets to an accuracy of one minute of arc (1/60°). Over 20 years, Tycho built up a huge record of the positions of all the planets that was remarkably accurate by the standards of the time.

Tycho himself could not accept the heliocentric model. He seems to have thought that the idea of the Earth moving was simply inconceivable. He also knew that the stars showed no parallax shift – even with his excellent instruments (Figure 10.11). But two observations led him to reject Ptolemy's model of crystalline spheres. The first came in 1572, when he observed a supernova – a very bright new star which appeared for a few months then gradually dimmed. This challenged the accepted view that the heavens were perfect and unchanging. Then in 1577, he observed a comet and studied it over several weeks,

Figure 10.11

At a time when most astronomers were using sticks and string to measure the position of stars, Tycho built a quadrant (like a large protractor) over 2.5 metres high. This enabled him to make very accurate observations.

making careful measurements to estimate its distance from the Earth. He concluded that its path took it through the orbits of several planets. This meant that the planets could not be mounted on crystalline spheres. Rather than switch to the Copernican model, however, Tycho proposed a compromise: the Earth at the centre, the Moon and Sun going round it, and the other planets going round the Sun. Tycho's model never really caught on. But the data he collected was later used by Johannes Kepler to make a significant breakthrough (see page 163).

Key term

Imagine 'lining up' a lamppost so that it looks exactly in line with a distant object. If you now move a step to the side, they no longer appear in line. The apparent motion of the lamppost (relative to the distant object) is due to **parallax** – you have moved to a slightly different position, so the line from you to the lamppost is in a slightly different direction.

The principle of parallax can be used to measure the distance to the Moon – by measuring the difference in its direction when viewed from two places on the Earth's surface at the same moment.

Galileo Galilei: challenging the authority of tradition

Galileo Galilei was born in Pisa in 1564 (the same year as Shakespeare). In 1592 he was appointed Professor of Mathematics at Padua, one of the leading universities of the day. Early in his career, Galileo did some work on motion that challenged the established ideas of Aristotle. Most natural philosophers at that time were convinced Aristotelians. So he had many professional critics and opponents.

Experiment not authority: the basis of modern science

Ancient Greek philosophy was based on reasoning. If you wanted to understand something, you read books by learned people and thought hard about it. Galileo's view was that you had to carry out experiments to find out what really happens, and base your explanations on that. This seems an obvious approach to us today, but in Galileo's time it was revolutionary. Because he was the first to argue this way, Galileo is regarded as the 'father' of modern science.

Galileo seems to have been convinced, from his early days, of the heliocentric view. He was also a sincere Catholic and became increasingly concerned that the Church was in danger of making a major error in supporting the geocentric view. At the time there was a rather uneasy truce between astronomy and the Church. Astronomers were allowed to use the Copernican model as a calculating device, but not to claim that it was a picture of how things really were. Galileo was convinced that evidence would eventually show that the heliocentric view was correct, and the Church's support for the wrong view would fatally undermine its authority. He expressed his views very confidently, sometimes appearing rather arrogant. He was also a person who did not suffer fools gladly. He called some of his Aristotelian opponents 'dumb idiots'. As a result he antagonised opponents and made enemies of some of them. It particularly angered him that some of his opponents brought theological issues into the argument, rather than keeping it a purely philosophical matter.

Galileo's first telescope

In 1609 Galileo obtained one of the first telescopes. (They were invented in Holland around 1608.) He immediately saw how useful it could be and made a better one to observe the night sky. Between 1609 and 1615 he observed mountains on the Moon and sunspots. These observations showed that heavenly bodies were not 'perfect' spheres. He also observed the moons of Jupiter, which showed that objects could orbit around a centre other than the Earth, and he observed that Venus (like the Moon) had phases (Figure 10.12).

Figure 10.12

The phases of Venus (similar to the phases of the Moon) showed that Venus was sometimes on the opposite side of the Sun (as the heliocentric model predicts). If it was always between the Earth and the Sun (as the geocentric model predicts), we would never see it 'full'.

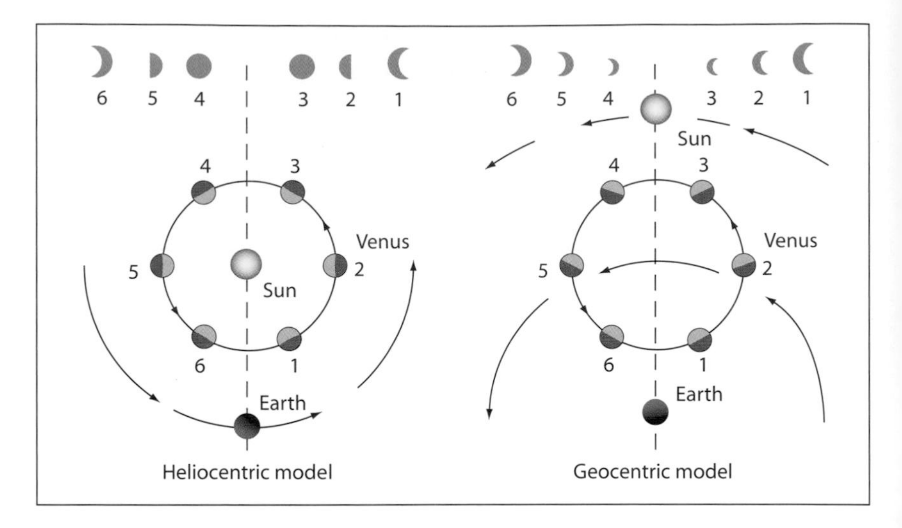

Question

10 Did Galileo's observations provide evidence against the geocentric model? Did his observations provide evidence for the heliocentric model? In both cases, explain why.

In 1615, following an after-dinner discussion with his patron, the Grand Duchess Christina of Tuscany, Galileo wrote her a long letter setting out his views on the Solar System. In it he crossed an important line by arguing that the heliocentric model should be regarded as 'fact'. He also argued that science and theology are not in conflict as they have different aims: 'the intention of the Holy Ghost is to teach us how to go to heaven, not how heaven goes'.

Galileo's opponents were quick to draw this letter to the attention of the authorities. Cardinal Roberto Bellarmine, leader of the Jesuits and probably the most powerful man in the Catholic Church at the time, censured Galileo, instructing him that the heliocentric model should be thought of as a calculating tool, which does not claim to tell us how things are, until there is evidence of the Earth's movement. Galileo went to Rome in 1616 to clear himself of these rumours of heresy and to argue his position. He was cleared, but the tribunal also confirmed Bellarmine's instructions. In fact, Galileo seems not to have been too put out by this, believing that he could collect convincing evidence that would show that the Earth moved.

Technology and science

Sometimes advances in scientific knowledge lead to new technical developments and new devices. But it can also work the other way round. The telescope was developed by craftsmen, before the scientific principles it is based upon were understood. It made it possible to see things in the sky that are invisible to the naked eye, and so enabled our understanding of the Solar System to advance.

The trial of Galileo

By 1630, Galileo was ready to write his great book, the *Dialogue Concerning the Two Chief World Systems*. He believed that his theory of the tides was clinching evidence of the Earth's movement, though we now know that it was completely wrong. Rather than write in Latin, he chose Italian, which everyone could read, and constructed the book as a dialogue between three characters. Galileo put his own views into the mouth of Sagredo, while Simplicio was a caricature of the Aristotelian philosophers who opposed Galileo. The third character, Salviati, was open-minded and willing to listen to all arguments. The reaction of the Church was rapid; the printer was ordered to cease printing and all copies were seized. Galileo, now almost 70 years old and rather ill and infirm, was summoned to stand trial in Rome before the Inquisition in April 1633. Only after being shown the instruments of torture did he give way. Dressed in the white robes of a penitent, he was required to kneel and publicly renounce his Copernican views. Galileo remained under house arrest for the last years of his life, cared for by his daughter, but forbidden to teach or publish.

The trial of Galileo brought to an end a highly creative period in Italian science. Four centuries later, in 1992, Pope John Paul II acknowledged that the Church had been wrong to deal with Galileo as it did. But what exactly went wrong? What was the dispute really about? It is often portrayed as a clash between science and religion. But Galileo was an ardent Catholic all his life. In part the dispute was about authority: if there is a conflict of ideas, should the decision about which to accept be based on evidence or on tradition? The Church was threatened by the challenge to its authority, and wanted to draw clear boundaries between what could and could not be questioned. Over time, however, this has proved to be an impossible line to draw. Another key aspect of the Galileo case lies in the personalities involved. In 1616, when Galileo was first summoned to Rome, Cardinal Maffeo Barberini was one of his supporters. By 1632, he had become Pope Urban VIII. Galileo made the serious tactical error of putting one of Urban's favourite arguments into the mouth of Simplicio, thus making Urban look naïve and foolish. This made Urban less likely to respond positively to the *Dialogue*. Feuds and power struggles within the Church may also have played a part. Historians still argue about the motivations of those involved.

With the benefit of hindsight we can now see that the *Dialogue Concerning the Two Chief World Systems* is one of the most important books ever written. It is a classic of science and of literature, and a milestone in the struggle for freedom of thought and expression.

Johannes Kepler: a model that fits the data

Johannes Kepler was born in 1571 in Weil der Stadt near Stuttgart. He came from a 'problem family' and was a rather solitary child with few friends, but his abilities were recognised at school and he went on to the local university in Tübingen. His original aim was to become a Lutheran minister, but he also studied mathematics and astronomy with a teacher who was one of the leading supporters of the Copernican view. Encouraged by his university teachers, Kepler's career quickly moved towards astronomy.

Kepler seems to have been convinced of the heliocentric model from an early age. He thought that it made it possible to **explain** the movements of stars and planets, rather than just **predict** them. He wrote, for instance, of the 'whirling effect' of the Sun keeping the planets moving in their orbits round it.

On 1st January 1600, Kepler fled from Austria to escape from conflict between Catholic and Protestant factions, and took a job as Tycho Brahe's

Questions

11 Explain what it means to say that the heliocentric model was a hypothesis which 'saves the appearances'.

12 Some powerful figures in the Church opposed Galileo's efforts to spread the heliocentric view. Why might they have been opposed to it?

13 The title of Galileo's book refers to the 'Two Chief World Systems'. Explain what these systems were.

14 What advantages do you think there were for Galileo in writing his book in the form of a dialogue? Were there any disadvantages?

Key terms

The usual unit for measuring angles is the degree. A full circle is 360°. To measure smaller angles, we use **minutes** and **seconds**:

- 1 minute is 1/60 of a degree
- 1 second is 1/60 of a minute (or 1/3600 of a degree).

Sometimes we use the term 'a minute of arc' or 'a second of arc' to make clear that we are talking about an angle and not a time measurement.

Questions

15 What data did Kepler use for his work? Did these data agree with the Copernican model of the Solar System, or falsify it?

16 What was the most important new idea that Kepler introduced, which enabled him to explain the motion of Mars?

17 Could we say that Kepler 'deduced' his model of the Solar System from his data? Explain your answer.

18 Kepler's work finally established the heliocentric model. But Kepler did not collect any new data. So how did his work lead to the acceptance of the model?

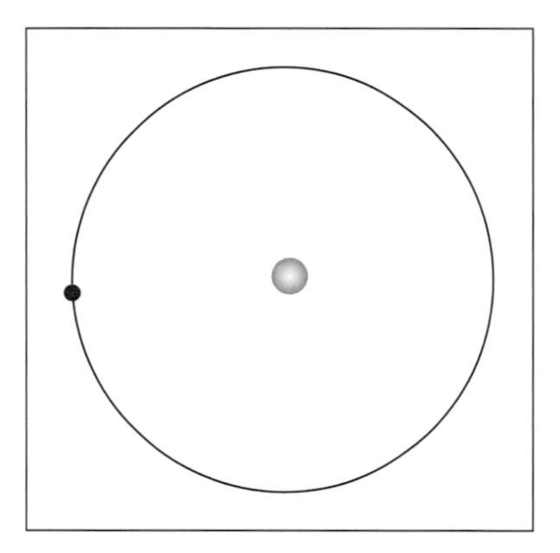

Figure 10.13

An ellipse looks like a slightly flattened circle. You should not think, however, that the orbits of the planets are pronounced ellipses. This is an accurate drawing of the shape of the orbit of Mars. It is an ellipse, but it looks like a circle. It is not hard to see why it took so long for Kepler to realise that the orbit was elliptical.

assistant in Prague, where he was then based. When Tycho died rather suddenly in 1604, Kepler inherited his data. Shortly before his death, Tycho had set Kepler the task of working out the orbit of Mars. Kepler said that he would do it in eight days – in fact it took him eight years. His problem was that he could not get Tycho's data to agree exactly with the predictions of the Copernican model. They were out by six minutes of arc. This is a tiny amount, which many scientists would have ignored, but Kepler had such complete trust in the accuracy of Tycho's measurements that he knew the Copernican model wasn't right.

After several years of painstaking calculations, trying to fit different kinds of circular orbits to the data, he came to the conclusion that the orbits of the planets were not circles, but ellipses. This was an enormously imaginative leap, as circles had been the basis of models of the Solar System for over 2000 years. Using elliptical orbits, Kepler drew up a large set of tables of predicted planetary positions (the Rudolphine Tables). Unlike the predictions of Ptolemy and Copernicus, these really did agree very closely with observation.

The statement that 'the orbits of the planets are ellipses' is known as Kepler's first law of planetary motion (Figure 10.13). Kepler also proposed two further laws:

- Each planet moves round its orbit so that the line joining it to the Sun sweeps out equal areas in equal times;
- The time it takes for each planet to orbit the Sun (T) is related to its average distance from the Sun (R) by a simple rule: the number R^3/T^2 is the same for every planet.

An explanation for the planetary motions

Kepler had proposed a model of the Solar System that fitted the data very accurately. But could the motions of the planets be explained? Galileo had done some important work on motion. Isaac Newton, who was born in 1642 (the year that Galileo died), built on this to develop his famous laws of motion.

In 1664, just after passing his examinations at Cambridge, Newton came home to Lincolnshire to escape from the plague. This gave him time to think about questions that interested him. One question was why does the Moon orbit around the Earth at a constant distance from it, unlike objects near the Earth which fall towards it? He argued that the Moon is constantly falling towards the Earth (Figure 10.14). Without the force of the Earth's gravity, it would move in a straight line. The gravity force acting on the Moon makes it deviate from that straight line, towards the Earth – and this keeps it in an orbit. Gravity does not act only on objects on Earth – it is a universal force of attraction between any two objects.

To check this idea, Newton worked out how strong the gravitational force of the Earth should be at the distance of the Moon, compared with its strength at the Earth's surface. The Moon is 60 times as far from the centre of the Earth as an object near the Earth's surface (384 000 km compared with 6400 km). If the gravitational force falls off as the square of the distance (an inverse-square law), it will be 1/3600 times weaker at the Moon than it is on Earth. So the Moon should 'fall' 1/3600th as far every second as an object on Earth. The value of the radius of the Moon's

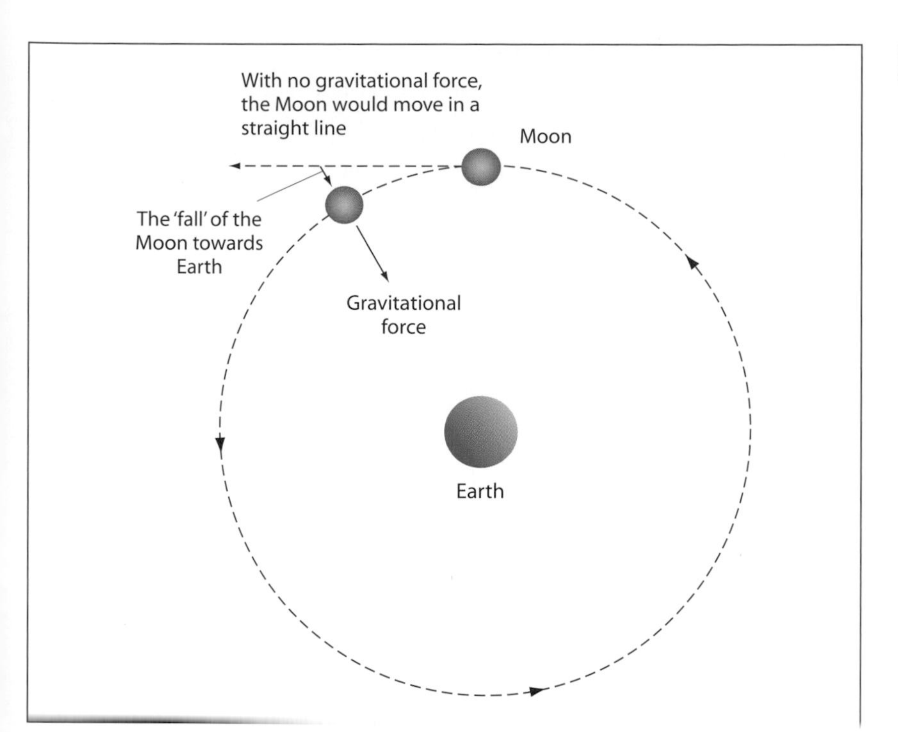

Figure 10.14

The fall of the Moon.

orbit which Newton used for his calculation is one that we now know was inaccurate – so his figures did not tally exactly. As a result, he did not immediately publish his work. It was eventually published in 1687, in his most famous book, the *Principia.*

One reason why Newton proposed an inverse-square law of gravitation was that it led straight to Kepler's three laws – they could be deduced from it. The 'natural' path of an object orbiting a larger one, due to an attractive force between them that obeyed an inverse-square law, is an ellipse. Kepler's other two laws can also be deduced. Newton's ideas explained why the planets move as they do.

Questions

19 What was the problem that led Newton to develop his idea of universal gravitation – an attractive force between any two masses?

20 Newton did not collect any new data on the Solar System. So why does his work increase our confidence in Kepler's model of the solar system?

Gravity

Newton's idea of a force that acted between two objects a distance apart ('action at a distance') was very controversial. It seemed rather like magic – two objects exerting an influence on each other across empty space. But there are other examples of forces like this. One is the force between two magnets; another is the force between two electrically charged bodies.

Gravity is the force of attraction between any two masses. Between everyday objects, the force of gravity is very weak. It is only significant for very large masses such as planets, stars and other objects in the universe. The strength of the force is proportional to the masses of the two attracting objects. This means that if you double either one of the masses, the force between them is twice as big, and so on. It also depends on the distance between the two masses. If you double the distance, the force falls to a quarter of its previous size. It is said to be inversely proportional to the square of the distance.

Questions

21 Explain why the invention of photography would have made it easier to discover the planet Uranus.

22 William Herschel was not the first person to observe the celestial object we now know as Uranus. So is it reasonable to claim that he 'discovered the planet Uranus'?

Discovering new planets

The discovery of Uranus

Until 13th March 1781, astronomers thought the Solar System consisted of the Sun and six planets. Then, quite unexpectedly, an amateur astronomer in Bath, working with a home-made telescope, discovered Uranus. William Herschel was looking for comets at the time. Some thought his discovery a matter of luck, but in fact he was engaged in a systematic and detailed survey of the sky. Herschel had a fantastic visual memory (remember that there was no photography in 1781) and so was able to spot changes in the patterns of stars. Uranus had, in fact, been observed before this but had mistakenly been thought to be a star. Herschel estimated the distance to Uranus and found it was far beyond Saturn. At a stroke he had doubled the size of the Solar System.

As astronomers observed Uranus, however, they began to encounter difficulties in working out the details of its orbit and predicting exactly where it should be at different times. It seemed to stray off its predicted course. Some astronomers suggested explanations that were compatible with the accepted model of the Solar System, based on the work of Kepler and Newton: perhaps Uranus had an invisible moon, or was struck by a comet just before it was discovered. But some began to question the accepted model: perhaps, as the Astronomer Royal, George Airy, suggested, Newton's law of gravitation does not apply at longer distances.

The discovery of Neptune

In 1843 John Couch Adams had just completed his degree at Cambridge in mathematics. He was exceptionally able and in his final exams was awarded double the mark of the next best student in his year. Adams thought that the irregularities in the motion of Uranus might be caused by another (as yet unknown) planet, even further away (Figure 10.15). After two years of painstaking calculations, he wrote to the director of the Cambridge Observatory, John Challis, in 1845 predicting where this

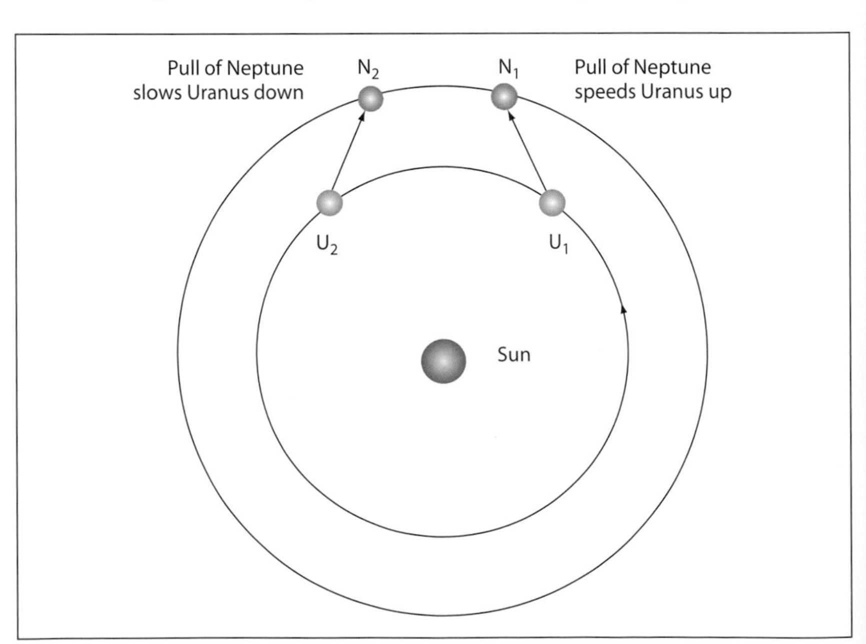

Figure 10.15

The gravitational attraction of another (as yet undiscovered) planet beyond Uranus could explain the 'wobbles' in its motion.

new planet should be. Had anyone looked they would have found it, within two degrees of the position that Adams predicted. But no one did!

Instead Challis sent him off with a letter of introduction to George Airy, the Astronomer Royal at Greenwich. Airy was in France, so Adams returned a month later, again without making an appointment. This time Airy's butler told him that Airy was at dinner and could not be disturbed. Adams left a copy of his calculations and went back to Cambridge. When Airy eventually got round to looking at these, he sent Adams some technical questions, relating to his own idea that Newton's law of gravitation broke down at large distances. Adams's earlier treatment had made him rather fed up with Airy, so he didn't reply – and Airy then assumed his questions had shown up flaws in Adams's calculation.

Meanwhile a French astronomer, Urbain Le Verrier, was working independently on a similar calculation. By June 1846 he too could predict the position of the unknown planet and, like Adams, he could not persuade any French astronomers to look. However, his paper was noticed by Airy, who compared the prediction with Adams's and found they were almost identical. This made him think it was worth looking. Ironically the best telescope was Challis's, in Cambridge, so Airy asked him to start a search. Instead of going straight to the predicted point, Challis embarked on a rather slow and painstaking search of a large area of sky around it. This would certainly find the planet if it existed, but was slow. Twice in August Challis actually observed the unknown planet – but did not recognise it. On 29th September he found one 'star' that he thought might be the planet, but decided to wait till the next night before looking at it again with higher magnification. By the next night, the skies were clouded over.

In any case, he was already too late. On 23rd September, a letter from Le Verrier arrived at the Berlin Observatory. The assistant astronomer, Johann Galle, persuaded the director to begin a search immediately. With a telescope, a planet shows up as a disc, while a star remains just a point of light. Galle hoped to see that one of the 'stars' in the predicted region was a disc, but was initially disappointed. Then a student present suggested they check off the stars they could see against a recently published chart of that region of the sky. Checking through them one by one, they found a new 'star' that was not on the chart. In one night they had done what Challis had failed to do in two months – they had found a new planet.

Le Verrier was generally acclaimed as the discoverer of this new planet, now named Neptune. Although he knew Adams had predicted it first, Airy did nothing to get Adams the credit he deserved. However, John Herschel (the son of William and also a leading astronomer) made Adams' contribution known and arranged for Adams and Le Verrier to meet at his house in Kent. Both are now recognised as the joint discoverers of Neptune.

As a result of this episode, people were even more convinced that the heliocentric model of Kepler, together with Newton's idea of universal gravitation, was right. It had successfully withstood a serious challenge.

Questions

23 Observations of the orbit of Uranus did not agree with predictions based on Kepler's model and Newton's laws. Did this prove that the model and laws were incorrect and needed to be replaced?

24 Adams and Le Verrier did not question Newton's law of gravitation (though Airy did). Which of these was the more reasonable viewpoint at the time? Explain why you think so.

25 Galle found Neptune by spotting a 'star' that was not on his chart. What other method might he (or Challis) have used to identify a planet?

26 Did the discovery of Neptune make astronomers more, or less, confident about their model of the Solar System? Explain why it had this effect.

An understanding that works

Perhaps the most convincing way to demonstrate that your understanding of anything is correct is to make something that depends on it, and show that it works. If so, the Voyager missions provide strong evidence that our model of the Solar System is correct. They are among the most remarkable technical achievements of all time.

The Voyager missions were planned by scientists at NASA (the US National Aeronautics and Space Administration) to take advantage of a rare arrangement of the four outer planets in the late 1970s and early 1980s. All four happened to be lined up on the same side of the Sun, making it possible for a spacecraft to follow a path which would pass close by each. Even better, the gravity field of each planet could be used as a 'slingshot' to accelerate the spacecraft on its way. Voyagers 1 and 2 were both launched in the summer of 1977. Voyager 1 passed close by Jupiter and Saturn. Voyager 2 made an even more amazing journey, passing close to all four outer planets, and several of their moons (Figure 10.16). It took 12 years and travelled 7128 million kilometres to reach Neptune. On 25th August 1989, it passed just 5000 km from the surface of Neptune, within 100 km of the planned distance. This is equivalent in accuracy to sinking a golf putt of 3630 km.

Figure 10.16

The journeys of Voyagers 1 and 2.

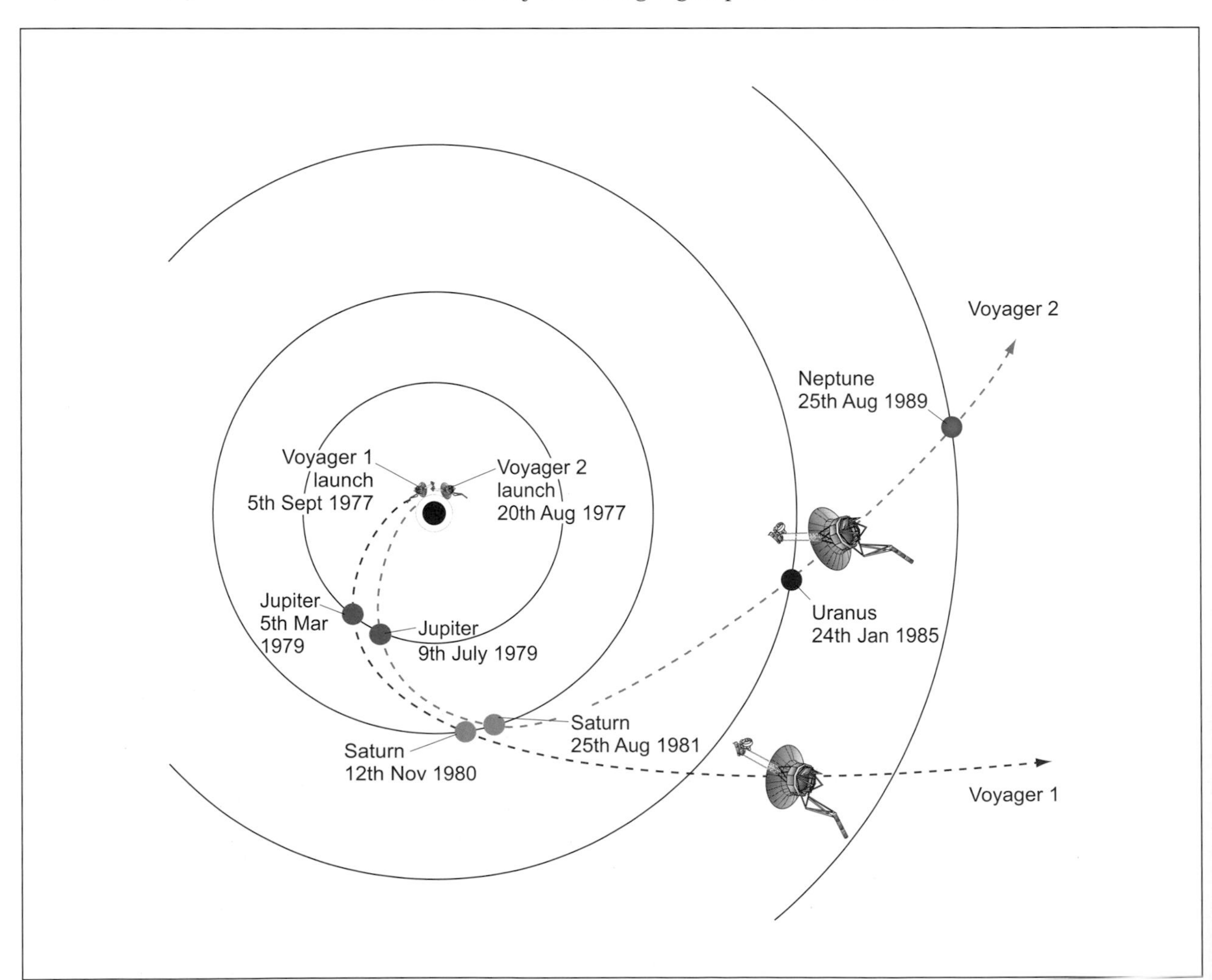

The Voyagers sent back a mass of data and some beautiful photos of the outer planets and their moons (Figure 10.17). Remarkably, the spacecraft are still working and sending back data for astronomers to analyse.

Figure 10.17

Neptune and its moon Triton as seen from Voyager 2. Just try to imagine what was involved in getting a camera to the position where it could take this photograph. The very existence of this photograph is compelling evidence that our model of the solar system is correct.

How far are the stars?

During the 1640s and 1650s, the heliocentric model of the solar system gradually took hold as the accepted view. The simplicity of Kepler's model based on elliptical orbits – and the very accurate predictions it made – were what convinced astronomers. But what lies beyond the solar system, and how far away are the other objects we see in the night sky, in particular the stars?

If the Earth is moving in a large circle round the Sun, then we would expect to have to look in a different direction to see any given star in (say) January and July – six months apart, when the Earth is at opposite sides of its orbit. If the stars are different distances away, we would expect to see changes in their patterns in the sky (Figure 10.18). When Copernicus first proposed his heliocentric model, astronomers looked for this stellar parallax but found none. Some thought that this showed the heliocentric

Key terms

'Stellar' means 'of the stars'. So **stellar parallax** means parallax of the stars.

One **light year** is the distance light travels in one year. The speed of light is 300 000 km per second, so in 1 year light travels 9 470 000 000 000 km (that is, almost 9.5 million million kilometres). The kilometre is not a convenient unit to use for these huge astronomical distances, which is why the light year is used instead.

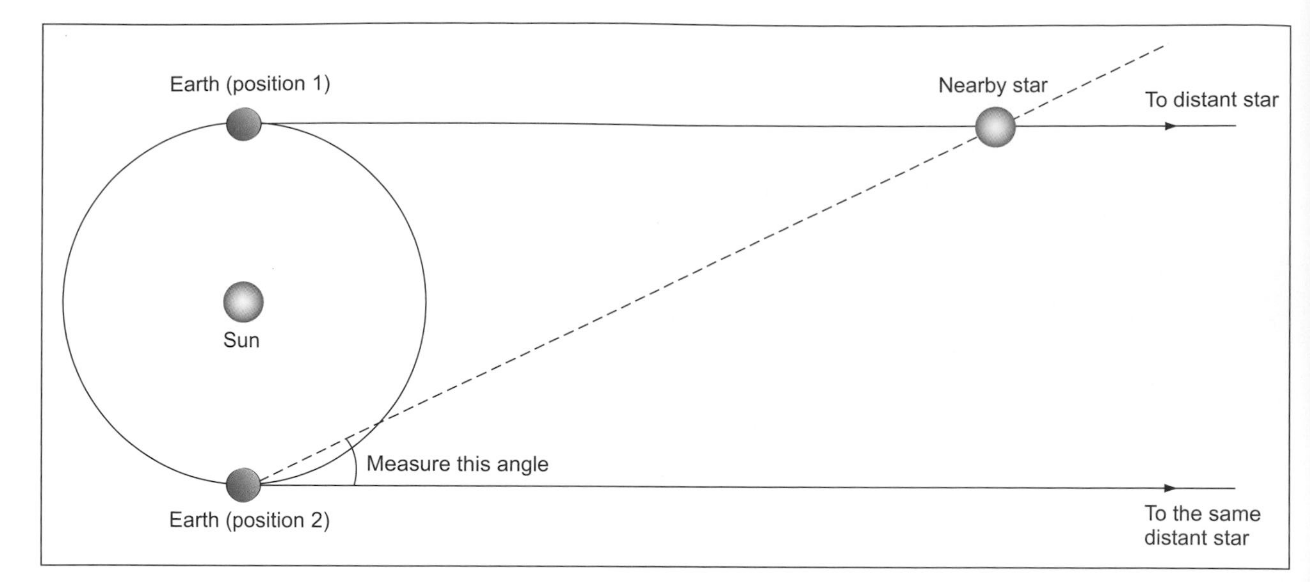

Figure 10.18

Imagine looking at a nearby star when the Earth is at position 1, and noting that it is in the same direction as another very distant star. Six months later, when the Earth is in position 2, the direction to the distant star will be the same, but the nearby star will then be in a slightly different direction. This effect is called parallax. By measuring the change in direction of the nearby star (the angle shown in the diagram), its distance from the Sun can be estimated. (Note that the closest star to the Sun is much further away than this diagram suggests, relative to the diameter of the Earth's orbit – so the angle is much smaller.)

model was wrong – others that it showed the stars were very far away. As the heliocentric model became generally accepted, most people accepted the second explanation and the search for stellar parallax was on.

Parallax observed

One of the problems in searching for stellar parallax is to know which stars to look at (Figure 10.19). Which ones are closest to us? One guess is that the closest ones will be bright. Astronomers used this and other 'hunches' to select stars to measure accurately. In 1838, Friedrich Wilhelm Bessel succeeded in measuring a parallax angle of 0.31 seconds for the star 61 Cygni. From this, he calculated that its distance is 11.2 light years.

Shortly afterwards, several other astronomers measured similarly small parallax angles for some other stars. The closest stars are over four light years away – about 1000 times further from the Sun than Neptune. Our solar system is like a tiny island in empty space.

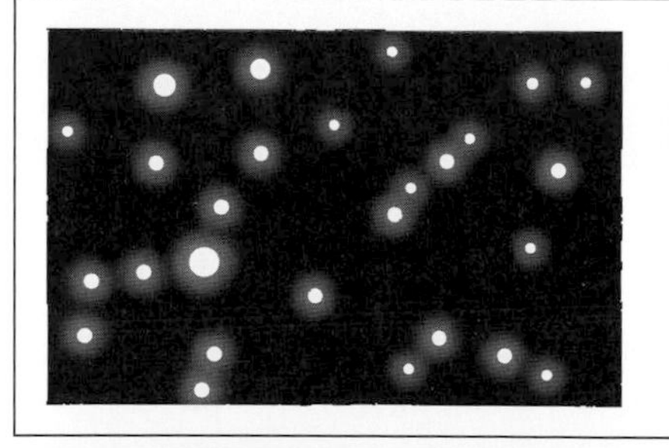

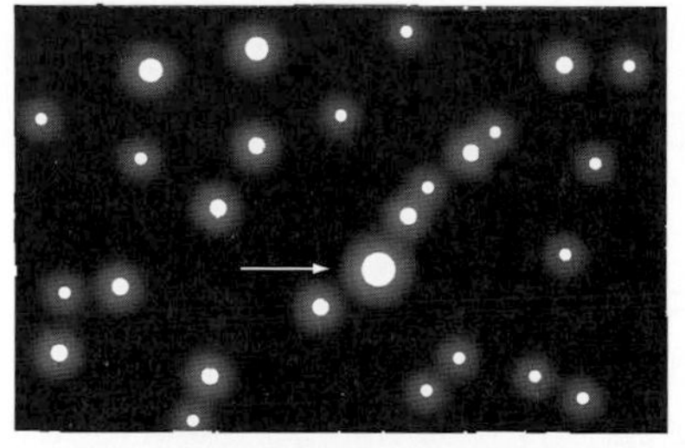

Figure 10.19

These two images, from a modern telescope, show stellar parallax. The bright star in the lower left-hand part of the first photo is further to the right relative to the other stars. By measuring the angle it has moved through, relative to the more distant stars in the background, we can work out how far away it is.

Measuring tiny angles

To get a feeling for how small these angles that astronomers were trying to measure actually are, consider the dot below. At a distance of 1 kilometre this would have an angular width of 1 second.

Bessel measured an angle which is about one-third of this. This was only possible because of the improvements in telescopes by the 1830s. With modern telescopes we can measure parallax angles down to about 0.01 second.

Question

27 Roughly how far away from you would a person be if the angle you measured between their eyes was 1 second? How easy would it be to measure this angle accurately?

New techniques

Two technological developments in the nineteenth century played a key role in the growth of knowledge of the universe. The first was the invention of photography. The first photo were taken in the 1820s, though the process only became widely available in the 1840s and 1850s. As photographic techniques improved, astronomers began to see its potential. The great advantage was that observations could be recorded and studied later. This made it much easier to spot changes. Also, by using long exposure times, it was possible to study objects that were too faint to be observed in 'real time', even with a telescope. The second development was spectroscopy.

Spectroscopy

The scientific study of spectra began with Isaac Newton. Using a prism, he showed that sunlight is made up of a continuous band of colours, from red to violet. Around 1800, Thomas Young showed that two light beams produced interference, thus convincing scientists that light travels as a wave. Subsequent work showed that the colour of visible light is related to its wavelength. Light at the red end of the spectrum has longer wavelengths. Light at the violet end has shorter wavelengths.

Figure 10.20

The spectrum of sunlight, showing the dark lines observed by Fraunhofer.

In 1814, Joseph Fraunhofer noticed that the spectrum of the Sun is crossed by many dark lines (Figure 10.20). These were not explained until 1859, following studies by two other German scientists, Robert Bunsen and Gustav Kirchhoff. They found that the spectrum of a glowing gas consisted of bright lines of different colours, with a pattern that was characteristic of the element present (Figure 10.21). Kirchhoff realised that the lines Fraunhofer had seen in the Sun's spectrum were due to light from the Sun's surface passing through the slightly cooler vapour surrounding it – and that he could use this to discover the composition of the Sun. He published his findings: the Sun is composed mainly of hydrogen and helium, and the gases surrounding it contain other heavier elements found on Earth. So he had evidence for the first time that celestial objects are made of the same elements as the Earth.

Figure 10.21

The emission spectrum of helium produced when the atoms in the gas are excited by heating or by electricity. The lines are a unique fingerprint of the element.

Questions

28 Use the general model of radiation, spreading out from a point source, to explain why a star that is far away from us looks dimmer than a similar star that is closer.

29 You might think that the closest stars are the brightest ones? Explain why this conclusion might not be correct.

Key terms

Intrinsic brightness is a measure of how bright a star really is.

Apparent brightness means how bright a star looks to us.

Intrinsic brightness and apparent brightness are not the same, because stars are different distances away. A faint star might have a low intrinsic brightness, but it could also be an intrinsically bright star which is very far away.

Beyond the closer stars

Measurements of stellar parallax depend on the fact that some stars are relatively close. But only a few stars are close enough for their parallax to be measured. As telescopes have improved, this number has increased, but we can still only detect parallax for a tiny fraction of all the stars we see. How then can we measure the distance to the other stars and to the other objects we see in the night sky?

In fact it took astronomers many years to work out some indirect ways of estimating these distances. The difficulty is that there is not very much that astronomers can observe about a star. They can measure accurately its position in the sky, how bright it is and (using spectroscopy) the composition of its light. But its brightness does not show how far away it is. One star might look dimmer than another because it really does emit less light, or because it is further away.

Around 1890, a group at the Harvard College Observatory in the USA decided to compile a catalogue of the spectra of stars. This was a huge undertaking, involving hours of painstaking work, with no obvious goal other than to collect reliable data. The work was deemed to be too routine for professional astronomers, so the director of the observatory employed a team of women, who were not permitted at that time to obtain a degree at Harvard, let alone be on the staff. Between 1890 and 1924 they steadily built up a huge catalogue of spectra of stars, eventually listing 225 000 stars.

One member of the team, Annie Jump Cannon, decided to try to classify stars according to their spectra. She began with seven classes, ranging from bluish stars, through yellow and orange, to red. With subsequent improvements and additions, this rose to 13 classes. In 1914, using the data in the catalogue, an American astronomer, Henry Norris Russell, tried plotting the brightness of a star against its spectral class. If you do this using the apparent brightness, there is no pattern. But if you do it for stars whose distances are known from parallax measurements, so that you can base it on their intrinsic brightness, then a pattern emerges (Figure 10.22).

This was an important breakthrough. It meant that, by observing the colour of a star, astronomers could estimate its intrinsic brightness. By comparing this with its apparent brightness, they could then work out how far away it is. This allowed them to estimate distances to stars that are too far away to be measured by parallax.

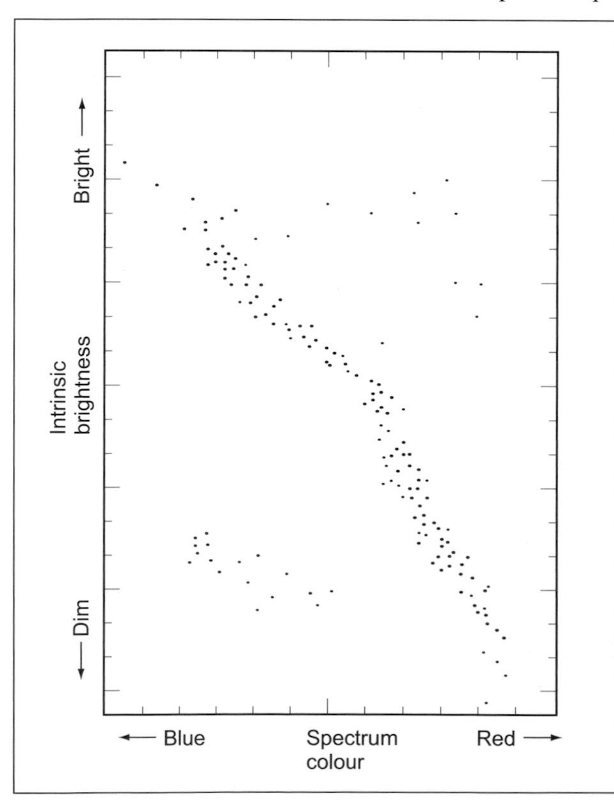

Figure 10.22

This graph shows the relationship between the colour of stars and their intrinsic brightness. It is known as a Herzsprung–Russell diagram. The band running from top left to bottom right, on which most stars lie, is called the 'main sequence'.

Cepheid variables

The work of one woman in the Harvard team led to a discovery that was to prove particularly important. Henrietta Swan Leavitt was interested in variable stars known as Cepheid variables (Figure 10.23). These are stars which pulse in brightness. Their brightness rises and falls regularly over a cycle lasting several days. She chose to study stars in one of the Magellanic Clouds, two clusters of stars that are visible only in the southern hemisphere. The reason for doing this was that the stars in the Magellanic Cloud are all about the same distance away. This means they can be compared directly because their apparent brightness is not affected by how far away they are. Leavitt's data took the form of photographic plates from the telescope at Harvard University's Observatory in Peru. She identified 1777 variable stars, and measured the periods of 16 of them. In 1912, she noticed a pattern: the brighter a variable star, the longer its period.

Figure 10.23

Henrietta Leavitt at work.

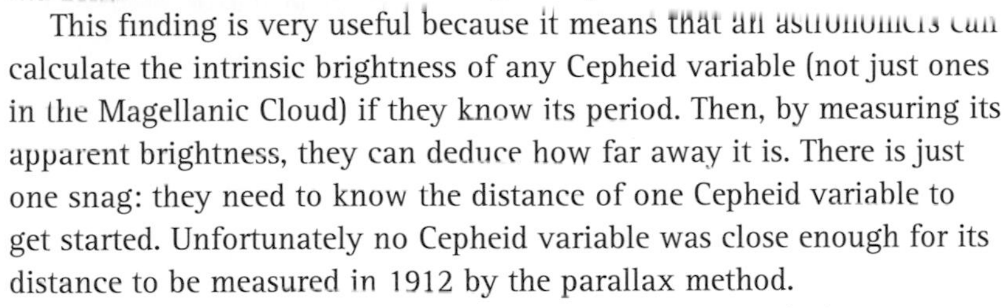

This finding is very useful because it means that astronomers can calculate the intrinsic brightness of any Cepheid variable (not just ones in the Magellanic Cloud) if they know its period. Then, by measuring its apparent brightness, they can deduce how far away it is. There is just one snag: they need to know the distance of one Cepheid variable to get started. Unfortunately no Cepheid variable was close enough for its distance to be measured in 1912 by the parallax method.

However, Harlow Shapley, the director of the Harvard Observatory from 1921, used a number of additional assumptions to estimate the distance to one Cepheid – and this then allowed the distances to all other Cepheids to be estimated.

Key terms

Most stars are constant in brightness. But some vary in brightness, for different reasons. These are called **variable stars**.

One group of variable stars rises and falls in brightness in a steady regular pattern. The first of these to be noticed was Delta Cephei (in 1784), and so the type are known as **Cepheid variables**.

The time it takes for a full cycle from maximum brightness to minimum, and back to maximum again, is called the **period** of the star.

The structure of the universe

Nebulae

Most of the objects we observe in the night sky are stars. But there are also many fuzzy, or cloudy, objects which nineteenth century astronomers called nebulae. Some are visible with the naked eye. The most prominent is a hazy band that runs across the sky, called the Milky Way. The Magellanic Clouds, visible only from the southern hemisphere, are two more. Another smaller one is visible in the constellation of Andromeda. These had puzzled astronomers for centuries. One view was that they were clouds of gas and dust, relatively close to us. Another was that they were clusters of stars, a very long way away.

An early supporter of the second view was the philosopher Immanuel Kant. He speculated (correctly, as it turned out) that the Milky Way was really our edge-on view of our own galaxy – and that the nebulae are other galaxies beyond our own. In 1755 he wrote that 'the nebulae are systems of stars lying at immense distances'. He called them 'island universes'. This was, however, just armchair theorising, and the question of what nebulae are was to rumble on for another 160 years.

Questions

30 What is special about Cepheid variable stars?

31 Explain why it was important that Henrietta Leavitt studied stars in the Magellanic Cloud rather than Cepheid variable stars in general.

Key term

Originally the term **nebula** was the name for any cloudy or fuzzy patch of light in the night sky. The plural is nebulae. The word comes from the Latin for 'a cloud'. Now the term is used to describe a cloud of dense gas and dust in space or surrounding a star.

William Herschel, who discovered Uranus in 1781, also observed and documented nebulae of different kinds (Figures 10.24 and 10.25). Some appeared to be stars surrounded by a halo of gas, others he took to be distant clusters of stars. He proposed a classification of five types. Later discoveries led to changes in this, but Herschel's basic idea was right – the nebulae are not all the same kind of thing. The development of spectroscopy reinforced this conclusion: some nebulae have spectra like that of a star, whilst others have spectra characteristic of a glowing gas. At the beginning of the twentieth century, a common view among astronomers was that nebulae are stars at different stages in the process of formation.

Figure 10.24

The Tarantula nebula. This nebula is nearly 1000 light years across. It is located 180 000 light years from Earth, in the Large Magellanic Cloud. Two bright clusters of stars are seen here. The star cluster at the centre is only 2–3 million years old. Its massive, young, hot stars are shaping the dust of the nebula.

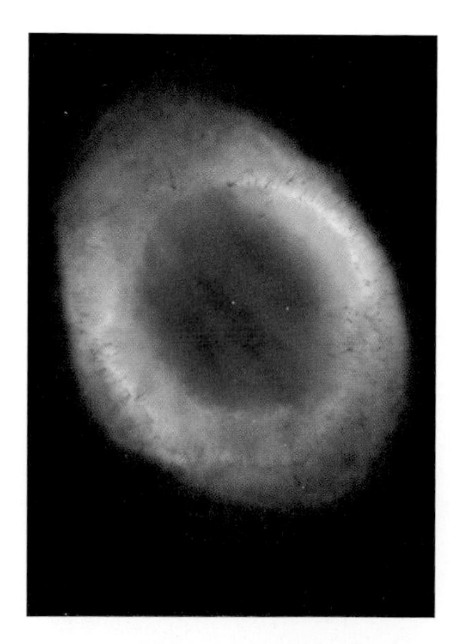

Figure 10.25

The ring nebula in the constellation of Lyra. The nebula is a shell of gas ejected from a central star. The distance to this nebula is over 2000 light years.

Globular clusters and the Milky Way

Leavitt's discovery of the link between the period of Cepheid variables and their intrinsic brightness provided a tool for tackling the problem of what the nebulae really were. With powerful telescopes, individual stars in some nebulae could be observed, and some of these were Cepheids. With great enthusiasm, the American astronomer Harlow Shapley embarked on a programme to measure distances to many of these star clusters. In just a few years, he made some remarkable discoveries. He found that the clusters seemed to form a huge sphere, whose centre was far away in the direction of the constellation Sagittarius (visible from the southern hemisphere). In a daring imaginative leap, he suggested (and astronomers still think he was right) that the centre of these clusters is the centre of the galaxy – and hence that the Sun is nowhere near the centre of the Milky Way (Figure 10.26). Just as Copernicus and Kepler had forced us to accept that the Earth

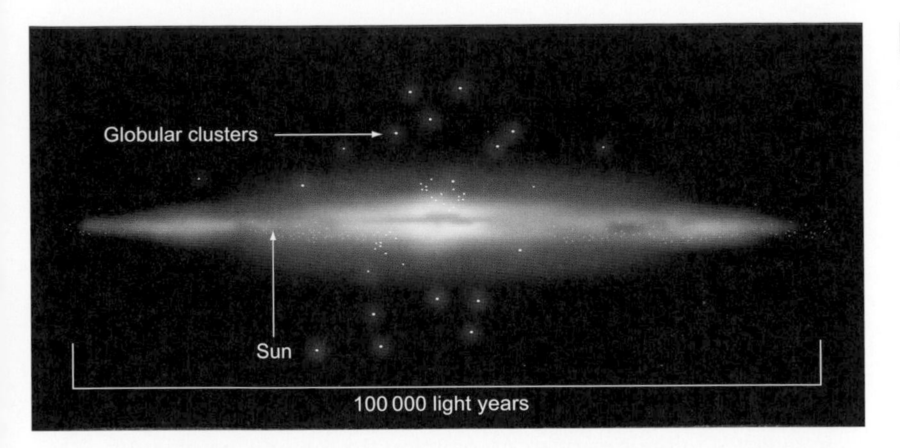

Figure 10.26

Astronomers believe the Milky Way is a huge disc of stars, with a central bulge, surrounded by some globular clusters of stars. The Sun is towards the outer edge of the disc. The galaxy is thought to have a spiral structure, though the evidence for this is not yet conclusive.

is not at the centre of the solar system, we now (in 1920) had to recognise that the Sun and its solar system are not at the centre of our galaxy.

Not all of Shapley's conclusions turned out to be right, however. His distance measurements, using the Cepheid variable method, were all considerable over-estimates, for reasons only discovered much later. As a result, his estimate of the size of the Milky Way galaxy was more than three times the size we now believe it is (around 100000 light years). Because he thought the Milky Way was so large, he concluded that it was the whole universe and that everything was within it – including all the nebulae.

Not everyone agreed. Another American astronomer, Heber Curtis, argued that some nebulae were galaxies, far beyond the Milky Way – the 'island universes' view. The issues came to a dramatic head in a debate between Curtis and Shapley at the National Academy of Sciences in Washington on 26th April 1920. Although Shapley was widely thought to have had the better of this debate, it is Curtis's view that is now believed to be correct. The issue was resolved not by debate, but by evidence. The astronomer who provided this evidence was Edwin Hubble.

Questions

32 Would it be reasonable to say that Shapley had proved that the centre of the galaxy lay in the direction of Sagittarius and that the Sun was nowhere near the centre? Explain your answer.

33 Astronomers in the nineteenth century were not sure what nebulae were. What two hypotheses were there? What evidence eventually helped to resolve this uncertainty?

34 What is meant by the 'island universes' idea? What is the alternative viewpoint?

Island universes

In October 1923, Hubble was studying the Andromeda nebula (as it was then known) (Figure 10.27). The magnification of his telescope was

Figure 10.27

The Andromeda galaxy. Our present estimate is that it is 2.2 million light years away (more than double Hubble's estimate). Despite this huge distance, it is our nearest neighbour among the galaxies. It is one of the 'local group', along with our galaxy, the Milky Way. It is the most remote object you can see with the naked eye. If anyone asks you how far you can see on a clear day (or night), the answer is 2.2 million light years.

Questions

35 What view of the universe was held by:
a) Harlow Shapley
b) Heber Curtis?

36 How was this difference of view among astronomers finally resolved?

37 On the basis of our estimates today, how many times is the Andromeda galaxy further away:
a) than the nearest star?
b) than the radius of the solar system (the distance from the Sun to Neptune, which is about 4500 million km)?

sufficient to enable him to detect a few individual stars in the nebula. In one pair of photos taken a few nights apart, Hubble noticed that one of these stars had varied in brightness – and a few more observations told him it was a Cepheid variable. That meant that he could use Leavitt's results to work out its intrinsic brightness, and from that he could work out how far away it was. His answer was an astonishing one: the star was 900 000 light years away – many times further than any star in our galaxy.

Over the next year, he found some more Cepheid variables in the Andromeda nebula and checked his calculations. The results were consistent. The Andromeda nebula was really the Andromeda galaxy – a separate 'island universe', far beyond our own galaxy. Hubble's results were reported to the American Association for the Advancement of Science on New Year's Day 1925. Everyone present knew the debate about the structure of the universe was over. Our galaxy is not the whole of the universe. It is one of many galaxies that make up the universe.

An expanding universe

At a meeting of the American Astronomical Society in 1914, well before the nature of nebulae had been sorted out, an American astronomer, Vesto Slipher, reported an unexpected discovery. He had found evidence that several spiral nebulae seemed to be travelling away from us at incredible speeds – up to 600 miles per second. The evidence came from measurements of their spectra and a phenomenon called the Doppler effect. Their spectra were 'shifted' towards the red end.

The reason for the red shifts Slipher observed were a mystery at the time. Others, however, confirmed his observations. And Slipher himself measured the velocities at which other galaxies were moving away, including one of 1100 miles per second.

Relativity and cosmology

Around the same time, astronomers were arguing about the implications of Einstein's theory of general relativity. Published in 1915, this suggested that gravitation is due to the curvature of space by all objects that have mass. One prediction of the theory is that light should bend by a certain amount as it passes a massive body. Observations of stars close to the Sun during a total eclipse in 1919 showed that the amount of bending was exactly as Einstein had predicted.

Einstein was aware that his theory had implications for cosmology. His equations led to the conclusion that the universe must either be expanding or contracting. He was unhappy about this and modified the equations to produce the result that the universe is static – as Einstein believed it must be. He later said this was 'the biggest blunder of my life'. In 1922, a young Russian meteorologist, Aleksander Friedmann, published a paper which showed that Einstein's original equations led naturally to the conclusion that the universe is expanding at a changing rate.

Key term

Cosmology is the scientific study of the formation and evolution of the universe. It is a branch of astronomy.

The Doppler effect
Figure 10.28

Almost everyone has noticed the way the sound made by a moving object changes when it passes you. A train whistle or horn shows the effect; so does the siren on a police car or an ambulance. As the vehicle approaches you, the sound is higher in pitch, and as it goes away from you, the pitch falls. The reason is that sound is a wave motion. As the vehicle approaches you, this makes the wavelength of the sound it emits 'bunch up'. A shorter wavelength means a higher frequency and so a higher pitch. Once it has passed and is going away, the sound waves are 'stretched out' a little. Longer wavelength means lower frequency so a lower pitch. The effect is named after the Austrian physicist who first explained it (in 1842), Johann Christian Doppler.

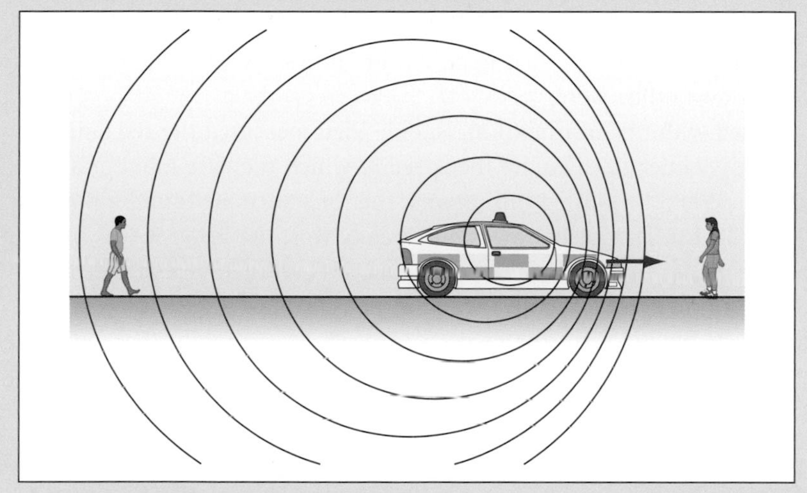

The apparent wavelength of the waves is shorter when the sound source is approaching you, and longer when it is going away from you.

The red shift
Figure 10.29

The Doppler effect also applies to light waves. If the source is moving towards you, the wavelength of the light is decreased. This makes it appear more blue than usual. If it is moving away, the wavelength is increased, making it look redder. If the light is from a glowing gas, its spectrum will be a series of lines. Coming from a receding object, these will have longer wavelengths than the lines in the spectrum of the same element in the laboratory. In other words, they show a red shift.

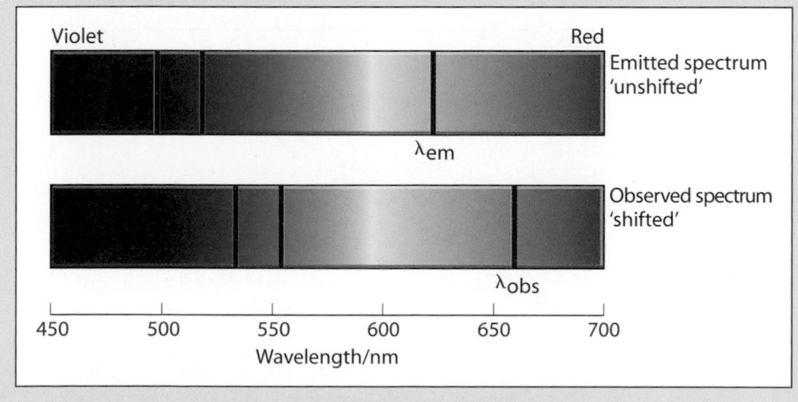

If a luminous object is moving away, its spectrum will be shifted towards the red end.

Edwin Hubble's theory

The evidence for this expansion was provided by Edwin Hubble. Using the same method that he had used to measure the distance to the Andromeda galaxy (based on Cepheid variables), Hubble estimated the distance to some of the other closer galaxies. He was not able to use the method for galaxies further away because the Cepheid variables in them were too dim to see clearly. He could, however, see the brightest stars in these more remote galaxies. So he used a roundabout method.

First he worked out the intrinsic brightness of the brightest stars in each galaxy. This enabled him to confirm that the brightest stars in every galaxy had roughly the same intrinsic brightness. He then observed the brightest stars in the more remote galaxies and assumed that they too would have the same intrinsic brightness as those in the nearer galaxies. By comparing this with their apparent brightness he was able to work out how far away they were.

Armed with this information, Hubble then measured the red shifts of these galaxies to calculate the speed at which they are moving away from us. When he plotted this against their distance, he found a pattern (Figure 10.30). The further away the galaxy was, the faster it was moving away from us. The distance to other galaxies can then be estimated by measuring their red shift, and reading their distance off Hubble's graph. The bigger the red shift, the faster it is receding and the further away it is.

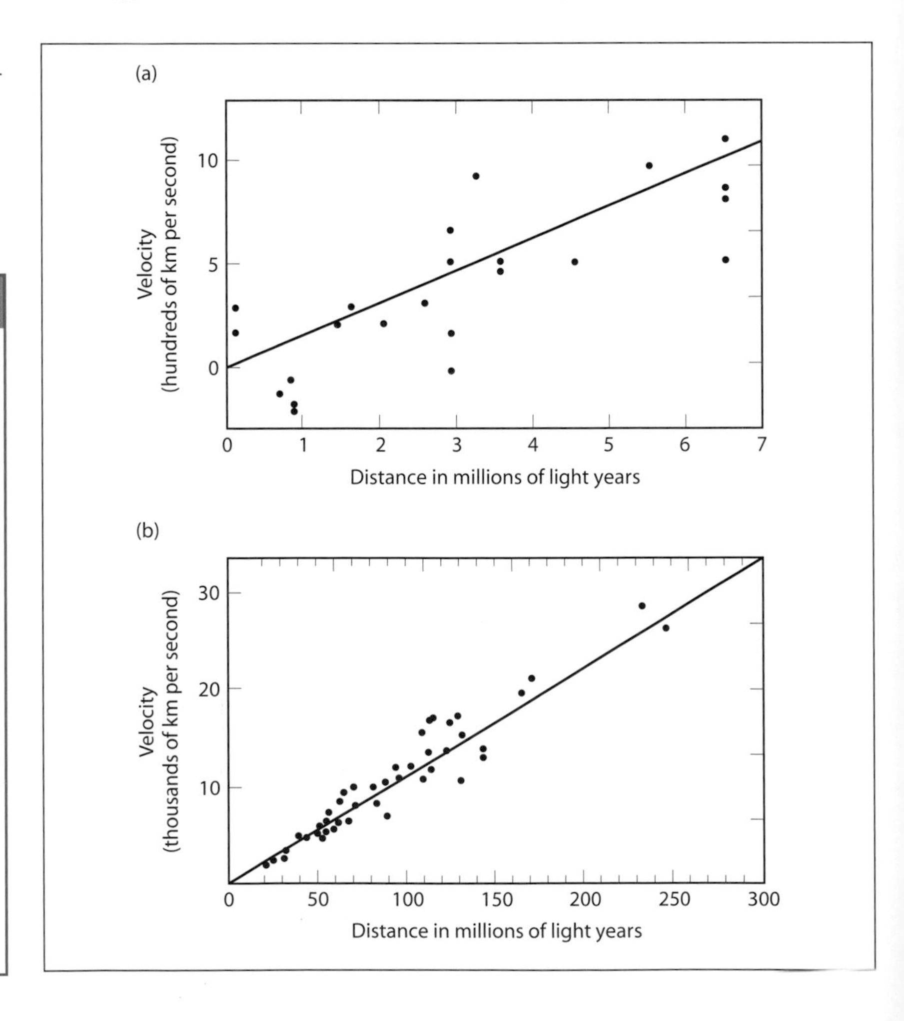

Figure 10.30

(a) Hubble's measurements of distance and velocity for the galaxies he could measure; (b) a distance and velocity graph for more distant galaxies, based on modern measurements.

Questions

38 Does Hubble's graph (Figure 10.30a) provide strong evidence to support his claim that there is a relationship between the distance of a galaxy and its speed?

39 Why do you think the points in Hubble's graph (Figure 10.30a) do not all lie exactly on the line?

40 By measuring the red shift of a galaxy, we calculate that it is moving away from us at a speed of 15 000 km per second. Use Figure 10.30b to work out the distance to this galaxy.

41 A friend has read that the universe is expanding and asks you how we know this. How would you explain the idea to him or her?

The life history of stars

Kirchhoff's work on spectra enabled him to deduce that the Sun consisted mainly of hydrogen and helium, with other heavier elements in its atmosphere. In the mid-twentieth century, scientists used their growing understanding of radioactivity and nuclear reactions to work out how stars generate their energy. It comes from nuclear fusion reactions in the star's core, in which hydrogen atoms are fused together to form helium.

Stars form when huge clouds of hydrogen and other gases are pulled together by their gravitational attraction. As the matter collapses together, its temperature rises enormously, starting the nuclear fusion reactions off. The total mass determines the temperature it reaches – and where the star appears on the 'main sequence' (look back at Figure 10.22, page 172).

When the hydrogen is used up, the core of the star collapses, but its atmosphere expands and the star becomes a red giant. Other nuclear fusion reactions involving helium keep it going until the helium is all used up, and it becomes a white dwarf star. After that, no further nuclear reactions occur and the star gradually cools down. This will happen to the Sun, though it will not reach the red giant stage for another 5000 million years.

Stars more massive than the Sun may continue with further nuclear reactions after all the helium is used up, forming heavier nuclei. They may then explode (a supernova) – and the central materials will then collapse under their own gravitational attraction to become a tiny, very dense neutron star. (A 1p coin made from neutron star matter would have the same mass as 1000 oil supertankers.) It may even become so small and dense that the gravitational pull at its surface is too large even to allow light to escape. (Remember that Einstein's theory of general relativity predicts that light is bent by a gravitational field.) It has then become a black hole.

A big bang

Hubble's work showed that the universe is expanding. This suggests that the universe began with a big bang. At first, some astronomers resisted this idea because it implies a 'creation' at one moment in the past. It also implies that the first stage in the evolution of the universe cannot be explained by the normal physical laws. These astronomers argued instead for a 'steady state' theory, with new matter being produced all the time as the universe expands. The appearance of the universe would then remain constant, even though it is expanding.

One problem with the big bang theory was its estimate of the age of the universe. Using Hubble's measurements of the speed at which galaxies were moving apart and his measurements of their distances, he was able to work backwards and estimate how long ago the big bang occurred. His answer was 2 billion years. However, in 1929, the estimate of the age of the Earth was 3.4 billion years. It is not possible for the Earth to be older than the universe. This anomaly was not resolved until 1952, when it was shown that Hubble had observed a type of Cepheid variable different from those studied by Henrietta Leavitt. This led to a correction in his estimates of distances to the galaxies – which doubled the estimate of the age of the universe to 4 billion years. Later work has increased this estimate further, to the present estimate of around 15 billion years.

The 'echo' of the big bang

Astronomers who championed the big bang theory during the 1930s and 1940s tried to deduce some of its theoretical implications. They showed that Einstein's theory of general relativity implied that the universe had initially been confined to a small space, perhaps even an infinitely small pinpoint, or 'singularity'. They also deduced that there should be an 'echo' of the big bang still around – left-over radiation from the bang itself. Because of the expansion of the universe, this would now appear as microwave radiation, of the wavelength that would be emitted by a body at a temperature of around five degrees above absolute zero (5 K, or −268 °C). The astronomers who reached these conclusions, however, were not leading scientists, and the predicted radiation would not have been easy to detect – so no one tried looking for it.

In 1964, two radioastronomers, Robert Wilson and Arno Penzias, were working with a radio antenna at Bell Laboratories in New Jersey in the USA. They were studying a supernova remnant, trying to work out the temperature of the radiation it was emitting. However, they kept getting other radio noise, which they did not want. This was constant whatever direction they looked in, so it could not be coming from a source on Earth. They thought it might be a fault in the antenna, perhaps caused by the droppings of the pigeons nesting in it. So they tried to get rid of the pigeons, but the radiation persisted. It was microwave radiation corresponding to a temperature of around 3 K. They were reluctant to publish this finding because it did not make any sense to them.

Meanwhile, unknown to Penzias and Wilson, two Russian astronomers had recently published a paper arguing that something must be wrong with the big bang theory because the background radiation it predicted would surely have been observed by antennae such as the one at Bell Laboratories. A theoretical group at Princeton University had also predicted the existence of radiation from the big bang – but were not aware that there was an antenna just down the road that was the best in the world for detecting it. So they started to build their own. In January 1965, Penzias happened to mention the radiation he and Wilson were observing in a phone call with another astronomer colleague, who happened to have heard from another friend of a talk by one of the Princeton group. The penny dropped – and in May 1965 Penzias and Wilson published a paper on their results, alongside one by the Princeton group which interpreted the findings. They had found evidence of the big bang.

Penzias and Wilson were awarded the Nobel Prize for Physics in 1978 – for finding something that they weren't looking for, and not realising what it was when they had found it (Figure 10.31).

Questions

42 What observations did Penzias and Wilson actually make? What was the explanation for these observations?

43 Does Penzias and Wilson's work prove that the universe originated in a big bang?

Unresolved questions

Although our knowledge of the universe has grown enormously since the days of Copernicus and Galileo – and even within the past century – there are still many things we do not fully understand, and some of the observations we can make with modern instruments raise new puzzles and questions.

Figure 10.31

Arno Penzias and Robert Wilson discovered cosmic microwave background radiation, using this horn receiver at Bell Laboratories.

Dark matter

One question which immediately arises from the discovery that the universe is expanding is: will it go on expanding for ever, or will the expansion gradually slow down, stop, and perhaps reverse, with the universe ending in a 'big crunch'? You might think that it is impossible to know, but in fact the answer depends on the density of the matter in the universe. Every bit of matter in the universe attracts every other bit – because of the gravitational force between any pair of masses. If the average density of the universe is above a critical value, which can be calculated, then the gravitational force of attraction will eventually reduce the speed of expansion, stop it, and start to make the galaxies move closer together again. But if the universe's average density is below this value, it will go on expanding for ever.

This naturally made astronomers keen to estimate the average density of the universe and the amount of mass it contains. Some of their measurements, however, were puzzling. Spiral galaxies (like the Andromeda galaxy shown in Figure 10.27, page 175) are rotating. Astronomers can estimate the mass of the matter in a galaxy by measuring its speed of rotation. In the early 1970s an American astronomer, Vera Rubin, did this for the Andromeda galaxy and several others – and found that their mass was almost 10 times bigger than it needed to be to account for the amount of radiation they emit. This suggests that as much of 90% of the matter in these galaxies is invisible to us – so called **dark matter**.

Earlier evidence, from studies of the motion of clusters of galaxies by Swiss astronomer Fritz Zwicky in 1933, had pointed to the same conclusion. The motions we observe seem to require more mass in the system than we can actually see. More recent studies have also corroborated the dark matter hypothesis. Even astronomers who are convinced of the existence of dark matter are unsure what it might be. One possibility is that is consists of the familiar elementary particles (protons, neutrons, electrons); another is that it is made of more unusual fundamental particles, not previously known. The name 'dark matter' might be seen as an indication of our lack of understanding, rather than of our understanding.

Questions

44 'Over 90% of the mass of the universe is dark matter.' Would you regard this statement as a fact, an observation, a hypothesis, a law, or a theory? Explain why.

45 What evidence led astronomers to propose the existence of dark matter in the universe?

46 Which can we be more certain about, the existence of dark matter or of dark energy? Explain why you think this.

Dark energy

If dark matter is a mystery, then **dark energy** is an even bigger one. Its existence was first proposed in 1998 to account for a surprising and unexpected finding. Two independent teams of astronomers were studying distant supernovae (stars that had exploded) to try to measure the rate at which the expansion of the universe was slowing down. But their observations suggested that the expansion was in fact speeding up – and had been for the past 10 billion years. They suggested that this was caused by some unknown sort of gravitationally repulsive material which they named 'dark energy'.

This, they suggested, was the 'driving force' pushing the universe apart at ever increasing speeds. Although this idea has been taken up by many astronomers – as it seems the only way to account for the evidence that the expansion of the universe is increasing in speed – other more recent studies have cast some doubt on the dark energy idea. Some astronomers question whether the evidence is strong enough to justify theories based on completely new laws of nature and on matter of a kind that has never been observed in other contexts. They wonder if there might be a simpler, explanation for the results based on more generally accepted physical laws. It is fair to say that 'the jury is still out'.

Perhaps what these recent developments really tell us is that, despite all the advances in knowledge, our understanding of the universe in which we live is always likely to be incomplete – and new observations may require us to question our current understandings.

Review Questions

47 Both Copernicus's model and Ptolemy's model led to predictions that were roughly equally accurate. How, then, might an astronomer have decided which is the better model?

48 Is it possible to deduce the heliocentric model of the Solar System from the data? If so, why did no one do it before Copernicus? If not, how did Copernicus arrive at it?

49 Kepler rejected a model based on circular orbits because of a discrepancy of just six minutes of arc. Is it reasonable always to expect a scientific model to agree exactly with the data? If the agreement is close but not exact, does this mean that the model is wrong? What other explanations might there be?

50 'The Galileo affair tells us more about political power and rivalry, and about Galileo's personality, than about the relationship between science and religion.' List the evidence which supports this statement and that which casts doubt on it. What is your own view?

51 Which was more important in the development of our understanding of the Solar System: collecting more accurate data, or using imagination to devise possible ways of explaining the data? Give some examples to support your viewpoint.

52 Anomalies in the motion of Uranus did not make astronomers immediately give up the accepted heliocentric model and Newton's law of gravitation. Were they being good or bad scientists in holding on to these, and looking for other ways to explain the anomaly?

53 On the evidence in this chapter, what seems to be necessary to make scientists adopt a new explanation?

54 'The most convincing way to demonstrate that your understanding of something is correct is to make something that depends on it and show that it works.' Suggest some examples (from this chapter, or elsewhere in this book) that either support or challenge this statement.

55 Building large telescopes on Earth and putting the Hubble Space Telescope into orbit cost millions of pounds. List the arguments for and against spending large sums of money on astronomical research. Do you think this area of scientific work should be supported by governments?

56 'The Sun is an average-sized star, towards the outer edge of a medium-sized galaxy. The diameter of this galaxy is around 100 000 light years. The next nearest galaxy is 2.2 million light years away.' How should you regard these statements? Are they facts, or guesses, or hypotheses, or theories, or what? How strong is the evidence supporting them?

57 How different is our knowledge of the universe from that of people 500 years ago? Do you think our knowledge of the universe affects the way we live?

Chapter 11

Are we alone in the universe?

The issues

We have begun to understand how small are the bits of space and time that have been occupied by life on Earth now that scientists have discovered more about the size of our universe and the evolution of life on Earth.

There are so many more galaxies in space and they have been developing for such a long time that it seems very likely that life may have evolved elsewhere as well as on our planet. Such speculation has been the basis for much popular science fiction in the last 50 years.

The science behind the issues

We now know that there are many stars that are in a similar stage of development to our own Sun. It is just becoming possible to detect planets travelling around some of these stars. Our knowledge of the evolution of life on Earth gives us clues about ways that life might have evolved on other worlds.

There is the intriguing possibility that intelligent life forms are as clever or cleverer than we are. If intelligent life elsewhere is looking at the Earth how could we know? For many years now groups of scientists involved in the Search for Extra-Terrestrial Intelligence (SETI) have been analysing radio signals from outer space, looking for a pattern of signals that might match communications coming from an intelligent living world like ours.

Figure 11.1

Head and torso of a fictional alien on an autopsy table. This is an exhibit at the International UFO Museum and Research Center in Roswell, USA. The town has tourist attractions related to the theme of UFOs. It was near Roswell on 2nd July 1947 that UFO sightings were reported during a thunderstorm.

What this tells us about science and society

People are very interested in the possibility that we are not alone in the universe. This is shown by the popularity of books and films about contact with beings from other worlds (Figure 11.1). If we actually found life elsewhere in the universe, how would we feel and what should we do about it?

Modern astronomy uses expensive equipment. To see further, astronomers have to design and build telescopes in places where the atmosphere is clear of modern pollution or they need to put telescopes on spacecraft outside our atmosphere. This type of research is so expensive that it is largely done by groups of scientists collaborating internationally. In the UK our government has to decide how much we can and should contribute to these international efforts.

The search for life beyond Earth

Searching for signals from intelligent life

The universe has existed for long enough for some scientists to think it probable that life elsewhere has had time to evolve 'intelligence'. These scientists speculate that beings in other parts of the universe may be using waves of electromagnetic radiation to communicate in the same way that we do. These alien beings may even be trying to detect us or send us messages.

Groups of scientists working in this area are using radio telescopes to scan for radio waves that contain patterns of information which seem unnatural; waves which show the sorts of the patterns that we radiate from Earth because we use radio waves.

Figure 11.2

An array of radio telescopes in the Cascade Mountains of California. The telescopes are being used by the SETI Institute in the search for life in the universe as well as for cutting-edge radio astronomy research.

Since 1927, we have been sending radio waves out from Earth. These waves have since travelled over 80 light years out into the universe. That means that they have already reached many hundreds of potentially habitable planets. So if there is intelligent life out there, they already know that we are here.

Searching for habitable planets

Of all the life that has evolved on Earth, only humans have developed the intelligence to use radio for communication. Simpler life forms are much more common. Scientists face the challenge of devising ways to detect forms of life which have no capacity to communicate at a distance.

Over the last half century groups of scientists have begun to build upon our increasing knowledge about how our universe and life on Earth evolved to speculate about what simple life could be like elsewhere. This science is multidisciplinary (Figure 11.3). Biologists, astronomers and many others work together to predict the type of environment necessary to allow living organisms of some sort to evolve. They can then design methods to detect those environments on distant habitable planets. They search for signs of that life.

Figure 11.3

Fish swimming among giant tube worms by a hydrothermal vent in the deep ocean. The fish are a species that was previously unknown. The tube worms are a key part of the ecosystem of deep ocean hydrothermal vents. They can tolerate high temperatures.

The search for habitable planets by UK scientists

Scientists in this country are working with international teams on the search for 'Earth-like' planets. Many scientists are involved. Each contributes to a small area of research. For example, some research explores life in habitats that could shed light on the types of living things that could exist on another planet in extreme conditions. Examples include the study of bacteria living in salt water between the cracks in Antarctic ice at −20 °C and of microorganisms that survive at 120 °C beside volcanic 'vents' deep under the ocean.

Another contribution to the overall picture comes from the analysis of data from satellites which measure the concentrations of gases in the Earth's atmosphere. Living things add to the range of gases that can be detected. Exploring our atmosphere indicates the chemical signs to look for in other atmospheres that might indicate the presence of life elsewhere.

Briefing the science minister

In 2007 the government chief scientific adviser invited a group of the UK's top scientists to a meeting with the science minister. All of the scientists who attended were working on some aspect of the search for habitable planets at the time.

This was not a meeting to make policy. Instead the aim was to give the minister and senior civil servants an opportunity to:

- discuss the challenge of identifying other solar systems and Earth-like (or habitable) planets within them
- recognise areas of research that had already proved successful
- understand how much we can learn about such planets using current technology.

Seven scientists attended the meeting. There were journalists present too.

One of the journalists present was Roger Highfield, who was the science editor of the *Daily Telegraph* at the time. He wrote an article about the meeting which his paper published the following week (Figure 11.4). The article shows how remarkable it seemed that the meeting happened at all.

If life is common elsewhere, how will we find it?

Astronomers believe that life is most likely to have evolved where the conditions have some similarities to those on Earth. This means that they are looking for rocky planets at temperatures and pressures which mean that water is usually liquid. They assume that the temperature extremes need to stay within the limits that microbes can survive. The planet's environment must also be stable over the millions of years that evolution takes.

Scientists also predict that life is more likely to flourish if there are other bigger, heavier planets in an orbit further out from the star. The gravitational field of these outer planets protects the inner planet from other hazards such as being struck by comets or asteroids.

Question

1 Suggest two examples (other than those in the text) of habitats where the study of life on Earth might help to inform the search for life on other planets.

2 Imagine that you had been a civil servant briefing the minister for the meeting. Write three questions that you would have suggested that he might ask the scientists at the meeting exploring the issue: 'Is there life out there?'.

3 Why do you think that journalists were invited to the meeting?

4 Suggest why the meeting was limited to talking about science in the UK.

5 On what time scale does the pace of evolution on Earth seem like 'indecent haste'?

6 What do scientists know that justifies their belief that life is common in the universe?

7 Why are Mars and Europa two places where scientists might be able to detect life in our own Solar System?

Key term

The **habitable zone** is the region of space around a star where the surface conditions of any planets could be favourable for life to evolve.

Aliens really do exist!

By Roger Highfield

A decade ago, my question about ET would have been greeted with nervous laughter, curling toes or embarrassed silence. Perhaps even a contemptuous sneer. So it was extraordinary to find myself sitting before a respected huddle of planet hunters with the Government's Science Minister as they declared they believe in aliens.

Seven leading British astronomers were briefing the minister in the bowels of the Department when I put to them two questions: Is there alien life? And if there is, do you think it is intelligent?

Until recently, the assembled scientists would (quite rightly) have pointed out that there is no universally accepted definition of life, let alone of intelligence. Any talk of smart extra-terrestrials belonged in science fiction.

Today, our understanding of the cosmos has changed so much that for the first time searching for signs of life in other Solar Systems is not just a philosopher's dream but on the list of planned human endeavours. All the scientists I questioned agreed that alien life is inevitable and ubiquitous.

And all but one believe it could be intelligent – that is with the faculty of reasoning and understanding.

Some even think we will find some form of life on our doorstep. Professor John Zarnecki of The Open University bullishly told the minister: 'My position is very simple. We will find extinct or some life in the solar system.' He believes that primitive bugs, or their remains, will be found on Mars. After all, once upon a time Mars was a warmer, wetter and more comfortable place. And he has high hopes for a future mission to the icy moon Europa, the heart of which is warmed by tidal forces as it orbits Jupiter. 'We shall find life on Europa in 2023,' he told us with determination.

Why so confident? Our own planet proves that life can flourish quickly. The first creatures evolved on primordial Earth with indecent haste, only half a billion years or so after its birth 4.5 billion years ago, and microbes have now been found in the most inhospitable nooks and crannies, able to dine on wet rock or thrive without sunshine.

This prompted Dr Ian Stevens of the University of Birmingham to speculate that other planets must support life, too: 'My guess is life is common out there.'

Figure 11.4

Roger Highfield's article from the Daily Telegraph.

Planets where these conditions might be found lie within the habitable zone around a star.

One of the scientists at the meeting was astronomer Dr Ian Stevens from the University of Birmingham (Figure 11.5). He wrote a briefing paper for the participants, in which he presented his views about the prospects for finding life in other parts of the universe (Figure 11.6).

Figure 11.5

Dr Ian Stevens is an astrophysicist whose interests include the search for planets outside our solar system.

Searching for planets round distant stars

Using telescopes on Earth

In 2007, Suzanne Aigrain's research was aimed at detecting planets outside our Solar System (Figure 11.7). In her contribution to the meeting she pointed out that in general we cannot see these planets directly, even with the best available telescopes. However, some powerful telescopes can detect the very slight wobble of a star caused by the gravitational pull of a large planet in orbit. The stars are so far away that it is extremely hard to spot the disturbance to the star's motion. Other clues come from the slight shifts

Figure 11.6

Memo by Dr Ian Stevens.

Will we find life on other planets?

The discovery of an Earth-like planet orbiting another star with detectable life signs will be a long-term activity, taking perhaps 10–15 years at the most optimistic and perhaps rather longer if easily visible life signs are rare (which is not the same as life being rare).

The outcome of this search (either positive or negative) will have a profound influence on the long-term development of humanity and our place in the universe, and will represent a major accomplishment for humanity.
We can divide the subject into two phases.

1 The discovery phase of Earth-like planets (which is just commencing). This will enable us to characterise some very basic properties of Earth-like planets – their masses, perhaps their radii, how common they are, what range of orbital periods they have, how many are in the habitable zone of their host star.
2 The study of nearby Earth-like planets and the search for life signs. This will only happen when we can take spectra of the Earth-like planets and see the composition of the atmospheres.

My personal view is that life is common in the universe, but that intelligent life is much rarer. Life began early in Earth history – we see some evidence of life stretching back nearly 4 billion years, and the Earth is only 4.5 billion years old. Geological evidence is very scant for this period and we see life just about as soon as it is possible to see it. An obvious conclusion from this is that given a basic set of conditions (nice planet surface, water, heat, mix of chemicals) life will emerge. This may be a bit trite, but seems a decent working hypothesis.

What is more difficult is seeing evidence of life very remotely. It is tough enough on Mars where at least we can send probes. For extra-solar planets we are definitely 'remote' observers, and we can only hope to see life if it has substantially remodelled the atmospheric composition of the planet. This has only been the case for about half the Earth's history (about 2 billion years). Observing the Earth prior to this, we may have concluded the Earth was dead, whereas it did contain simple bacterial life.

Questions

8 What assumptions are scientists making when they predict that life has only evolved in the habitable zone of a star?

9 Sketch and label a diagram to show which of the planets in our Solar System appear to be in the habitable zone of the Sun.

10 Why is spectroscopy a suitable technique for studying the atmosphere of planets?

11 If life is common in the universe, why are easily visible signs of life rare?

Figure 11.7

Dr Suzanne Aigrain is an astrophysicist at Exeter University. Here she is standing outside the ESO telescope in Chile.

in the lines in the spectrum of light from the stars. These can indicate regular movement towards and away from us (see Chapter 10). Despite the difficulties, scientists had discovered several hundred planets by 2007.

The planets that can be detected by these means have to exert a very much larger gravitational effect on their star than the Earth does on the Sun. Also it is hard to be sure whether a change in a star's motion is due to a smaller but nearer planet or a larger planet that is more distant. Most seem to be planets that are the size of Jupiter or bigger but orbiting closer to their star than Jupiter is to the Sun.

In early 2007, a team of astronomers led by Stephane Udry of the Geneva Observatory announced the discovery of Gliese 581c. This planet is in orbit around a star about 20 light years away. At that time Gliese 581c was the smallest planet to have been discovered around a distant star. The planet is about 15 times closer to its star than Earth is to the Sun but the star is about 50 times dimmer than the Sun. This means that the planet could just be in the habitable zone.

In her own research to detect new planets, Suzanne Aigrain uses a method that observes the brightness of stars. In some cases a star varies in brightness when a planet moves in orbit to partially eclipse the star as seen from Earth. This is the 'transit method' for detecting planets (Figure 11.8).

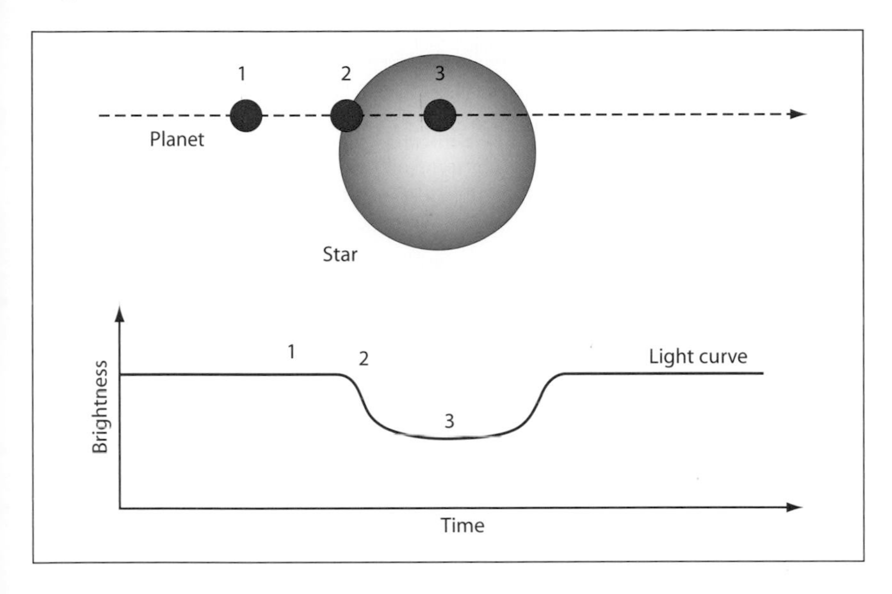

Figure 11.8

As a planet moves in front of its star, the brightness of the light detected falls. This is the basis of the transit method of detecting planets outside our Solar System.

Questions

12 How would you explain to someone who knows little science why it is possible to discover planets with telescopes that can only see stars?

13 Suggest reasons why popular news media were keen to report the discovery of Gliese 581c in 2007.

14 A journalist reporting on Gliese 581c wrote that it was found at 'Goldilocks distance' from its star. Why do you think that the journalist used this term?

15 COROT can observe the same star continuously for 150 days. Why can this not be done by a telescope on Earth?

Using telescopes on spacecraft

Unfortunately the most powerful telescopes on the Earth's surface can only detect giant planets. In order to detect rocky planets which might be habitable, astronomers need telescopes outside the Earth's atmosphere that can observe one star for longer than is possible from any point on the Earth's surface. This is why Suzanne Aigrain is part of the European team involved in the COROT project, which has a telescope on a satellite. COROT was launched at the end of 2006.

The COROT satellite is able to detect the 1 in 10 000 changes in intensity of light from a star as it is eclipsed by an Earth-like planet. Over the life of the project the satellite will collect data from about 60 000 stars. Suzanne Aigrain confidently expects that a number of rocky planets and several hundred new giant planets will be detected using data from COROT by 2010 (Figure 11.9).

Figure 11.9

The COROT satellite, which is designed to keep its telescope pointing precisely at one star continuously for up to 150 days. It has a baffle to give almost complete protection from light reflected by the Earth.

Detecting life on Earth-like planets

Discovering Earth-like planets is just the first step towards the big prize of detecting life elsewhere in the universe. This is only possible with telescopes in space so that there is no interference from the atmosphere of the Earth.

Living things leave traces of their existence because of their biological

Key term

A **signature of life** is any evidence from observations and measurements that life has existed in the place that is being studied. Just by respiring, for example, life on Earth affects the composition of the atmosphere.

activity. In particular, they change the composition of the atmosphere. On Earth, for example:

- plants produce oxygen by photosynthesis
- in sunlight some of the oxygen turns into ozone
- all organisms give out carbon dioxide as they respire
- organic matter that rots away in the absence of oxygen releases methane into the atmosphere.

Astronomers use spectroscopy to investigate the chemical composition of distant planets and stars. They can analyse the light reflected from a planet through its atmosphere to look for signatures of life. Astronomers can claim that they have found evidence of life on a planet if they detect methane and oxygen in its atmosphere, or water and ozone.

Seeing abundant oxygen would be encouraging, since oxygen reacts rapidly and gets destroyed. If found with other gases, notably ozone, nitrous oxide and methane, it would suggest a 'biosphere' teeming with creatures.

The amount of light coming through a planet's atmosphere is tiny compared to the amount coming directly from its star. To see a dim planet around a bright star is like looking for a candle flame next to a searchlight. To solve this problem, scientists have developed a way of combining the signal from several telescopes in such a way that the light from the central star is cancelled out, leaving the much fainter planet easier to see.

The European Space Agency (ESA) is planning a mission that will search for Earth-like planets around other stars and use spectroscopy to analyse their atmospheres for the chemical signature of life. The project team will launch a cluster of telescopes on spacecraft by 2018 (Figure 11.10). This Darwin mission will consist of three space telescopes, each at least 3 metres in diameter, and a fourth spacecraft to serve as communications hub (Figure 11.11).

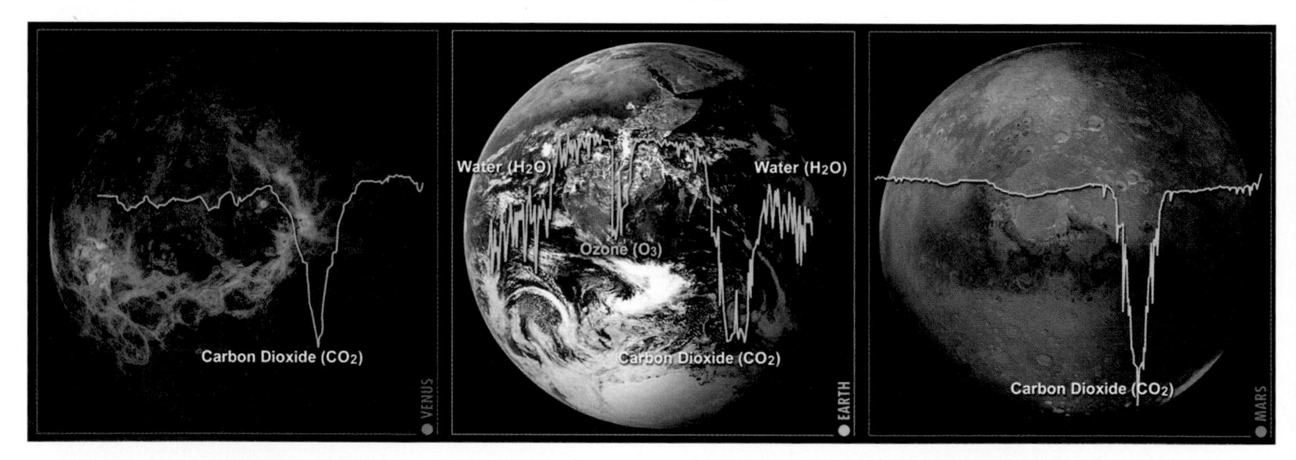

Figure 11.10

Spectra from three rocky planets with atmospheres in our solar system. These are the kinds of spectra that astronomers look for when searching for Earth-like planets in other solar systems.

Professor Glen White of the Open University is a member of the Darwin project's science team. In 2007 he told the meeting of scientists with the science minister that the mission will be able to scan 500 stars over five years within a distance of 60 light years and study the light from 50 alien planets to seek signs of life.

Professor White was pretty sure that, if there is life out there, the Darwin scientists are going to find it. He predicted that by around 2020 we will have very definitive answers to these fundamental questions that impact on our understanding of our true place in the universe.

Figure 11.11

An artist's impression of the European Space Agency Darwin project. Darwin will consist of a flotilla of free-flying spacecraft to search for signs of life on distant planets.

Questions

16 If there is intelligent life studying Earth from a distant planet, what evidence might convince the aliens that there is life on our planet, including intelligent life forms?

17 Why will the Darwin telescopes have found very strong evidence of life if they detect ozone, liquid water and carbon together on a planet?

What kind of life do the astronomers expect to find? It is most likely that alien living things are microbes. But no one can rule out the possibility of more complex life forms.

Next steps

Dr Andy Longmore summarised the meeting with the minister. In his view, scientists can be very confident that rocky planets exist in habitable zones (Figure 11.12). In other words, we can be certain that there are habitable planets around stars other than the Sun, probably in very large numbers. This shows that there has been a remarkable change of scientific opinion in a short time. Only 10 years previously there had been great controversy when astronomers announced that they had detected a planet with about the mass of Jupiter.

According to Dr Longmore, scientists can now be much more certain about their predictions because they have observational techniques that are sensitive enough to measure the tiny wobble of a star affected by the gravity of orbiting planets. He cited the recent discovery of a rocky planet orbiting the relatively small star Gl 851 as another important step in turning ever stronger evidence into a certainty.

Dr Longmore outlined for the science minister some of the likely next steps in the search for habitable planets and the contribution that UK scientists might make to this work. In his view the next key steps were to:

- Establish firmly the number and variety of habitable planets in the universe. He expected this to be done over the following seven years, mostly using existing facilities or facilities that were then being built

Figure 11.12

Dr Andy Longmore, of the Royal Observatory in Edinburgh.

Questions

18 Suggest reasons why it was important to brief the minister about the possible contributions that UK scientists and engineers might make to future international projects.

19 If you had charge of limited funding for a search either for simple life on habitable planets or for the search for intelligent life, explain which you would choose.

for international projects like COROT. These facilities are mainly designed to find planets that are so far away from the Sun that they could never be studied by direct imaging techniques. However, knowing the statistics is essential for planning future projects such as the Darwin mission.

- Directly image the nearest examples of habitable planets and analyse the light from these planets to estimate their size and surface properties.
- Start the search for life itself by determining the physical and chemical conditions on the surface and in the atmosphere of planets. This requires measuring the spectrum of each imaged planet to detect specific molecules.

Dr Longmore explained that the major facilities required for the most challenging breakthroughs would be international because of their cost. UK scientists need rapid access to the data from these projects. They can earn this access either by bringing a special contribution to the science programme or by helping to engineer the technology that makes the science possible.

Review Questions

20 Give at least one example from this chapter, or elsewhere in the course, to illustrate each of these ideas about how science works:

a) Scientists test an explanation by seeing if specific predictions based on it are in agreement with data from observations.
b) It is important that a new theory is consistent with existing theories that are well-established and generally accepted.
c) The topics that scientists choose to work on and the methods they use are strongly influenced by previous work of other scientists.
d) When data conflict with a well-established explanation, scientists are reluctant to reject the old explanation until a better explanation is available; but this can sometimes result in the initial rejection of ideas that later become accepted.
e) The interests and concerns of society influence the direction of scientific research and the extent of funding for work in different areas.
f) The popular media play a part in providing information and influencing opinion on issues involving science and technology.
g) Decision makers are influenced by the mass media, by special interest groups and by public opinion as well as by expert advice.

Chapter 12

Reading and writing about science

This chapter is for you if you are following a course leading to the GCE AS level *Science in Society* qualification. To gain this qualification you have to submit two written reports for Unit 2. One is a *Critical account of scientific reading* and the other is a *Study of a topical scientific issue*. You should refer to the specification for details of the assessment criteria and the marking scheme. You can download the specification from the qualifications section of the AQA web site (www.aqa.org.uk).

The *Critical account of scientific reading* gives you the opportunity to read some popular science writing. You are asked to write an account which shows that you have developed an understanding of the scientific ideas involved, can express your personal response to the issues and assess its appropriateness for the intended readership.

The *Study of a topical scientific issue* gives you the chance to find out more about an issue which is not covered elsewhere in the course – maybe it has hit the news since this book was written, or is of more particular local or personal interest. You are asked to seek out the necessary information and write a report to show that you have developed your understanding of the topic and the science involved.

In each of the other chapters in this textbook you were asked to consider the issues, the science behind the issues, and what this tells us about science and society. You will find it helpful to bear this in mind throughout your own writing for *Science in Society*. These are the areas that your supervisor and the examiners expect you to consider.

Tip

Choose your reading and issue carefully – this may be the single most important thing you do. If you get this right the rest will follow and you will gain the marks you deserve.

Starting points

Choosing your reading

You may be able to choose reading that has caught your eye because you have a special interest. Alternatively, your supervisor may provide a list of sources that are easily available locally or particularly suitable for you (Figure 12.1).

Whether you are making your own free choice or choosing from a more limited selection of recommendations, it is important to seek a piece that allows you to demonstrate the skills which gain marks in the final assessment. You are free to review any piece of popular science writing – fact or 'hard science fiction'. However, it needs to be a substantial piece of writing (a minimum of 40 pages

Figure 12.1

Books shortlisted for Royal Society Prizes for Science Books 2007. Check for lists and reviews of previous winners and the latest news at www.royalsoc.ac.uk/sciencebooks.

or so of a typical paperback). You also need to bear in mind that you have to demonstrate that you understand the science in your reading, so you need to be careful to choose writing that has some clear scientific content.

Be particularly careful if you are considering selecting a piece of science fiction, as you will need to be able to identify and explain some of the real science from which the fiction is developed. Agree your choice with your supervisor before you start reading.

Tip

Avoid a book or topic if there is little available scientific explanation or evidence – even if it looks interesting. Topics involving the supernatural fall into this category. The student who chose 'Spontaneous human combustion' as an issue found that she spent many more hours than she had planned tracking down rare scientific data, with a disappointing result.

Choosing your topical issue and working title

You need to find a topic that interests you, allows you to do everything needed to meet the assessment criteria, and is well documented. There has to be a range of source material available.

It is a good idea to investigate an issue which you find interesting and will enjoy, but if you are already familiar with most of the arguments and have made up your mind on the issues it can be more tricky to gain marks for seeking out and organising information. There is a danger that you will leave out the science behind the issue or fail to present opposing points of view. On some issues there may be heated debate, but that debate may not really involve much science. For example, students have failed to gain high marks for studies of social issues such as the right to abortion.

Once you have a topic and have agreed it with your supervisor, create a working title. It is a good idea to make the working title a question. Then in the rest of your work you set about providing an argued answer and judge the relevance of any information you come across accordingly.

Suppose you choose 'diabetes' as your topic, your working title might be, 'What is the recommended treatment for diabetes and is it available to everyone on the NHS?' The question gives an immediate direction and point to your enquiry. As your work develops and you gain more understanding of the issue, you can extend or change your working title.

Organising your work

Tip

The advice from students who have completed this course already is 'plan ahead'. Give your work the time it needs and work steadily and evenly over the weeks you are given. Your deadline will come all too quickly, and the amount of other coursework you have to do in all your subjects builds up even more quickly. Don't be the one who says, 'I wish I hadn't left it till...'.

Make sure you know your final deadline and any important intermediate dates. Your supervisor will want you to present a summary of your work at each stage and will specify a last date on which you can hand in a draft for advice. You may also have subsidiary deadlines, such as a presentation of your work for functional skills, to include in your planning. Create a time line for your work. Plan to spend about half the time you have available on your reading or the research on your issue and the other half on writing your reports.

In addition to marking your work, your supervisor is required by the Awarding Body to provide information about their oversight of your work as you progress. It will help you, and it will help your supervisor, if you keep a brief diary (in a small notebook or online) noting the date of reading and making notes on a particular chapter, the date you undertook a piece of research, the dates you discussed drafts with your supervisor and so on. If you do all your work on your own without any advice and simply hand over your final reports it will be hard for your supervisor to give all the marks you might deserve.

World Wide Web

There is a huge amount of authoritative information quickly and easily available on the Internet through websites and, for many issues, newsfeeds and forums. The Internet is certainly an excellent starting point in finding information and giving you an idea of what is available. The resources section of the website for this course (www.scienceinsocietyadvanced.org) can help get you started, with links to many relevant websites related to topics in the specification. If you have chosen a topic not included in the course, use search engines to locate appropriate data. Remember that you should evaluate sources. Sites like Wikipedia, which can be altered by anyone, may provide very valuable introductions to a topic and links to authoritative sources but are rarely useful on their own.

School/college library

It is likely that your own school or college library will stock, or have subscriptions to online versions of books and journals (such as *New Scientist*) that relate to the course. Ask your librarians what sources they would suggest.

Local public libraries

Your local library has a different range of books from that in your school or college library. A local library can easily order books for you from other libraries. The librarians also have access to a variety of online databases. Discuss your research needs with them. They will usually be very pleased to help you.

Specialist libraries

These are libraries which specialise in particular topics. Some are public libraries, others may be linked to a manufacturer or association. They often have restricted access or may be geographically inconvenient for you, but some will deal with enquiries by phone or have their catalogues available online. Others can help if approached through public librarians.

TV/Video

Make notes from programmes which are about your topic, including the national and international news. Keep a note of the details of the programme, channel, date, so that you can give full references if you decide to include the information in your report. Check ahead in the listings in magazines or in newspapers for programmes relevant to your topic so that you do not miss them.

Newspapers

You will find issues to do with the public understanding of science in newspapers every day. Weekend newspapers, in particular, often carry excellent features and are very well presented. Keep an eye out for items related to your issue. Most newspapers now have archives which are accessible through the Internet and some allow you to set up alerts for news on a particular issue.

Magazines and journals

Useful articles can be found, particularly in specialist magazines related to your topic. Magazines like *New Scientist* and *Scientific American* will certainly include articles relevant to many *Science in Society* issues. Indices are available on the web but it is often worth searching out the hard copy version of articles you are interested in – for copyright reasons web versions often contain just the text without supporting pictures and illustrations.

Museums and exhibitions

Many topics may relate to exhibits in a local or national museum. Visit the museum website so that you can plan your trip and make good use of your time. Ask museum staff for help. They will be able to answer questions and to direct you. If travel is difficult, many museums will respond to specific queries sent by e-mail or letter.

Interviews

Consider interviewing an 'expert' in the field of your topic or someone who holds strong views/opinions/ideas. Always have a detailed list of questions ready and be ready to take notes or to tape-record the interview.

Letters and e-mail

You may seek information by letter or e-mail from organisations with an interest in your topic. Always be quite specific about what you are asking the reader of your message. Before you write check that the information is not available on a website. A request for 'anything to do with the environment' will either be ignored or will encourage the reader to send unwanted general information. Only well-written letters or messages are likely to get a useful reply and a stamped self-addressed envelope or return e-mail address will help. Always write in a polite tone – never demand information. When writing your letter, say: who you are; what you are doing; why you have chosen to write to them and, as clearly as possible, what you hope to get in response. Don't rely on getting a reply. It may take a long time before you get a response, particularly from busy voluntary organisations, and they may not have the time or money to reply at all. If there are a number of possible places you could write to, send all the letters at once and give yourself a good chance of getting at least some replies quickly. As long as you have clear, brief and specific questions, many academics will respond to an e-mail query. You can usually find e-mail addresses on their university website.

Telephone

If you simply want an advertised leaflet or, perhaps, to find out whether it is worth writing or who to write to, then you may prefer to telephone. Prepare exactly what you are asking for and be sure to give your name and address. Keep a note of the date of your call and who you spoke to.

Family and friends

Discuss your ideas with people you meet outside school or college. You may get suggestions as to how and where your research could develop. Or you may discover 'experts' among these people – their hobbies, their jobs or their personal history could be relevant.

Surveys

Some topics lend themselves to surveys of knowledge, attitudes or opinion. Take advice when devising your questionnaire. You are rarely going to achieve better data than a larger survey reported by a reliable source but you might be able to add a local comparison or update. Don't spend longer on a survey than your results will justify.

Figure 12.2

Sources of information.

Once you have begun research you may find that you are led to visit places and to read articles you had not considered before. This is all part of the excitement of research. It is similar to the job of a detective: tracking down information, asking questions, keeping notes and going on to find out more. So always be ready to change your plan as you go along, but do start with a plan. Consider the research needed for each part of your issue report, and check opening times of places like libraries and museums. You do not want to be surprised by a holiday closure just when you intended to work.

Keeping notes

As you read and research, keep notes. These should contain a reference to all the things you want to write in your final report. Keep a file of any leaflets and smaller bits of paper such as news cuttings, compliments slips and photos. Organise your notes so that you can easily refer to relevant records when writing. For example, you could number all the pages in a notebook and have a contents page at the beginning, or use it as a diary so that it follows a time order. As a minimum, you should keep an up-to-date record of:

- the important content of your reading
- all the people/organisations you have contacted, together with dates, addresses, telephone numbers, and so on, as well as the dates of any replies you may have received
- all the books, magazines, journals, websites, and so on that you have consulted – keeping full records of sources as you go along will save you hours of time when completing the list of references in your report (see Figure 12.7, page 205)
- any other sources you have used (Figure 12.2).

Some people find it helpful to note references on index cards or in a computer file. This aids sorting later. Check with your supervisor, who may also wish to see a sample printout page from any website that you are going to use in your work.

> **Tip**
>
> Don't be the student who is saying to their supervisor on the deadline, 'I've got to put the name of the book I used on the form but I've forgotten – it was something like 'Genetics for Now' and big and red and on the top shelf in the library, but it's not there now.'

Evaluation of sources

For your account of reading you probably only have one source – your chosen book – but for your study of an issue you are expected to seek out a range of sources. In both cases you should evaluate the sources (the authors and/or publishers) so that you can make appropriate judgements about the content.

This section suggests ways of responding critically and reflectively to articles in newspapers or magazines, or when listening to reports on the television or radio. You do not have to adopt these approaches but you may find them helpful.

It is always a good idea to separate the 'facts of the matter' from the conclusions and explanations that people put forward.

Looking critically at data

Although data are usually more reliable than explanations and predictions, you should not just assume that the 'facts' are right. If an article or media report presents some data, then Figure 12.3 shows some questions you might ask about this information.

- Is the data reliable? Have measurements been repeated? If so, do the results agree?
- Has the data been checked by anyone else?
- Is the data reasonably easy to obtain, or is it a tricky thing to observe or measure?
- Do different scientists involved agree on the data?
- If the data comes from a sample of some kind, is this sample big enough? Is it a suitable sample to have chosen?
- Is it real measured data, or has it been calculated from measurements of something related? Or from a computer simulation?

Figure 12.3

Questions to ask about data in a report.

Looking critically at explanations

Even if the facts are generally agreed, it is still perfectly reasonable to question the claims in the article about explanations and predictions. Figure 12.4 suggests some useful questions to ask.

- Is the explanation based on generally accepted scientific ideas?
- Are there other data and evidence that support this explanation?
- Is this the only explanation, or are there other competing explanations?
- Are the predictions about situations similar to those already investigated? Is it reasonable to extend the results to these new situations?

Figure 12.4

Questions to ask about explanations and predictions.

Who to trust

Often articles quote some of the scientists involved. Radio and television reports may include interviews with these scientists. How do you decide whether to accept these claims? Indeed, many reports present differing views from two or more scientists. How then do you decide whom to agree with? There is no infallible 'recipe' for making these decisions – but Figure 12.5 highlights some of the things you might bear in mind as you reflect on news stories about scientific issues. You may not always feel you have the background knowledge to 'score' for all these aspects, so concentrate on those where you feel you can make a judgement.

Try using Figure 12.5 to assess an expert's comments in newspaper or magazine articles related to your study, or to the conclusions being expressed by the author of your reading. If you award a score (from 5–1) for each aspect, and add these, does it help you to decide which views to take more seriously? What are the weaknesses of a 'scoring system' like this?

Thinking along these lines will help you when it comes to writing your reports and evaluating your findings.

Preparation

Preparing for your account of reading

Read and assess your text, bearing in mind the intended readership. For books you can often get some indication about the author's intentions from the information on the cover, or in an introduction or preface.

For long articles from a periodical you need to judge who buys it. Your librarian may be able to help with some advice.

Score	Aspect				
	Theoretical ideas involved	**Nature of the data**	**Status of the scientist**	**The scientist's institution**	**Personal affiliation**
5	Core science – agreed by all	Reliable and agreed experimental or observational data	A recognised authority in this field	A famous university or scientific research institute, or a major company	Works for an official regulatory body with responsibility for this area
4	Agreed by many, but still contested by a few	Experimental or observational data that is challenged by some	A professional scientist working in the area – but not a top name	A known, but less prestigious, institution or company	Has no direct personal or professional interest in the issue
3	There are several competing explanations in this field – and this is one of them	Data that is agreed to be sketchy and uncertain	A respectable scientist but whose expertise is in a different field	An institution or company with a more doubtful reputation	Has been involved in these issues for a time and is known to hold a particular view
2	A new field in which there is no agreed theory as yet	Data calculated from computer models, or projected from other data	A relatively junior scientist with no established reputation	An institution or company which few people have heard of	Has known views or contacts that might bias views
1	A fringe theory accepted only by the author and his/her friends	Data little more than an educated guess	A known maverick (or crank)	Not employed in an academic or scientific research institution	Works for a company with a direct interest in the issues

Figure 12.5

Assessing how far to trust an 'experts' views.

Tip

Try reading some short book reviews in magazines or newspapers to see how experienced writers assess a book critically in a few words. Many book reviews are only about 250 words long.

You might try finding your book's page at an online bookstore. They often add reviews contributed by readers and you can gain a picture of the readership from them. (You might also gain other ideas – but remember that those reviews are not written with an examination in mind, it can be hard to evaluate them as sources, and you must be careful not to copy ideas directly.)

If you are reviewing a few chapters of a book, it is a good idea to at least skim the introduction and conclusion of the book and the other chapter titles so that you are fair to the author. The chapters you are reading may be much easier to understand if you'd read the chapter before, or it might fall into place if you read the following chapter too. Of course there is nothing to stop you – and everything to gain – from reading the whole book anyway and just using examples from one section in your account.

Key term

In this chapter the word **research** mainly refers to the type of web, library or text study that you will undertake for your own writing. This is similar to the work of a science journalist 'researching' a story for the press, radio or TV. It is distinctly different from the scientific research which has been described in Chapters 1 to 11.

Preparing for your report of a topical issue

With your final deadline in mind, plan your research so that you have plenty of time both for considering the results and for any further research that has become necessary to clarify a point.

Divide your topic into sections for research based on your working title. Work out what questions you are going to try to answer in each section. For each question you are working on, consider the best places to search for answers. For example, you might best find out what diabetes is from a medical text or from the web, but finding out how it feels to receive certain treatments might require an interview with a diabetic. Be prepared to follow up any new questions your research raises.

Figure 12.2 lists a number of possible sources where you can begin to search for information. Few will use all of the suggestions – some topics will rely more easily on some sources of information than others. You are required to use a 'range' of sources. Many of these may be from the web but, if so, you should have a range that covers a number of different aspects of your topic from different viewpoints.

Drafting and writing your reports

Remember your audience. This is a *Science in Society* course, so you should write a report that explains your understanding of your reading and your issue to a 'general reader'. Specifically, though, you are writing for your supervisor, who will read and check your work, and an examiner appointed by the examination board, who will also read your report.

When you are writing your report, make it your own. You are expected to be using sources of information – indeed you are credited with marks for finding those sources and picking appropriate parts to add to your arguments. You are expected to be explaining ideas that you have gained from your reading. It is virtually impossible to get good marks if you do not use information from somewhere else. You must, however, avoid any possibility of being judged guilty of plagiarism, which is representing someone else's work as your own (Figure 12.6).

Do not assume your reader has any previous knowledge of the subject you are writing about. This will encourage you to write clearly. Imagine someone else in your class is going to read it. Will they understand your account?

Read and re-read it. Write and rewrite. Draft your sections and ask your supervisor to comment on them. Do not waste time presenting everything beautifully just to have your supervisor suggest a change at the top of the first page. Leave details like page numbering until the end. Your supervisor will probably prefer to see a second or third draft rather than your first rough notes, but you do need to provide evidence of your developing work so that your supervisor can confirm that it is your own.

Figure 12.6

Plagiarism and how to avoid it

Plagiarism

'plagiarise, to steal from the writings or ideas of another.'

Chambers English Dictionary, 2002.

When you hand in your work for assessment you sign a *Candidate record form.* You should have included all your references and sources as appendices to each of your pieces of writing. If you left anything out, there is the space to add it here. The form reminds you that: 'To present material copied from books or other sources without acknowledgement will be regarded as deliberate deception'. You also sign to say you have read a *Notice to Candidate* which says: 'The work you submit for assessment must be your own. If you copy from someone else or allow another candidate to copy from you, or if you cheat in any other way, you may be disqualified…' It is not just in examinations, but also in academic work generally, that the crime of plagiarism is taken very seriously. Plagiarism is not just copying directly and not acknowledging that you have done so; it is also following someone else's idea or argument too closely.

Here are three ways to avoid plagiarism:

- When you are ready to write the first draft of each section of your work, look through any books, notes and other material you have accumulated. Then put them out of sight and write that draft from your head. Do not worry if you cannot remember the odd detail or want to put in a quote, just carry on. When you have finished the section, go back to your books and notes for quotations and missing details which you can now add to give more substance to your work.
- If you copy the actual words from your source, put the words in quotation marks ("…") and give a reference (see Figure 12.7). Similarly, if you copy pictures or diagrams you must give a reference. If you slightly alter a diagram to make your own point, then write that it is 'adapted from…' and give the reference.
- If you outline a particular idea based closely on one source, or use specific information from a source, then give the reference immediately. If you have used a number of ideas from several sources, then give the references at the end of the paragraph or at the end of the section.

If you work in this way you should find that the pattern and thread in your writing is your own, and that you are interweaving ideas and information from different sources (both of which get you more marks). You will be giving clear references which are indications of your research and selection of material (which gets you marks). You then avoid the risk of being accused of plagiarism.

Writing your account of reading

Your final account must be short, in the range 500–800 words. The best accounts really are short and succinct. You should aim to write no more than 800 words. You will not lose marks by exceeding the limit slightly but you will not gain marks by doing so.

The specification suggests that your account should include:

- a precise reference to the material you have read
- a critical discussion of the effectiveness of the text for its purpose, its style and language and the values or attitudes it conveys to the reader and related to any social or moral issues – set in a wider context, if that is appropriate
- your personal response to the ideas and explanations which arise
- an explanation of the understanding of science or how science works that you have gained.

There is no fixed style which you have to use for your account. You do not have to stick rigidly to the advice below and your supervisor may well have suggestions to make. Apart from the introduction and the conclusion, you do not have to write the components of your account in the order given below.

Reference

At the top of the page, or as part of the title, give the details of what you have read – including the title, author, edition, publisher, publication date and page numbers.

An introduction

Aim to open your account with two or three sentences that engage your reader and show why your chosen text is interesting and has something worthwhile to say. At the start you might set the passage in context and identify the intended readership.

Explanation of scientific ideas

You should then write a few paragraphs explaining the main science ideas about which you have read and any issues which arise. Remember that this is a short review and you are aiming to demonstrate that you have understood your text. So use your own words to explain the main ideas clearly, as if to a fellow student. If there is a lot of detailed science in the reading you have chosen, then you should select one or two ideas which stand out as good examples.

You may find that you need to look up some points in another text in order to understand your reading or to be able to make a comparison. If so, you should make this clear but do be careful not to start full-scale research into the issue in other sources. This is supposed to be a succinct review of one piece of reading. The place for wide research on an issue is in the other piece of work.

Add a few paragraphs commenting on any issues which arise from the topic of your text and comment on any social or moral issues which may arise.

Your personal response

You need to add a paragraph giving your personal response to the reading. Be honest (favourable or unfavourable) and give reasons and examples to illustrate the points you make. Remember to use appropriate language for a critical review. Do not lapse into very informal language just because you are stating your own opinions.

Critical discussion

You should finish with a paragraph or two in which you assess the effectiveness of the writing. It is perfectly appropriate to consider issues such as the type of illustration, the cost, and so on, as well as the actual style and language of the writing.

This is the place to include any comment on particular views or opinions the author gives, unless you have already made that a part of your personal response.

Conclude with a sentence summarising your view of the effectiveness of the text as an explanation of science ideas for its intended readership.

Writing your study of a topical issue

Unlike the account of reading, your report on your study of an issue is expected to be in the set style of a formal research report:

- a title
- an abstract of the report (50–150 words)
- the main points of evidence with full citation of sources
- a discussion of the evidence, including an evaluation of the sources
- a full list of references used in the study.

Plan the structure of your report so that you know how you are going to use your material and roughly how much you will need to write for each section. The total length should be 1500–2000 words – no more.

Your final report may have even fewer words if you are going to present a lot of your information in tables, graphs, flowcharts or labelled diagrams. Illustrations help a reader by breaking up the text, but in your report there is a very limited place for pictures that are just there to 'make it look pretty'.

Plan how and where your report is going to do justice to all your research and thinking. As you are writing, make sure each sentence and paragraph is contributing to your marks, not only by being relevant to your argument and either presenting detailed knowledge or contributing to your overall evaluation, but also by making the stages in your work clear. For example, in your introduction to your report, 'Diabetes care for different ethnic groups', you might write:

'There are two main types of diabetes, which are defined as...'

This is fine and would begin to contribute to your marks for 'Content'. However, if you read several different leaflets and book chapters on diabetes before finding a clear definition that you found useful, you can write:

'Among a number of publications, the best definition of the two types of diabetes I came across was...'

In this way you are showing the examiner the breadth of your research as well. You are gaining credit for finding material and making a sensible decision to leave it out. It is very important to ensure that evidence of all your hard work appears in the report.

Tip

Be prepared to edit your writing – and do not be afraid to 'chop and change' it.

Your title

Your working title was the question that guided your research. You can use the same title for your report but you need not do so. Often a short title is better, but consider adding a subtitle. Choose a title which tells your readers what your report is about rather than a jokey title with no meaning. Remember that the content of your report is marked according to how relevant it is to your title and to what you say your report is about in your introduction – so make sure they match.

Your introduction

Your introduction should be a brief 'way in' to your report. It should arouse your readers' interest and outline what they are going to get out of reading your work. Include in your introduction a few sentences to describe why the topic is of importance to you or to the public. For example, you might explain a personal connection.

Consider starting with an opening sentence to grab your readers' attention and make them want to read more. For example:

> 'Harlequin stood on the platform, nerves tingling, as he watched for the pull of the lever which would send water pouring along the tubes behind him and launch him through the trap-door into an explosion of smoke above.'
>
> [from a study of the use of new technology in the rebuilt Sadlers Wells Theatre]

Other features of your introduction might be:

- A few sentences to explain the focus of your report.
  For example, if your report is called 'Alternative Medicine', you might explain that you actually have chapters on homeopathy and hypnotism, because you had discovered that these were popular forms of alternative medicine amongst patients in your own GP's practice.
- A few sentences to describe the structure of your report so that your readers know where they are going.
  For example, in a study on animals in scientific experiments you might indicate that, after a chapter defining cruelty and pain, you have four sections giving examples of the use of animals in experiments, in which you choose to show the possible range from necessary to unnecessary, and that you then have a section on the current law, before you give your opinion as your final conclusion.
- A final sentence which states the question that the rest of your report answers.

When you have written the rest of the report, return to your introduction to check that it accurately reflects what you've written. You might even consider leaving the final drafting of the introduction until after you have finished the bulk of your writing.

The main body of your report

The main points you have selected from your research will make up this central, and longest, part of your report. You should already have created, on a separate paper, an outline of what you are going to write and in what order. Writing over a thousand words can be a daunting task but it will be more manageable if you split your work up into smaller parts.

Organise your work into groups of related areas, and present these as sections with sub-headings. Do make sure that there is a clear logic to the order. Ideally each new point should be related to the previous one in some way. Each section should be titled and listed at the beginning of your report in a contents page.

Many pages of continuous text can be hard to follow and unattractive to the reader. Select relevant diagrams, graphs, labelled pictures or photos with captions to summarise information. A labelled diagram is not only a useful and attractive visual aid but it can also often save you a great deal of writing. However, do not include illustrations for their own sake. Make sure you refer to them in your writing and always give each a caption.

Remember that you must give a precise reference to the source of each piece of information that you present.

Your discussion

In this section of your report you review the work you have carried out and justify your findings. This section might be quite short and concise, but it is essential to demonstrate that you really have understood and thought about the issue you have studied. It is possible, and sometimes preferable, to write some of this comment in your other sections as you go along, but it is generally a good idea to have a separate section as well. It keeps your thoughts clear and makes it easier for the examiner to give you the marks for 'evaluation'. This part gives your personal opinions, so it can be appropriate here to write in the first person ('I want to argue that...' rather than 'It is argued...'). When expressing your own opinion, always justify your statements with reasons and examples. When analysing information, look at the subject impartially and give alternative solutions. Then base your conclusions on the facts you have discussed.

You should comment on the main points you have presented, for example:

- If different sources disagreed, how did you decide which to follow?
- If several references gave diagrams, how did you choose which one to adapt for your use; or have you used parts of one for clarity and another for completeness?
- If you have had to make judgement between different opinions, what questions have you asked yourself and how have you weighed the evidence?
- If you have come across any controversial questions (or maybe your whole report is about a controversial question), how have you distinguished between facts and opinions?
- If you have paid more attention to some sources than others, was it because some were more recent or in a more prestigious journal – or because they matched your opinion?
- How did you decide what material to use and what to leave out?

You are expected to have a balance between discussion of the issues and the underlying science. Do not make the mistake of concentrating on one aspect at the expense of another.

Only include work which is directly relevant to your chosen issue and your working title. If you cannot bring yourself to leave out something interesting but only marginally relevant, consider including it at the end of the report in an appendix.

Distinguish between: describing, explaining, informing, comparing and contrasting, and opinion. It is muddling to the reader if you mix too many of these things in one paragraph. If you find yourself adding a lot of description and explanation, it probably belongs in the previous section.

Your conclusion

Your conclusion should include a clear statement of your own views on the issue you have been studying. It should leave your readers satisfied and absolutely clear about what you are saying.

Consider including:

- a reference to the opening sentence in your introduction which grabbed the attention of the reader – it rounds a piece off nicely to refer back to your introduction by giving your explanation or answer to the question or issue you started with
- a summary (almost a list) of the most important points you have made and your main opinions or conclusions
- a few sentences which indicate what further research needs doing.

Your conclusion should not include new information. You should already have given all the facts and your evaluation of them in earlier sections.

Your abstract

Write your abstract last. This may repeat phrases or sentences from elsewhere in your report, if appropriate. You need one or two short sentences to describe the range of points on your issue which you cover, followed by a further sentence or two to describe the main points and conclusions you have come to. The aim of the abstract is to tell readers, briefly and clearly, what they are going to find in the report. It goes in a separate section at the front of your report, but should always be written last since it must be an accurate reflection of your final piece of work.

References and sources

You have to cite references in a separate section at the end of your report. You will need to include all the sources that you have cited in your text but also add any more general sources that you used. Don't be tempted to list other sources that you came across but did not actually read in the end. Number your references or list them in alphabetical order. Remember to name anyone who has helped you. If you have discussed your work with anyone or been given any advice, acknowledge this help. The only exception to this rule is that you do not have to credit your supervisor.

It is very important to have kept a clear record of your sources as you went along so that you can make your list of references accurately and quickly (see Figure 12.7).

You must list:

- any books, periodicals or other media you have quoted in your report
- any books, periodicals or other media from which you obtained background information
- any people, other than your course supervisor, who helped you
- any computer software packages you used (e.g. to prepare graphs).

Key term

Citation provides all the information needed to identify a publication. This includes information such as the author, title, page number, and date about a book, journal article or other format (see Figure 12.7, page 205).

There are several different ways of citing references. For example, you might list your references in the order they occur in your report, number them and give the numbers in your report as you go along. Alternatively, you might cite author and page number and make your full reference list in alphabetical order by author. It doesn't matter which style you use, but be consistent and if you are unsure try the following:

Books and periodicals
In your text citation put the name of the author and the date in brackets. For example:

> 'Even Ludwig Boltzman, physicist to the core, saw Darwin's theory as the most significant achievement of nineteenth century science' (Haw 2007).

Then, at the end of your report, add a list of detailed references in alphabetical order by author.

For books give the author (last name then initials), date of publication, title of the book and the publisher. For example:

> Haw, M. (2007), *Middle World: the restless heart of matter and life*, Macmillan.

Notice that the title of the book is in italics.

Often you will find that you are using an edited collection on a particular topic or reading about a piece of work in a textbook. For a piece in a book with an 'editor', give the author, date and title of the actual part you are using, and then give the full reference for the book. Use the abbreviation 'ed.' to show that the book has an editor rather than an author. For example:

> Dawkins, R. (2006), 'Intelligent Aliens' in Brockman, J. (ed.), *Intelligent Thought: science versus the intelligent design movement*, Vintage.

A similar format is used for articles in periodicals but you need to add the volume number or precise date. For example:

> Vines, G. (2007), 'The beaver: destructive pest or climate saviour?' in *New Scientist*, 22nd August 2007.

For something quoted from another source give the reference to the book you actually used. For example:

> King, D. quoted in Gore, A. (2006), *An inconvenient truth*, Bloomsbury.

You may be using pamphlets or leaflets with information from companies or charities. Sometimes these do not have details of the author or date on them. Do the best you can, using the name of the organisation as the 'author'.

Electronic media
For websites give the title of the site or forum, the full web address of the page you are referencing and the date or dates on which you accessed the information. For example:

> Royal Institution (2007), *Inside Out Discover Space: Big questions*, http://insideout.rigb.org/ri/space/big_questions.html (6th September 2007).

For references to other media (online databases, DVDs, TV programmes and so on) you need to give information which is similar to that used when referencing books and periodicals. It should be possible for your readers to find the original source you used so that they can follow it up themselves in more detail if they are interested (and, of course, your supervisor and examiner may need to check on how you interpreted the source). For example:

> *The Science Behind Medicines* (CD-ROM 2005), GlaxoSmithKline.
>
> *The Cosmos: A beginner's guide: Episode 4: Exploring the Cosmos* (BBC2 2007), broadcast 28th August 2007.

People and places
For people who have helped, give their names, their jobs or relationship to your work, and a few words to say how they helped. For example:

> Professor Ike Eberstein, Center for Demography & Population Health, at the Florida State University program in London, allowed me to interview him about epidemiological issues after I met him on a visit to Broad Street, March 2006.

For places you have visited, give the name of the place and a few words to indicate what use you made of the visit. For example:

> Wellcome Collection, Euston Road, made notes on anatomical exhibits and bought the postcards used for illustrations at 'The Heart' exhibition, visited 7th September 2007.

Figure 12.7

Citing references and sources.

A final check

Before you make your final draft and hand in your reports, you should make a final check that your report contains everything asked for in the specification for the course. Read through your work one more time, making sure as you go that nothing is missing, and make any necessary last additions or amendments.

- Use double spacing for word-processed or handwritten reports as it is so much easier to read. Always number your pages.
- It is good practice to set your reports out as a booklet rather than as a long essay. This means that you will have a title page, contents and so on.
- Consider including a glossary of any special terms as an appendix.
- Do not add leaflets, booklets, print-outs or photocopied documents to your final reports. There are no marks for bulk or excessive length, or for work done by other people.

Index